Policing and Special Units

❖

PETER W. PHILLIPS

Editor

The University of Texas at Tyler

M.L. DANTZKER

Series Editor

PEARSON

Prentice
Hall

Upper Saddle River, New Jersey 07458

Library of Congress Cataloging-in-Publication Data

Policing and special units / Peter W. Phillips, Mark L Dantzker, editors.
 p. cm.—(Policing and . . . series)
 Includes bibliographical references and index.
 ISBN 0-13-048211-0
 1. Police—Special weapons and tactics units. 2. Police administration.
I. Phillips, Peter W., 1940– II. Dantzker, Mark L., 1958– III. Prentice Hall
policing and . . . series.

HV8080.S64P65 2004
363.2'3—dc22

2004013303

Executive Editor: Frank Mortimer, Jr.
Assistant Editor: Korrine Dorsey
Production Editor: Anita Ananda, Integra Software Services
Production Liaison: Brian Hyland
Director of Manufacturing and Production: Bruce Johnson
Managing Editor: Mary Carnis
Manufacturing Buyer: Cathleen Petersen
Acquisition Editor: Frank Mortimer
Design Director: Cheryl Asherman
Senior Design Coordinator: Miguel Ortiz
Cover Designer: Carey Davies
Cover Image: Paul Edmondson/Getty Images
Printing and Binding: Phoenix

Pearson Prentice Hall™ is a trademark of Pearson Education, Inc.
Pearson® is a registered trademark of Pearson plc
Prentice Hall® is a registered trademark of Pearson Education, Inc.

Pearson Education LTD
Pearson Education Singapore, Plc. Ltd
Pearson Education, Canada, Ltd
Pearson Education–Japan

Pearson Education Australia PTY, Limited
Pearson Education North Asia Ltd
Pearson Educaçion de Mexico, S.A. de C.V.
Pearson Education Malaysia, Plc. Ltd

10 9 8 7 6 5 4 3 2 1
ISBN 0-13-048211-0

Contents

❖

Acknowledgments		v
About the Authors		vi

PART I
SPECIAL UNITS IN CONTEXT

Chapter 1	Special Units in Policing: An Overview *Peter W. Phillips*	1
Chapter 2	Special Units in Policing: Functionality v. Legitimacy Maintenance *Lorie L. Rubenser*	24
Chapter 3	An Examination of the Compatibility between COP and Special Units *Stan Shernock*	54

PART II
PROBLEM-ORIENTED SPECIAL UNITS

Chapter 4	Elder Abuse Units *Allen Sapp and Carla Mahaffey-Sapp*	78
Chapter 5	Gang Units *Jeffrey P. Rush and Gregory P. Orvis*	97
Chapter 6	Bias Crime Units *Joseph B. Vaughn*	119
Chapter 7	Volunteer Units: A Case for the Use of Citizen Volunteers in Sundry Police Tasks *Peter W. Phillips*	141

PART III
METHOD-ORIENTED SPECIAL UNITS

Chapter 8   From SWAT to Critical Incident Teams: The Evolution
                of Police Paramilitary Units   162
                Robert W. Taylor, Stephanie M. Turner, and Jodi Zerba

Chapter 9   Specialized Drug Enforcement Units: Strategies
                for Local Police Departments   181
                David Olson

Chapter 10   Horse Mounted and Bicycle Units   200
                Mitchel P. Roth

PART IV
CAVEATS AND CONCLUSIONS

Chapter 11   Federalism, Federalization, and Special Units: An Issue
                in Task Force Management   226
                Peter J. Nelligan, Peter W. Phillips, and W.A. Young, Jr.

Chapter 12   Summary and Conclusion   244
                M.L. Dantzker

Acknowledgments

❖

Few publishing projects are without occasional frustrations, and this one was no exception to that generalization. No volume editor, however, could have asked for or received greater support than I received from Prentice-Hall's *Policing and . . . Series* Senior Editor, Mark Dantzker. For his unwavering support, guidance and encouragement, I thank him profusely.

To Frank Mortimer and Korrine Dorsey of Prentice-Hall, please accept my expression of sincere gratitude for your support of this project.

For tolerating the many hours that I remained sequestered with this project and neglected certain "honey-dos," and for quelling my occasional temper tantrum when the "stupid computer" lost pages or worse, I thank my dear, dear wife, Bobbi.

—Pete Phillips
Tyler, Texas
November 2003

About the Authors

❖

Carla J. Mahaffey-Sapp, Ph.D., is the President of Preferred Futures, Inc., a consulting firm specializing in Long-term Health Care and Criminal Justice issues. Her educational background includes degrees in Nursing, Social Gerontology, Criminal Justice, Human Services and the Doctor of Philosophy in Health Care Management. She is a Registered Nurse and a licensed Nursing Home Administrator. Dr. Sapp specializes in Long-term Health Care Crisis Management and focuses her efforts in the area of quality care for long-term care clients.

Peter J. Nelligan (Ph.D., University of Hawaii) is an Associate Professor of Criminal Justice, Coordinator of the Criminal Justice Program, and Director of the Master of Arts in Criminal Justice Degree Program at California State University, Stanislaus. Prior to taking this position, Dr. Nelligan was Associate Professor of Criminal Justice at the University of Texas at Tyler where he chaired the Department of Social Sciences from 1990 to 1996 and held the Bart Brooks Endowed Professorship in Ethics and Leadership. Dr. Nelligan teaches criminological theory, program evaluation, victimology, and research methods. His current research interest is in the area of victim impact statements in capital murder cases.

David Olson is a Professor of Criminal Justice at Loyola University, Chicago, a member of Loyola's graduate faculty, and a Senior Scientist at the Illinois Criminal Justice Information Authority. Prior to his appointment at Loyola, Dr. Olson was the Director of the Illinois Statewide Drug and Violent Crime Control Strategy Impact Evaluation Program, where he oversaw the evaluation and monitoring of federally funded drug control efforts in the state. He has also served as staff to the Illinois Governor's Task Force on Crime and Corrections, the Illinois Legislative Committee on Juvenile Justice, and the Illinois Truth-in-Sentencing Commission. David has managed more than $5 million in research and evaluation grants, has published more than 75 articles, research reports, and research bulletins, and has presented more than 50 papers at professional conferences, training symposia, and governmental hearings. David's most recent research has been published in *Law and Policy*, *The Journal of Law and Economics*, *Justice Research and Policy*, and *The Justice System Journal*. Dr. Olson received his B.S. in Criminal Justice from Loyola University Chicago, his M.A. in Criminal Justice from the University of Illinois at Chicago, and his Ph.D. in Political Science/Public Policy Analysis from the University of Illinois at

Chicago, where he was the recipient of the Assistant United States Attorney General's Graduate Research Fellowship.

Gregory P. Orvis is an Associate Professor of Social Sciences at the University of Texas at Tyler, a position he attained after practicing law for several years in New Orleans. He has earned a Juris Doctorate from Tulane School of Law and a Ph.D. in Political Science from the University of Houston, and is a member in good standing with the Louisiana Bar Association. His research interests focus primarily on public administration, due process rights and organized crime, and he has published several book chapters on these subjects, as well as publishing in such prestigious journals as *Police Quarterly*, *Justice Quarterly*, *American Journal of Criminal Justice*, and *Harvard Law and Policy Review*.

Peter W. Phillips (Ph.D., Florida State University) is currently an Associate Professor of Criminal Justice at the University of Texas at Tyler. Dr. Phillips' career in law enforcement includes service as a deputy sheriff in Texas, police officer in New York, and special agent and resident agent-in-charge of the Bureau of Alcohol, Tobacco & Firearms. At the Federal Law Enforcement Training Center (Glynco, GA), he was Assistant Director for Police Training and educational policy advisor to the director. He also has held positions as Director of the Criminal Justice Institute, Sam Houston State University, and Director of Criminal Justice Programs, Utica College of Syracuse University. He has published recently in *The Justice Professional*, *Police Forum*, *Police Quarterly* (with Greg Orvis), and *The Police Chief* (with William Peverly). He reviews regularly for *Police Quarterly*, *The Journal of Criminal Justice Education*, and *The Justice Professional*. Dr. Phillips is a member of the Academy of Criminal Justice Sciences and a founding member of the ACJS Police Section in which he currently holds the elected position of Executive Counselor. Dr. Phillips' teaching and research interests include criminal justice policy, police management, and tribal policing.

Mitchel P. Roth is an Associate Professor of Criminal Justice at Sam Houston State University. He is the author of seven books and a number of journal articles on a wide variety of historical and criminal justice issues. He received his Ph.D. in history at the University of California, Santa Barbara.

Lorie L. Rubenser received her Doctor of Philosophy Degree from the University of Nebraska at Omaha. Presently, she is an Assistant Professor of Criminal Justice at Sul Ross State University, Alpine, Texas where her course load focuses on police services and police management. She also serves as an instructor at the university-based law enforcement academy. Dr. Rubenser has presented numerous research papers at professional conferences in topics ranging from issues relating to the Mexican Federal Judicial Police to psychological and organizational issues in American policing.

Jeffrey P. Rush is UC Foundation Assistant Professor of Criminal Justice in the School of Social and Community Services of the University of Tennessee at Chattanooga. A former police officer and juvenile probation officer, Dr. Rush is currently a Reserve Deputy Sheriff in Jefferson County, AL and the author or co-author of three books, numerous journal articles and many professional presentations and training sessions. A Certified Gang Specialist, Dr. Rush is a faculty member of the Tennessee Basic Police Academy at Cleveland State Community College, is certified as an instructor in Verbal Judo, OCAT and Defense

Without Damages and as a Law Enforcement Instructor by the State of Alabama. He is also a supervisor/trainer for an event security company and served as the Faculty Coordinator for the Southeastern Command and Leadership academy. His areas of interest include juvenile justice, gangs, law enforcement and violence. A past president of the Southern Criminal Justice Association, he is a board member of the juvenile justice and police sections of the Academy of Criminal Justice Sciences and of the Criminal Justice section of the American Society of Public Administration.

Allen D. Sapp, Ph.D., is a Professor of Criminal Justice at Central Missouri State University. He has been the recipient of several awards for outstanding research and scholarship. He is a certified law enforcement instructor in several states and has been a training, research or management consultant to more than 100 law enforcement agencies. He has been a frequent lecturer at the FBI National Academy and served as the contract Behavioral Scientist-Futurist for the FBI Critical Incident Response Group for five years. Dr. Sapp is the author or co-author of more than 100 publications, including books, monographs, book chapters, and articles. He has made more than 350 presentations at international, national, and regional seminars and professional meetings.

Stan K. Shernock is Professor and Chair of the Department of Justice Studies & Sociology at Norwich University in Vermont. He received his B.A. from the University of California, his M.A. from Indiana University, and his Ph.D. from the University of Virginia. He has been President of the Northeastern Association of Criminal Justice Sciences, and served on different committees of the ASC and the ACJS and on the editorial boards of three criminal justice journals. He has published articles on numerous police topics, as well as other criminal justice and political sociology topics.

Robert W. Taylor is currently Professor and Chair of the Department of Criminal Justice at the University of North Texas in Denton, Texas. Previously, he was Professor of Criminal Justice and Public Administration, and Director of the Office of Research Services at the University of Texas at Tyler, Tyler, Texas. Dr. Taylor has authored or co-authored over 100 articles, books, and manuscripts. Most of his publications focus on international and domestic terrorism, police administration, drug trafficking, computer fraud, and criminal justice policy. His articles appear in numerous journals including *Defense Analysis* (University of Oxford, England, Press), the *ANNALS* (American Academy of Political and Social Sciences), *Crime and Delinquency*, and the *Police Chief* (International Association of Chiefs of Police). Dr. Taylor is co-author of the leading text, *Police Administration: Structures, Processes, and Behavior*, currently in its sixth edition with Prentice-Hall, 2005. His latest work is concentrated on international terrorism, decision-making crisis situations, leadership and management practices, and community policing.

Stephanie M. Turner is a Police Officer in the Fort Worth, Texas, Police Department and a graduate student at the University of North Texas where she is pursuing a Masters of Science in Criminal Justice degree. She has a keen interest in forensics, crime scene investigation, hostage negotiations and SWAT operations.

Joseph B. Vaughn is Professor of Criminal Justice at Central Missouri State University where he serves as the department's graduate program coordinator. He received his Ph.D. in Criminal Justice from Sam Houston State University in Huntsville, Texas. He also holds the

Education Specialist Degree in Public Services Administration, the Master of Science Degree, Bachelor of Science Degree, and Associate of Science Degree in Criminal Justice. While attending Sam Houston State University he was appointed as a Senior Research Fellow in recognition of his collaborative research projects with local, state, and federal agencies.

Dr. Vaughn was previously the Undersheriff of Camden County, Missouri. He serves as a member of the Central Missouri State University Department of Public Safety and as an investigator for the Missouri Rural Crime Squad. He has police officer certifications in both Missouri and Texas.

Dr. Vaughn frequently serves as a consultant to law enforcement and corrections agencies. He is an internationally recognized expert in the field of house arrest and electronic monitoring. He is certified by the State of Missouri as an instructor for law enforcement academies and has served as an instructor for the National Police Institute.

W. A. (Bill) Young, is currently Chief of Police in Deer Park, Texas. With 31 years in law enforcement, he previously held positions as Assistant Chief of Police in Houston, Texas, and Chief in Pueblo, Colorado, and in Tyler, Texas. Chief Young holds a BA degree in Public Administration from the University of Houston at Clear Lake. He has held adjunct faculty positions at San Jacinto Junior College and Pueblo Junior College and has been a guest presenter and trainer at numerous university seminars and conferences. Chief Young has published in the area of criminal investigations.

Dedication

❖

As with my other publications, I dedicate this volume to the memory of my father, LeRoy A. Phillips, the first Chief of Police for the Town of Queensbury, New York, after the township acquired first-class status in 1960. When I was a teen, contemplating college and a career, my father asked what I *really* wanted to do. My response was, "All I ever want to be is a police officer as good as you are." I don't know whether or not I yet have achieved that status because, as the old saying goes, "Those [were] mighty huge shoes to fill."

    Dad, I'm still trying.

—*Pete*

1

Special Units in Policing

An Overview

Peter W. Phillips

Special units have long been part of the organizational structure of police agencies, although some are of quite recent origin. Juvenile aid bureaus and vice squads may be among the oldest of these, while hate or bias crime units and child and elder abuse units may be among the newest. The multi-jurisdictional task force is the type of special unit now experiencing the most rapid proliferation, especially at the small- and medium-size police agency level.

Police special units vary in organizational outlook or mission focus and are present in two basic forms: one problem-oriented, the other method-oriented. Although there are no pure types, examples of problem-oriented special units include those that focus primarily on offender type, such as gang suppression units, and those that focus on a particular offense, such as vice and narcotics units. The tactics and strategies employed by problem-oriented units are not usually different from the broad range of police procedures employed in general law enforcement and criminal investigation. Chapters 4 through 7 focus on problem-oriented special units.

The organizational outlook or mission focus of method-oriented police special units, however, is different. They are organized based on a narrow or specific set of operational methodologies and may be deployed in a multitude of situations where the unique knowledge, skills, and attitudes (KSAs) of the officers—and often the specialized equipment they use—are essential to the unit's mission. The primary example of a method-oriented unit is the special weapons and tactics team, followed by bomb squads, and, to a slightly lesser extent, mounted, airborne, and marine patrol units. The selection of officers to staff method-oriented special units, their training, and their management tend to be different from that of problem-oriented units. Because of the contingencies to which they are so often called to respond, the potential liability for their actions also tends to be higher. Chapters 8 through 10 focus on method-oriented special units.

The purpose of this volume in Prentice-Hall's *Policing and . . . Series* is twofold: First, to explain the contemporary nature and use of police special units; second, to alert readers to the administrative, managerial, and supervisory issues involved. A review of the literature produces little in-depth analyses of the need for certain special units, with only an occasional case study of a particular special unit. At the same time, there is increased evidence that special unit operations make their parent agencies particularly vulnerable to criticism and legal liability, especially when these units suffer from inadequate supervision.

In selecting a sample of special units for in-depth analyses, this volume will provide critical insights regarding both operational and administrative considerations applicable to a wide variety of special units. The information in the following chapters will be useful to three specific audiences:

1. Students of policing, particularly in the academic community;
2. Front-line police officers, whether in the academy learning about the intricacies of the occupation or, later, making career choices regarding specialization;
3. Police administrators looking for a ready-reference on the subject when making decisions regarding development, implementation, and evaluation of special units.

Each chapter consists of original writing prepared especially for this volume. Authors were selected not only for their subject-matter expertise, but also for the practitioner-academic blends their collective backgrounds represent. Before analyzing particular special units and the management issues involved, it may be helpful to distinguish between the notion of specialization as division of labor and special units as programmatic elements within a police agency.

SPECIALIZATION VERSUS SPECIAL UNITS

Specialization Generally

Max Weber (1947/1987), father of the concept of bureaucracy, the organizational structure overwhelmingly the norm in American policing, noted that within organizations there must be a clearly defined division of labor (or division of work). Today, we understand this to mean specialization—but specialization is bi-directional. Viewing the police organization from a vertical perspective, work is divided hierarchically by rank. Although there may be some overlap, front-line officers have one set of tasks, sergeants have another, lieutenants another, and so forth. Viewing the police organization from a horizontal perspective, work is divided by time (different shifts), space (different beat areas), and function (different assignments; e.g., patrol, traffic, investigation). The functional division of labor may also include assignment of sworn personnel to different divisions within the police department. For a full discussion of the concepts of differentiation (the vertical element in organizational structure) and integration (the horizontal element), see Lorsch (1970/1987). The reader should note that there is a long line of research regarding the organizational structure of police departments; see, especially, Langworthy (1986) and Maguire (2003).

In the organizational structure of a contemporary police agency, one may find divisions, bureaus, or sections. These terms appear interchangeable as some departments prefer one to the other, and there is but modest conformity nationally at similar levels in tables of organization. Examples of divisions commonly found in police agencies include, but are not limited to, management services or administrative divisions, patrol divisions, investigative or detective divisions, traffic enforcement divisions, and community services divisions. Examples of sections or bureaus include records and information bureaus, planning and research sections, professional standards (internal affairs) bureaus, and training sections. As with any organizational element, the job-tasks of personnel assigned complement the mission of that division, bureau, or section. The tasks of a patrol officer, a detective, a crime analyst, or a training officer, for example, are generally well understood, and may even be considered specialized. That the same divisions, bureaus, or sections in some police agencies may even be called units, however, does not qualify them as *special units*. By whatever name known, they represent functions too common and too necessary in the administration of contemporary police departments, and, to one degree or another, are found in most.

Before discussing further the unique characteristics of special units, however, it may be well to say a bit more about specialization in general—something this author wishes to call the *specialization trap*. One function of management is to optimize the relationship between efficiency and effectiveness. Generally, as efficiency is driven upward, for example, handling ever more calls for service with the same number of personnel, the effectiveness with which each call is handled goes down. As Bayley (1996, p. 48) notes: "Police action without beneficial results is irresponsible. Efficiency without effectiveness is a false economy." When police management permits excessive specialization in a department, it risks losing the ability to coordinate the work (differentiation and integration), the overall organization loses effectiveness, and the apparent efficiency produced by so-called specialists becomes a "false economy."

Fyfe, Greene, Walsh, Wilson, and McLaren (1997, p. 138) aptly describe this syndrome as "*suboptimization*, meaning that the subgoals (specialties) of the organization appear to be more important than the overall goals of the organization" [emphasis in original]. Fyfe and his colleagues use the classic rift between plainclothes and uniformed officers as an example of suboptimization, where each side defines the organization in terms of its narrow perspective and coordination between these two major police divisions is often far less than optimal from a managerial perspective. The former is a malady tending toward the larger police agency. The smaller police department often becomes a victim of the specialization trap when it identifies specialists within its ranks without having adequate resources to fully support the specialization. More will be said about this in the section on personnel qualifications.

Special Units

Special units possess two unique characteristics. First, a special unit is created to address a special problem and does so with a particular programmatic focus. For example:

> Sometimes a police department is confronted by an inescapable operational challenge, like hostage-takings or Uzi-toting youth gangs, and develops SWAT teams or gang squads to deal with it. Sometimes problems that the department has neglected—domestic assault, child abuse, drunk driving—are nominated for increased attention by political forces in the

community, and the police respond with a new program. Sometimes a program to deal with a frustrating police problem such as drugs or domestic assault is shown to work, and other departments adopt it to remain on the cutting edge. (Sparrow, Moore, & Kennedy, 1990, pp. 198–199)

Special units, therefore, are designed to deal with particular problems. That one jurisdiction may have a particular problem does not mean that another jurisdiction has a similar one. Even if they do, the other jurisdiction may not have the same perception of need to address the problem or they may choose to address it quite differently.

When we speak of a special unit, therefore, we are speaking of an administrative decision to separate a number of officers from a broader range of policing responsibilities and assign them to a new organizational entity with a much narrower range of policing responsibilities. The range of responsibilities is defined by the nature of the problem to be addressed. Problems, however, generally emerge in two forms: *persistent problems* and *contingent problems.*

Persistent problems come to light when crime analysis shows that over time there are growing numbers of the same types of offenses and the ability to deal effectively with the increased numbers *exceeds the capacity of routine patrol or general investigation.* Both the magnitude and the persistence of the problem lead to an executive decision to create a special enforcement unit. Officers assigned to the unit are chosen based on appropriate criteria assessing prior performance and current interest, and are given specialized, advanced-level training in detecting and investigating the class of crime for which the unit was created to suppress. An example of a persistent problem is drug crime, and a drug enforcement unit is an example of a persistent problem special unit.

Contingent problems, as the term implies, are potential problems; problems that are clearly possible, perhaps probable, but not of an ongoing or continuous nature. A contingent problem, unlike a persistent problem, is not present "24/7," but when the problem emerges it is critical and demands immediate, if not emergency, response; the kind of problem for which the police agency should have a contingency plan. The highest level of selectivity, training, and evaluation must be exercised in assigning personnel to contingent problem-oriented special units as these present the police agency with the highest threats of liability for wrongful action. A barricaded armed robber with hostages is an example of a contingent problem and Special Weapons and Tactics (SWAT) teams (also referred to as SRTs, Special Response Teams, and ERUs, Emergency Response Units) and Hostage Negotiation Teams are examples of contingent problem special units.

This leads to the second unique characteristic of a special unit: It is optional. As pointed out by Sparrow et al. (1990), depending on the problems extant in a particular community, identified either by the police themselves or by local political forces, the initiation of a special unit *may* be an appropriate response. So might some other response, ranging from no action at all to changes in policies and procedures (e.g., directed or targeted patrols)—but short of establishing a special unit. Not every police agency needs or chooses to support an Elder Abuse Unit, a Gang Suppression Unit, a Bias Crime Unit, a Mounted Unit, or a Citizen Volunteer Unit. Encountered more frequently in city, county (including sheriff's departments), and state police agencies are SWAT and Drug or Narcotics Enforcement Units. Regardless of the extent to which they exist, however, even SWAT and Drug Units are optional. Through interagency agreements (often called MOUs, Memoranda of Understanding) and multi-jurisdictional task forces organized to make critical police functions available to the smallest departments, neighboring police agencies may practice shared resource management (also known as consolidation of resources). The question of whether or not an agency supports a special unit, therefore, depends on needs and resources, and these vary from jurisdiction to jurisdiction.

THE DECISION TO CREATE A SPECIAL UNIT

Whether or not a special unit is the best alternative for addressing a particular police problem usually requires considerable analysis. First, we must establish a common definition of what constitutes a police problem of sufficient magnitude as might justify organization of a special unit. Herman Goldstein (1990), the "father" of problem-oriented policing, provides a basic definition. A problem is:

- A cluster of similar, related, or recurring incidents rather than a single incident [and]
- A substantive community concern [and]
- A unit of police business (p. 66).

One occurrence (or several, but over a long period of time) does not rise to the level of a problem needing the organization of a special unit and the concomitant dedication of departmental resources for the management and operation thereof. On the other hand, frequency is not the sole criterion, as the criticalness of occurrences must be factored also. Although no precise formula for weighting frequency and criticality can be offered because too many local variables intervene in every community, some guidance may be derived from the literature on police training which suggests that criticality be given the greater weight (Peak, 2001; Phillips, 1987).

Whether or not a problem is substantive is itself problematic. As Goldstein (1990) notes:

> Because the police—as the "hired hands"—are available to deal with the unsavory aspects of life in the community, the citizenry tends to define substantive problems as *police* problems. But it would be more accurate to define substantive problems as *community* problems. (p. 34; emphasis in original)

However, here's the rub. Goldstein continues:

> But even when police are instructed in the meaning of "substantive" problems and then asked to focus on them, they are apparently so conditioned to thinking in terms of the problems of the organization that they frequently slip back to identifying concerns in the management of the organization. *Thus, focusing on the substantive, community problems that the police must handle is a much more radical step than it initially appears to be, for it requires the police to go beyond taking satisfaction in the smooth operation of their organization; it requires that they extend their concern to dealing effectively with the problems that justify creating a police agency in the first place.* (p. 35; emphasis in original)

What does Goldstein's admonition mean regarding the question of whether or not a special unit is the best alternative to addressing a particular police problem? In the next chapter, Lorie Rubenser asks a complementary question: Do police special units serve a functional purpose or are they instituted simply to provide legitimacy maintenance to the organization? The search for public support, or legitimacy status, has caused police departments to attempt a variety of new programs—often implemented as special units. Ultimately, it has been suggested for legitimacy purposes that it does not matter if the unit actually performs its stated objectives as long as the citizenry believe it is beneficial to them. This suggestion could lead police agencies to be more concerned with having their special units *appear* to work effectively rather than with actually working. Goldstein alerts police decision-makers regarding this pitfall and Rubenser (Chapter 2) demonstrates a decision-making process for assuring functionality.

The third element in Goldstein's (1990, p. 66) definition of a problem, "a unit of police business," seems rather straightforward. Criminal behavior surely is a matter of police business. Order maintenance may or may not be police business depending on the will of the community. Manning (1997) defines order maintenance in part as:

> . . . the provision of public services, intervention in disputes or disagreements, and control of the public streets. . . . Here the problem is not simply one of attempting to ascertain whether a law has been broken and what appropriate steps are required . . . but of deciding what often ambiguous code of public conduct has been violated, *in whose view*. (p. 249; emphasis added)

Police administrators should be wary of special interest group pressure to deal with order maintenance issues, investigating them fully to assure that bias is not the motive and that the police department does not unwittingly reinforce discriminatory practices in the community.

The important point here regarding the decision of whether or not to implement a special unit relates to the significance of the problem to be addressed:

- Its seriousness; the extent of harm, i.e., number of victims, loss to victims or society.
- The prognosis for its continuance at a high incidence rate or criticality level.
- The success (or lack thereof) of other policing alternatives to diminish the harm.

Drunk driving certainly is a crime, is of substantive community concern almost everywhere, and results in great loss to its victims and society. In certain areas of the country, "Stop DWI" programs provide external funding for police agencies to pay officer overtime (and certain other expenses) for "hunting" drunk drivers exclusively. These are not special units, however, and special units are not indicated to fight DWI because every patrol officer is trained to detect this violation and should be on the lookout for drunk drivers wherever and whenever probable cause to make a traffic stop exists. By contrast, as Allen Sapp and Carla Mahaffey-Sapp in Chapter 4 and Joseph Vaughn in Chapter 6 demonstrate with Elder Abuse Units and Bias Crime Units, respectively, these special units are indicated only under certain circumstances. The "circumstances" include the variables mentioned above; that is, substantive community concern with great loss to victims and society (assuming that the number of incidents follows a clear trend). The difference here is twofold. First, these crimes tend to take place in private places (with bias crimes, often in secluded places) where normal patrols cannot or seldom go proactively. Second, special investigative skills are often necessary, the development of which is as dependent on officers' personalities as on their technical knowledge. As noted later in this chapter regarding selection of officers for special units, it takes a "special person" to deal psychologically with certain crimes day-in and day-out without becoming cynical, authoritarian, burned-out, or even clinically depressed.

Goldstein (1990) may have considered departmental resources integral to his "unit of police business" mentioned earlier, because a department should not undertake responsibilities for which it lacks the resources to deal effectively; but this is not clear from his original writing. To make Goldstein's (1990) definition of a problem applicable to decision-making regarding special units, therefore, we must add a fourth assessment:

- Agency resources are adequate.

Most special units are intradepartmental. Those that are not, exemplified by the multi-jurisdictional task force, require the dedication of at least some departmental resources nonetheless. Special units draw their resources from the existing departmental programs to create a new program. This is known as a zero-sum game and will be discussed more fully later in the chapter. Before a decision is made to implement a special unit as an alternative response to a problem, decision-makers should undertake a systematic study of the problem and consider other feasible alternatives.

PLANNING AND DECISION-MAKING

There exists a direct relationship between planning and decision-making. As noted by DuBrin and Ireland (1993), planning is

> defined as a process that identifies objectives and the commitments, resources, and actions required for their achievement . . . When planning, managers make decisions in light of conditions they believe the [organization] will face in the future . . . [Astute managers] constantly think about what they should be doing in the present to assure the [organization's] future success. (p. 107)

Boone and Kurtz (1992, p. 123) say it a bit more succinctly: "Since plans specify the actions necessary to accomplish organizational objectives, they serve as the basis for decisions about future activities." Planning is not making tomorrow's decisions; rather, it is making today's decisions with tomorrow in mind.

A rather standard and well-established model exists for planning and making decisions leading to the implementation of policies, procedures, and/or programs (we are concentrating on programs, of course, that implement special units). Variously called the Rational Planning Model (Starling, 1998), the Police System Planning Model (Strecher, 1997), and the Synoptic Planning Model (Swanson, Territo, & Taylor, 2001), our interpretation of the model is shown below. The model consists of five major steps, each with two or more parts. (See also an excellent seven-step model proffered by Welsh and Harris, 1999.) Before we get to the model, however, consider the following scenario.

Assume that you are a mid-level manager in a police agency. A police executive has given you an assignment to study a crime problem that appears to be emerging in your jurisdiction. It could be any of the problems discussed in this volume (e.g., drug crime, gang crime, crimes against the elderly) or a different problem (e.g., internet crimes, child abuse, auto theft). The assignment is to thoroughly research the problem and present your findings to the chief (or sheriff) at a command-level meeting in two weeks. Your findings must include one of the following recommendations—in all cases explaining fully the facts discovered and your rationale for arriving at that particular recommendation:

(a) Do nothing. There is no significant problem, the problem is currently being addressed appropriately, or the problem appears to be of short duration and will self-abate.

(b) Implement a new departmental program or redirect (refocus or modify with some new elements, but not all) a current program to address the problem more efficaciously. Several alternatives can be proposed, including recombination of present programs, but not the implementation of a special unit at this point.

(c) Implementation of a special unit. After careful analysis of the problem and the development and consideration of all other feasible alternatives, you have determined that a special unit is the best alternative—and that is your recommendation.

To make this exercise even more exciting, assume that your next promotion (or demotion!) depends on the thoroughness of your research and the feasibility of your recommendation.

The Decision Model

The decision model outlined below consists of five steps or stages:

1. Define the problem,
2. Review the field,
3. Identify and analyze alternatives,
4. Develop an implementation plan,
5. Monitor and evaluate results.

1. Define the problem   Precisely what is the crime or disorder problem for which present police programs appear less effective than desirable? One must be careful to dig deep enough into the issues to assure that the right problem is being identified and that sufficiently probing questions have been asked to assure that the problem definition is neither superficial nor prejudiced by limited viewpoints (Strecher, 1997). Too often, there are knee-jerk reactions to certain crime problems and such "quick draw" pseudo-solutions seldom attack problems squarely and are usually wasteful of departmental resources. Re-evaluating and redirecting present police programs may be considerably more efficacious than implementing a new program.

As part of the first step in the decision-making process, there is also a need to *measure the significance of the problem*. The problem definition is not complete without a clear understanding of the impact the problem has on victims, the community-at-large (or a neighborhood, as the case may be), and the department. Harm is a fundamental element of criminal law (see, e.g., Territo, Halsted, & Bromley, 1998) and, as such, provides an appropriate measure of significance. Although quality-of-life issues are always important, the harm caused by the crime problem considered for targeted intervention should be measured quantitatively. How many incidents? How many victims? Over what period of time? What were the losses to the victims? What has the past-to-present trend been regarding this problem? What is a reasonable prediction of the future trend (six months, a year, five years hence) regarding this problem if no action is taken, that is, if the status quo in departmental response is maintained?

2. Review the field   This is an extremely important step for three reasons. First, by tracing the evolution of an alleged problem it may become clear that the real problem is different from the originally alleged problem. Maybe the right questions were not asked and the first impression of the problem was either "superficial or based on a limited viewpoint" (Strecher, 1997, p. 26). Second, in examining what other law enforcement agencies and communities have done to address the same or a similar problem, one may avoid, as the old saw goes, "reinventing the wheel." This applies to both the discovery of others' successes and of others' failures. When the philosopher/poet George Santayana said, "Those who cannot remember the past are doomed to repeat it," he very well could have been speaking about errors in police work. Of course, one must have learned about the past in order to remember it. Quite frankly, one of the more serious deficiencies in police administration today—especially affecting small and rural agencies—is a lack of knowledge regarding successful experiments, innovations, and emerging management trends. This derives from an ignorance of, and often a lack of interest in, the past successes and failures of policing and the insights that the contemporary literature on policing can provide.

Third, from an examination of the field and the possible discovery of others' successes in dealing with the problem under investigation, one may strike upon a particular kind of program—an exemplary program that, with only minor "tweaking," can be replicated directly in the home jurisdiction. In research, this is called a *conceptual framework*—a model, a system of organization, a typology—into which one can fit new data systematically. Goldstein's (1990) Scanning-Analysis-Response-Assessment (SARA) model is a conceptual framework for problem-oriented policing. The decision-making model we are building here is a conceptual framework. A methodologically sound, well-documented, successful police or police-community program in Jurisdiction A that diminishes the harm caused by the crime or disorder syndrome under consideration may provide a conceptual framework for abating the problem in Jurisdiction B.

The tweaking mentioned above refers to fitting local data into the conceptual framework. No two jurisdictions are exactly alike; one of the pitfalls, for example, is making comparisons between two or more jurisdictions based on single measures such as UCR data. In the author's opinion, however, a far greater problem exists when police agencies summarily dismiss innovations developed elsewhere simply because they were developed elsewhere. Despite the efforts of such organizations as the Community Policing Consortium to disseminate information regarding exemplary programs, there remains a tendency among police agencies to reinvent the wheel.

An organizational Napoleon Complex exacerbates this tendency, especially among small police agencies. The so-called Napoleon Complex is characterized by over-assertiveness, even aggressiveness, by certain persons of less than average body size as a form of psychological compensation—the behavior ostensibly making them feel proportionately bigger and symbolically more powerful. In this author's experience, the Napoleon Complex is observable in certain small police agencies (defined here as less than 25 full-time sworn officers) that implement special units or other programs or functions without having truly adequate resources to do so. Others believe that this is a prideful reaction, similar to the notion of "keeping up with the Joneses" where smaller police agencies simply want to emulate their larger brethren (Kraska & Cubellis, 1997; Kraska & Kappeler, 1997; Kraska & Paulsen, 1997).

In reviewing the field, one should be alert to descriptions of past and present proposals from credible sources regarding problem interventions. What proposals may have been made but not implemented; why weren't they? What programs currently are being proposed and what is being said about them by experts in the field? Have there been any formal or funded evaluations of those programs (by entities such as the National Institute of Justice, Police Executive Research Forum, or Police Foundation) or articles published in reputable, peer-reviewed (refereed) academic journals (such as *Police Quarterly*, *Policing*, *Policing and Society*, and *Justice Quarterly*) that appear relevant to the problem under investigation, especially any that may serve as a conceptual framework?

3. Identify and analyze alternatives   At the onset of this stage in the planning and decision-making process, it is important to develop all plausible alternatives. As mentioned in the "assignment," one alternative might be to do nothing. Police policy is public policy for the respective jurisdiction, and, as Dye (1995, p. 2) notes, "Public policy is whatever governments choose to do or not to do." The choice to do nothing is no less of a choice. Possibly the problem was misidentified—which does not always mean there is no problem, as we already have been alerted to the fact that research often reveals that the wrong questions were asked. If a different problem is identified, it is necessary to return to Step One of the model.

Research also might reveal that the problem is being addressed appropriately and no further resources need to be allocated at this time unless the nature of the problem changes. In this instance, the "do nothing" recommendation becomes "do nothing more."

Problem analysis (defining the problem *precisely* in Step One) might suggest a set of alternatives in which specific modifications to one or more police programs already in operation appear promising. Shifting resources to meet demands such as with targeted or directed patrols, stakeouts, decoys, and crackdowns or sweeps are examples of alternatives that may be identified and in the final decision-making phase weighed against the probable costs and potential benefits of other alternatives.

Still another set of alternatives might suggest combining present, modified, or new police programs with the efforts of other governmental or quasi-governmental organizations. Task force operations often fall into this category, where, using a child-sexual-abuse task force as an example, police officers from several agencies combine with social workers, mental health workers, and assistant district attorneys to investigate and prosecute this crime (Phillips, 1999). Task force operations are also discussed in Chapters 3 and 11 of this volume.

Once all plausible alternatives have been identified, it is time to reduce the number to those that are feasible. The distinction here is between those alternatives that are logically possible and those that are practical given the police agency's discretionary resources. It order to do this, it becomes necessary to develop decision-making criteria.

First, however, we must discuss the concept of a *zero-sum game*. At some point, there is a limit to everything and fiscal resources are no exception. Without money, a police agency cannot "buy" personnel (historically, about 80% of a police agency's budget is allocated to personnel costs, salaries, and fringe benefits), nor can it buy equipment or maintain facilities. Agency budgets are set in advance and the money to operate is appropriated by the funding authority (city council, county government). Most municipalities have been operating in a fiscally austere environment for 20–30 years. Within a particular fiscal year, it is sometimes possible for the agency administrator to request a supplemental budget appropriation—usually to meet an unexpected contingency, such as the total loss within a short period of several police vehicles from a small fleet. Sometimes an agency may be rewarded by sources outside local government by winning a federal agency or philanthropic organization grant to support a particular initiative. Agencies can never *depend* on additional revenue sources mid-fiscal year, however.

It also must be remembered that when agency heads present their budgets to the funding authority, they are in competition with one another. Resources are finite and what public works, fire and EMS, and the building inspector want compete with what the police department wants. This competitive situation is called a zero-sum game. As Starling (1998) describes this game:

> In other words, if you add all the wins and loses in a transaction, the sum is always equal to zero, since any wins for one party must cover the losses of the other. Another name for this situation is "win-lose." (p. 87, *n. 1*)

The zero-sum game is played intra-agency just as fiercely. As an agency's budget request to the funding authority is being finalized, division heads and bureau chiefs compete with one another for shares of a finite pie. Everyone wants a bigger piece of the pie for his or her share of the operation, for present programs and projects to be continued into the next fiscal year; and for newly proposed initiatives, for example, the implementation of a special unit. Now to the development of specific decision-making criteria.

Against what measures or limitations will each alternative be judged in order to determine which one is best? Important considerations include:

- Personnel,
- Equipment,
- Facilities,
- Time,
- Budget.

Finally in Step 3 of the decision-making process, the most preferable alternative must be recommended and selected.

- Recommendation and selection

Personnel How many officers will be required to adequately staff the alternative approach? Should the officers chosen for the alternative have particular aptitudes and competencies? Does the agency have the requisite personnel?

Assigning personnel to alternative responsibilities drains this most valuable resource from other departmental assignments, if only partially. Depending on the demand for special investigations and/or operations (significance of the problem), and the intensity with which officers must give of their time, devotion to a new program may be part-time or full-time. Even if officers are assigned to a program that activates only sporadically in response to a critical incident (e.g., SWAT/Hostage Negotiation), specialized training and regular drill requirements pull officers away from other police duties.

Qualifications of personnel are important here, as well. When considering the implementation of an alternative approach to a problem, particularly in the method-oriented category discussed earlier, extreme care must be exercised in assessing each officer's interests and qualifications for the assignment. Although qualifications may be enhanced by quality training initially, and rigorous, frequent drill and practice is given to assure that special KSAs are appropriately reinforced, what is critical is assuring that officers are physically and psychologically prepared to undertake the tasks of the assignment. For example, not every police officer is physically or emotionally fit to be a police sniper, has the personality even to be trained as a hostage negotiator, or, for that matter, the ability to deal day in, and day out with personal, intimate crimes such as in the several categories of abuse (spousal, child, child-sexual, elder). Misfitting officers to tasks can lead to disgruntlement, poor performance, high stress, and early burn-out. Officers in one or more of these mental states are apt to make poor decisions leading the agency to potential liability for wrongful action.

Whether or not a police agency has the requisite number of personnel to detach from regular duties in order adequately to staff, train, and supervise an alternative is one question. Whether available personnel are physically and psychologically capable of carrying out the functions of the alternative approach is another question of equal import. Both questions require the decision-maker to engage in a zero-sum game. In the first instance, one police function must lose resources for another to gain; in the second, a temporary departmental gain may turn into a permanent departmental loss.

Quantifying the exact number and kinds of personnel necessary to implement an alternative is one important factor in determining whether that alternative is the best solution for the problem under analysis.

Equipment   Will the alternative being considered for implementation require additional or special equipment? In establishing decision-making criteria, the availability of any additional or special equipment or the ability to make it available through purchase, lease, or grant award is an important consideration. This presumes that the decision-maker has excess equipment of the right type, has the fiscal resources to purchase or lease the equipment, or can acquire it through a known grant or interagency loan program. If none of these, but the equipment is already on the agency's inventory, once again the decision-maker must play the zero-sum game: take it from one departmental entity and give it to another.

Quantifying the types of equipment, exact numbers of units, and costs of each unit necessary to implement an alternative is the second important factor in determining whether that alternative is the best solution for the problem being analyzed.

Facilities   Will the alternative under consideration require additional or special space? Is there a need to separate personnel assigned to the function from other police personnel or from visible association with government, as is often the case to protect the identity of officers working undercover. Is there a need for particular accommodations within the space for special equipment, possibly because of size, configuration, security (e.g., sensitive case files, weapons, or seized property), or environment (e.g., special air conditioning or air filtering)? If present facilities are neither adequate nor desirable, does the required space exist or must it be designed and constructed, and are fiscal resources available to cover these costs? Are fiscal resources available to cover other costs as necessary, such as rent, utilities, and maintenance?

Quantifying the requirement for facility needs above and beyond those already part of the police inventory is the third important factor in determining whether that alternative is the best solution for the problem being analyzed. Even if present facilities appear adequate, the decision-maker must recognize the probability that operations may expand and outgrow space dedicated initially. When this happens, the decision-maker again may find him- or herself engaged in a zero-sum game—reducing facilities for one function in order to accommodate another or using scarce fiscal resources to provide facilities for one while withholding those resources from another.

Time   In establishing decision-making criteria, the variable "time" should be considered multi-dimensional. In relation to personnel resources, if officers assigned new program responsibilities are also required to perform regular duties, will they have time to do both sets of tasks well? In relation to overall departmental resources, given the ever-widening demands for police services, is there time to develop, implement, operate, and evaluate yet another program? Given the already determined significance of the problem being scrutinized, how much time is necessary in which to achieve a reasonable, measurable abatement in the harm it causes?

Possibly the most important time dimension relates to implementation of the selected alternative. In developing the decision-making criteria against which all alternatives must be weighed, a time line should be established that indicates particular milestones to be reached in the process of developing, implementing, operating, and evaluating the selected alternative. What steps must be completed in the process before another step logically can commence? What steps might overlap in time? What steps can be completed simultaneously? What events mark critical points between and among steps? How much time is reasonably required to accomplish each step and to fully implement the program?

Quantifying the time necessary to implement an alternative is the fourth important factor in determining whether that alternative is the best solution for the problem being analyzed. Time does not escape as a factor in the zero-sum game, either. Time spent on one task cannot be spent on another.

Budget   As noted earlier in this discussion, all resources are finite and all reduce to money. Once personnel, equipment, facilities, and time needs have been precisely defined and component costs determined, the decision-maker must assess each alternative against its total cost. What is the apparent cost–benefit ratio of each alternative presented?

Related to the notion of a zero-sum game is the concept of *lost-opportunity costs*. Lost opportunities occur when agencies fail to prioritize correctly the demands for their services or their needs for resources. If an agency had not spent scarce resources (or too much of its resource allocation) dealing with a particular problem (or small set of problems), might overall gains in crime and disorder prevention have been greater if the resources were directed elsewhere (or in different proportions)?

Recommendation and selection   The last phase of Step 3 in the decision-making model is the actual selection of an alternative. After all alternatives are evaluated and compared with one another, including the consequences of the alternative to "do nothing," the alternative with the highest probability of success should be recommended and selected. Success, however, means that results are predictable and measurable, and that potential losses have been cut to a minimum in the zero-sum game.

Success also means that predictable problems can be overcome. Two inextricably linked concepts in criminal justice are *discretion* and *accommodation*. Whenever discretion is exercised, that is, a decision to pursue one alternative course of action over another, there is a ripple effect. The ripple effect is observable as accommodation, the need for affected components of the system to make adjustments because of decisions beyond the affected components' control. The following is an example in macro-criminal justice perspective: A plan is implemented to hire 100,000 new police officers (discretion) who will make additional arrests. The arrests will increase the workload and burden on prosecutors and the courts (accommodation). The courts will convict more persons and send them to prison, increasing the burden on institutional corrections (accommodation). Prisons will become more overcrowded, requiring additional early releases (accommodation). Early release will increase the burden on community corrections' caseloads (accommodation) and on the security of society-at-large as increased recidivism (accommodation, again).

In micro-criminal justice perspective, one must examine the potential ripple effect in the jurisdiction in which the recommended alternative is adopted. Whenever organizational change is implemented, there will be a ripple effect. Consider change as a rock thrown into a pond. The magnitude of the splash and the concomitant ripples (waves, and sometimes tidal waves, if the rock is big enough or on top of other recently thrown rocks) is dependent on the magnitude of change. The shoreline of the pond is quite irregular and there are many people standing there, some very near the water's edge—the organization's own; others farther back—sister agencies, local government, and citizens generally; stakeholders all. Depending on where the rock hits the water, and, again, how big it is, some people will be splashed sooner than others and some with much greater force than others. As part of the decision-making process for any organizational change, the responsible administrator will scan the shoreline carefully to identify those standing there and will ask them how they feel about getting wet.

Identifying representative stakeholders and consulting with them regarding their perception of the planned change—how their job-tasks might be affected—is essential to program success. The potential error in problem definition caused by limited viewpoint, as noted earlier in this discussion, is equally valid here. The zero-sum game has a corollary: The success of current programs (or intra-agency cooperation and morale, or intergovernmental relations, or citizen satisfaction) should not be diminished by new ones. At the intergovernmental level, as Nelligan and Phillips demonstrate in Chapter 11 of this volume, dire political consequences can result from not anticipating where the ripples will go and who they might splash. As the alternative with the highest probability for success is chosen, potential problems in implementation—the ripple effect—must be predicted and contingency plans established for overcoming these problems.

4. Develop an implementation plan   The next step in the decision-making process includes the development of an action plan for the implementation of the selected alternative—the development of program budgets, project schedules, control procedures, and staffing plans.

Several scheduling models are available to facilitate the coordination of activities associated with program implementation. Three of the most often applied models are Gantt charts, the critical path method (CPM), and the program evaluation and review technique (PERT). A full description of these models is beyond the scope of this discussion, except to urge decision-makers not familiar with their use to become so. For those needing a refresher in these methods, we recommend a contemporary management text, a few of which have been cited previously (DuBrin & Ireland, 1993; Starling, 1998; Swanson et al., 2001).

In light of recent police scandals and public outcries involving special units in Los Angeles (Lait, 1999), New York (Kocieniewski, 1999), and Chicago (Mills & Hanna 2002), sadly to mention only the most spectacular of late, we feel that it is important to emphasize the development of control procedures as a critical component of any implementation plan. These will be used to monitor the operation, as explained further in the next section.

5. Monitor and evaluate results   *Monitoring*, according to Welsh and Harris (1999, p. 160) is "[a]n attempt to determine whether program or policy implementation is proceeding as planned. Monitoring is a process that attempts to identify any gaps between the program or policy on paper (design) and the program or policy in action (implementation)." Strecher (1997, p. 29), defines monitoring as "managerial oversight of the new activity, attention to the details of its daily action and productivity." Some of the questions Strecher recommends administrators constantly repeat include: "Are the intended outcomes happening? Are unintended side effects also happening? How are the staff taking to their new assignments? How are citizens reacting to the new police activity?"

With Strecher's definition of monitoring as managerial oversight, we return to the notion that control procedures are essential to the success of all police operations and critical to the success of special units. A full discussion of vicarious liability is beyond the scope of this volume and readers are referred to three especially well-written treatises on the subject, one by del Carmen and Walker (2000), the others by Kappeler (2001a, 2001b), as well as this author's admonition to stay current on case law. Suffice here to say that we recommend more than oversight, which may be understood by some simply as reviewing reports. We recommend direct managerial involvement in special unit operations—police managers should assume active participation in unit supervision. Because liability devolves up the chain of command, it would seem important to assure that police managers experience directly the successes and work personally to avoid the failures of some special unit operations. Monitoring involves constant assessment of progress and re-engineering of the implementation plan, should deviations from the best-intended course of action be detected.

Evaluation tests the mid- to long-term consequences or impact of police programs. Dye (1995) defines impact as effects on real-world conditions, including:

1. Impact on target situation or group;
2. Impact on situations or groups other than the target (spillover effects);
3. Impact on future as well as immediate conditions;
4. Direct costs, in terms of resources devoted to the program;
5. Indirect costs, including loss of opportunities to do other things (p. 321).

Dye's noting of spillover effects should remind us to "follow the ripples"; for direct and indirect costs, we should be reminded about the zero-sum game.

Bayley (1996) makes an important distinction between two indicators that measure police performance—direct and indirect measures:

> Performance is measured directly when the indicators reflect what police achieve by way of public benefits. Performance is measured indirectly when indicators show what the police are doing in terms of actions. . . . In other words, direct measures evaluate the ends of policing; indirect measures evaluate the means of policing. (p. 46)

It is the end product of police action that should be measured to determine the success (or failure) of a particular program. For example, has the program, over a reasonable intervention period, actually lowered the crime rate attributable to the targeted behavior or criminal element?

The above should not imply, however, that the means of policing are less important overall. If the police do not behave ethically and professionally, comport themselves civilly, and act Constitutionally, far more will be lost in public confidence than ever will be gained in crime and disorder reduction.

Bayley (1996) further divides both direct and indirect measures into the categories of "hard" and "soft" measures, with a substantial list of each (p. 47). One example of a direct/hard measure is the crime rate or the number of criminal victimizations; of direct/soft measure, the fear of crime; of indirect/hard measure, the number of police personnel; and of indirect/soft measure, police morale. Bayley continues: "Direct measures are better than indirect measures for determining the police's social value. Only direct measures show whether the police are effective." Further, "[s]ubstitution of soft for hard measures encourages police to become managers of opinion rather than of circumstances" (p. 48). In police program evaluation, the primary emphasis should be on direct/hard measures. Evaluation should strive empirically to establish police effectiveness.

Rigorous program evaluation serves several important ends. If the problem for which the intervention was designed reduces the harm to victims and the community, the results both enhance the police image in the citizens' eyes and provide validation for future fiscal appropriations. If the harm continues without significant abatement, the lesson is no less valuable because an opportunity exists to redefine the problem and amend the intervention strategy, thereby cutting further losses in terms of either the zero-sum game or lost-opportunity costs.

Many police agencies do not engage in rigorous evaluation of their programs not because they fear the results, but because they lack the know-how. For those without an internal program evaluation capability, resources are readily available. To start, we recommend three Police Executive Research Forum publications, two edited by Hoover (1996, 1998) and one by Brodeur (1998); and a specific text, albeit somewhat more statistically oriented, by Maxfield and Babbie (2001). We also respectfully recommend our colleagues in the academic criminal justice community with experience in policing and in program evaluation. Many excellent programs are ongoing in a multitude of police agencies. Some have been evaluated and reported, and are now available as conceptual frameworks— but more have not been evaluated and more should be planned with a firm evaluation scheme in place. University criminal justice faculty members are continuously on the lookout for opportunities to provide real-world research experiences for their graduate students. Police agencies and academics alike are urged to forge research and evaluation partnerships.

We have worked methodically through the five stages of the decision-making model: (1) Define the problem; (2) Review the field; (3) Identify and analyze alternatives; (4) Develop implementation plan; (5) Monitor and evaluate results. Of course, the results cannot be monitored and evaluated until the alternative you recommend has been adopted by the agency, but you still must propose the plan to do so. Having done the best job you are capable of doing, therefore, are you now ready to present your findings to the police chief (or sheriff) and recommend the alternative that is best for the community and the department? If your recommendation is to establish a special unit, can you justify your conclusion?

CONCLUSION

We began this chapter by explaining the authors' collective purpose in developing this volume. As part of the Prentice-Hall series *Policing and . . .* , we wish to explain the important role that special units play in today's police organization. This knowledge is important to students of policing in their understanding of the complexities of police organization and to police officers contemplating technical specialization in terms of career development. It is also important for police administrators and managers, present and future, to understand that the decision to implement a special unit should not be a knee-jerk reaction to political or community pressure, nor to the image of neighboring police agencies, possibly with greater or different resources. The decision must be based on: a precise definition of the problem to be addressed, an in-depth review of the field to increase probabilities of success and decrease probabilities of non-success, a thorough and systematic analysis of all reasonable alternatives to address the problem, an honest assessment of the agency's resources available under a set of decision-making criteria from which the optimum approach can be selected, and a methodologically sound plan to measure the true difference the police can make.

We have differentiated between specialization in policing and police special units. Whether special units are typified as problem-oriented or method-oriented, in the last analysis they all address crime or disorder issues that are problematic to the jurisdiction. Special units also are optional entities in any police organization because the kinds of problems for which they are designed to address may not occur in every community or may occur in varying degrees in relationship to other community characteristics. In either case, an alternative approach to the problem, other than a special unit, may optimize the police response.

In the process, we have discussed such management concepts as differentiation and integration of job-tasks, zero-sum game and lost-opportunity costs, and suboptimization.

Also, an effort was made to remind readers that special unit tactics often increase an agency's vulnerability and potential liability, all to reinforce the importance of direct managerial involvement in special unit operations.

CONTINUING THE THEME OF THIS VOLUME

Continuing with Part I—Special Units in Context—Chapter 2, by Lorie Rubenser, deals with a serious question about the real purpose of special units. As she will explain, two other investigators looking into the operations of particular special units, came to essentially the same conclusion: Special units are essentially "ceremonial," they simply create the appearance of effectiveness in order to satisfy public or political pressure for the police agency to "do something!" If true, then the outcome of that zero-sum game is obviously lost opportunity. Rubenser's program evaluation, on the other hand, paints a far more promising picture. The Omaha Police Department spent two years working through a planning and decision-making process leading to the implementation of a Nuisance Task Force—an intradepartmental special unit. In this case, the unit proved its effectiveness in abating the targeted problem and as a complement to the department's overall order-maintenance function. Following the decision-making model significantly increases the probability of special unit success by direct/hard measures.

In Chapter 3, Stan Shernock investigates another question posed quite frequently today about special units: Are they compatible with the premises of community-oriented policing? As he will explain in detail, a certain perception exists that special units are predominantly paramilitary, or at least zero-tolerance oriented, and that model is contraindicated in many treatises on community policing. Antagonists allege of community policing that it is "soft on crime." The protagonists' rebuttal is to point out quite forcefully that traditional police methods, for example, pro-arrest policies, are not abandoned or even diminished, nothing is subtracted; what is added is problem-solving (problem-oriented policing) and coproduction. Through a most in-depth review of the field (called "literature review" in the academic community), Shernock establishes that certain special units, despite a paramilitary or zero-tolerance perspective, are quite compatible with community policing precepts. Community policing clearly advocates getting the "bad people" off the streets so that there is more room for the "good people" (we just have to police each other to assure that the "removal process" is Constitutional).

Chapter 4 begins Part II of the volume in which problem-oriented special units are discussed. Again, we respectfully remind the reader that although all special units are intended to address community crime or disorder problems, a problem-oriented special unit is characterized both by its focus on offender type and that its methodologies are not usually different from the broad range of police procedures used in general law enforcement and investigative work.

Elder abuse units are the topic of Chapter 4 by Allen Sapp and Carla Mahaffey-Sapp. Elder abuse is a growing problem in America and one about which we are just recently taking cognizance. It is a growing problem because Americans are living longer—the phrase, "the graying of America," has become quite common in reference to this phenomenon—and the recent spurt in crimes against the elderly has brought the problem more to the forefront of attention. Elder abuse in the home is still grossly under-reported because, unlike child abuse where children are more often observed outside the home by neighbors, teachers, and others, the elderly, especially the infirm, just are not as easily observed by outsiders. Elder abuse in nursing homes, however, is being reported with increased frequency and the amount is shocking. The form of elder abuse most readily observable and most frequently reported to the police presently is financial abuse. Scams to defraud the elderly of their often meager savings run rampant in most communities. The Sapps explore crimes against the elderly, clearly demonstrating both the variety and extent to which they exist; predict that most larger police departments, at least, soon will be called upon to establish elder abuse units if they have not done so already; and provide a detailed plan for the implementation of such a special unit.

Chapter 5, "The Gang Unit," by Jeffrey Rush and Gregory Orvis, deals with several important issues. First, many communities are in denial of their gang problems, often prompted by ignorance because the symptoms are not like the stereotypical Los Angeles, which leads to delay in responding to the emerging problem and subsequently to the problem becoming entrenched. Second, Rush and Orvis provide guidance in defining the gang problem in order to facilitate its recognition when the signs appear. Third, they explain different gang unit structures and enumerate the special duties and responsibilities of gang unit officers. Lastly, these authors discuss gang unit operations in great detail. It is here that we gain certain managerial and operational insights regarding procedures that, if followed, likely will reduce agency vulnerability and potential liability for overzealous—and as we unfortunately know from the news, unconstitutional—police behavior.

Bias crime units are the subject of Chapter 6 by Joseph Vaughn. These are acts to which we also commonly refer as hate crimes. As Vaughn points out, however, the technical definition of a hate crime is one motivated by bias—bias must be established by the prosecution—thus bias crime appears to fit better in the legal lexicon favored by the police. The author discusses the evolution of bias crime statutes, including sentencing enhancement provisions, and the Congressional mandate for the FBI to gather statistics regarding this offense. As do most of the authors in this volume, Vaughn dedicates the greatest proportion of his work to a discussion of special unit management and operation. Within the section, he also provides a very useful summary of victim and offender characteristics, and modus operandi clues to alert investigators to the probability of a bias crime.

The subject of Chapter 7 is volunteer units. As Peter Phillips writes, volunteer units take many forms—from sworn personnel in reserve and auxiliary police units to disabled persons writing tickets for handicapped-zone parking violations; from volunteers performing clerical duties at headquarters and mini-stations to citizens recording traffic light violations at selected intersections. Not only do volunteer units contribute many hours of low-cost labor to police agencies, they also are important elements in a department's community policing plan. Because reserve and auxiliary units are discussed at length in Chapter 3, in this chapter Phillips concentrates on the wide range of contributions available from other types of volunteers. While identifying the management issues surrounding deployment of volunteers, it is this author's position that the benefits of well-organized volunteer units in policing far outweigh the liabilities.

Part III of this volume consists of three chapters that describe the operations of method-oriented special units. Here it is important to recall that method-oriented special units are organized based on a narrow or specific set of operational methodologies and may be deployed in a multitude of situations where the unique KSAs of the officers—and often the specialized equipment they use—are essential to the unit's mission.

Robert Taylor, Stephanie Turner, and Jodi Zerba are the authors of Chapter 8, "Critical Incident Teams and Hostage Negotiation Units." These authors conclude that SWAT units, early in their evolution and when at their best, were very important and highly specialized. As time passed, however, their use has become over-generalized to deal with less than critical problems and situations in which the highest levels of force may be deemed excessive. Where a genuine need for such a unit is established, Taylor, Turner, and Zerba strongly advocate the implementation of Critical Incident Teams (CITs) that highlight negotiation strategies over tactical solutions. In keeping with the plan of the book, the authors of this chapter provide detailed information regarding the selection of CIT members, their training, and their supervision. As is truer of certain special units than of others, CIT operations may significantly increase a police agency's vulnerability and potential vicarious liability, therefore making such variables as selection, training, and management most important.

Specialized drug enforcement units are the subject of Chapter 9 by David Olson. Because there is hardly a jurisdiction in the country without a drug problem, one might guess that drug enforcement units are prolific. As Olson demonstrates, however, this is not true. In 1997, only about 15% of all municipal police departments operated an "in house" drug enforcement unit with at least one officer assigned full-time. This may be explained in part by the demographics of policing in the United States—as of 1997, approximately 17,000 police agencies, 78% of which have less than 25 sworn officers and 52% have less than 10 (Reaves & Goldberg, 2000, p. 2)—where personnel resources may preclude officers' exclusive dedication to the drug problem. Olson discusses the structure of both in-house and multi-jurisdictional drug enforcement units and management responsibilities attendant to both forms. One important point about drug enforcement efforts is local dependence on federal funding and its continued availability (which may be dubious), another concerns issues related to asset seizure and forfeiture, both of which Olson discusses in detail.

Early in this chapter, we noted that little had been written about police special units compared to other police topics. In the sparse literature concerning special units, even less has been written about mounted units—horse, bicycle, motorcycle, airborne, and marine patrols. Mitchell Roth fills this void in Chapter 10 by providing an insightful analysis of horse-mounted patrols and bicycle patrols, the management and organization principles of which are clearly generalizable to other mounted units. Horses are most useful in crowd control, parking lot security, and tracking operations in rough terrain (with or without dogs, in search and rescue, and in prisoner apprehension). Roth documents the growth in the use of bicycle units in the past 20 years, particularly as an enhancement to other community policing efforts. Bicycles are especially valuable tools in drug enforcement, campus policing, neighborhood patrols, tourists assistance, and special operations. When special equipment becomes the raison d'être for a special unit, the procurement and maintenance of that equipment adds a fifth tier to the general management concerns for recruitment, selection, training, and supervision of the officers involved.

Part IV of the book contains two chapters. One focuses on the politics of police special units, particularly those in the form of multi-jurisdictional task forces. The other provides a summary of the preceding chapters and develops important conclusions regarding policing and special units.

In Chapter 11, Peter Nelligan, Peter Phillips, and Bill Young discuss two issues that are inextricably interwoven: federalization of the criminal law and its impact on local law enforcement. In recent years, the United States Congress has passed federal criminal laws that, in many respects, are mirror images of state criminal laws already well covered in the penal codes of the states and already well enforced by municipal and state police. Exacerbating this situation is the apparent move by the FBI to take control of these investigations through expansion of its violent crime task forces. Nelligan, Phillips, and Young document the adverse impact this movement has had on local police agencies. When presented with a proposal to join forces with a federal agency in any task force endeavor or to be *required* to "call out" the "feds" (unless exclusive federal jurisdiction is apparent, e.g., bank robbery), should not only "read the fine print," but should "read between the lines." This applies to any "contract," whether violent crime task force, fugitive task force, High Intensity Drug Trafficking Area task force, or other. Not to do so may result in serious personnel resource and command and control issues.

M.L. Dantzker, the Series Editor, is the author of Chapter 12, our capstone chapter. In it, Dantzker reminds us that both the patrol and investigations are general police functions in the majority of police agencies, and the officers assigned to these divisions, themselves, must be generalists. The role of special units in these departments, therefore, is to provide the specialists necessary to deal with the intricacies of particular cases or critical incidents. Not in every case must these be different police officers, only that they must have a particular expertise not generally shared by the average officer.

RECOMMENDED READINGS

General Police Management and Public Administration

FYFE, J.J., GREENE, J.R., WALSH, W.F., WILSON, O.W., & McLAREN, R.C. (1997). *Police administration* (5th ed.). New York: McGraw-Hill.

STARLING, G. (1998). *Managing the public sector* (5th ed.). Fort Worth, TX: Harcourt Brace.

SWANSON, C.R., TERRITO, L., & TAYLOR, R.W. (2001). *Police administration: Structures, processes, and behavior* (5th ed.). Upper Saddle River, NJ: Prentice-Hall.

Police Organizational Structure

MAGUIRE, E.R. (2003). *Organizational structure in American police agencies: Context, complexity, and control*. Albany, NY: State University of New York Press.

Planning and Decision-Making

STRECHER, V.G. (1997). *Planning community policing: Goal specific cases and exercises*. Prospect Heights, IL: Waveland.

SWANSON, C.R., TERRITO, L., & TAYLOR, R.W. (2001). *Police administration: Structures, processes, and behavior* (5th ed.). Upper Saddle River, NJ: Prentice-Hall.

WELSH, W.N., & HARRIS, P.W. (1999). *Criminal justice policy & planning*. Cincinnati, OH: Anderson.

Program Monitoring and Evaluation

BRODEUR, J-P. (1998). *How to recognize good policing: Problems and issues*. Thousands Oaks, CA: Sage.

HOOVER, L.T. (1996). *Quantifying quality in policing*. Washington, DC: Police Executive Research Forum.

—•—. (1998). *Program evaluation*. Washington, DC: Police Executive Research Forum.

MAXFIELD, M.G., & BABBIE, E. (2001). *Research methods for criminal justice and criminology* (3rd ed.). Belmont, CA: Wadsworth/Thomson Learning.

REFERENCES

BAYLEY, D.H. (1996). Measuring overall effectiveness. In L.T. HOOVER (Ed.), *Quantifying quality in policing* (pp. 37–54). Washington, DC: Police Executive Research Forum.

BOONE, L.E., & KURTZ, D.L. (1992). *Management* (4th ed.). New York: McGraw-Hill.

BRODEUR, J-P. (1998). *How to recognize good policing: Problems and issues.* Thousands Oaks, CA: Sage.

DEL CARMEN, R., & WALKER, J.T. (2000). *Briefs of leading cases in law enforcement* (4th ed.). Cincinnati, OH: Anderson.

DUBRIN, A. J., & IRELAND, R.D. (1993). *Management and organization* (2nd ed.). Cincinnati, OH: South-Western Publishing.

DYE, T.R. (1995). *Understanding public policy* (8th ed.). Englewood Cliffs, NJ: Prentice-Hall.

FYFE, J.J., GREENE, J.R., WALSH, W.F., WILSON, O.W., & MCLAREN, R.C. (1997). *Police administration* (5th ed.). New York: McGraw-Hill.

GOLDSTEIN, H. (1990). *Problem oriented policing*. New York: McGraw-Hill.

HOOVER, L.T. (1996). *Quantifying quality in policing*. Washington, DC: Police Executive Research Forum.

——. (1998). *Program evaluation*. Washington, DC: Police Executive Research Forum.

KAPPELER, V.E. (2001a). *Critical issues in police civil liability* (3rd ed.). Prospect Heights, IL: Waveland Press.

——. (2001b). *Police civil liability: Supreme Court cases and materials*. Prospect Heights, IL: Waveland Press.

KOCIENIEWSKI, D. (1999, February 15). Success of elite police unit exacts a toll on the streets. *The New York Times*, p. A1.

KRASKA, P.B., & CUBELLIS, L.J. (1997). Militarizing Mayberry and beyond: Making sense of American paramilitary policing. *Justice Quarterly, 14*(4), 607–629.

KRASKA, P.B., & KAPPELER, V.E. (1997). Militarizing American police: The rise and normalization of paramilitary units. *Social Problems, 44*(1), 1–18.

KRASKA, P.B., & PAULSEN, D.J. (1997). Grounded research into U.S. paramilitary policing: Forging the iron fist into the velvet glove. *Policing and Society, 7*, 253–270.

LAIT, M. (1999, December 26). 1999: The year in review. *The Los Angeles Times*, Record Edition, p. 1.

LANGWORTHY, R.H. (1986). *The structure of police organizations*. New York: Praeger.

LORSCH, J.W. (1987). Introduction to the structural design of organizations. In L.E. BOONE, & D.D. BOWEN (Eds.), *The great writings in management and organizational behavior*. New York: McGraw-Hill. (Original work published 1970.)

MAGUIRE, E.R. (2003). *Organizational structure in American police agencies: Context, complexity, and control*. Albany, NY: State University of New York Press.

MANNING, P.K. (1997). *Police work: The social organization of policing* (2nd ed.). Prospect Heights, IL: Waveland.

MAXFIELD, M.G., & BABBIE, E. (2001). *Research methods for criminal justice and criminology*. Belmont, CA: Wadsworth/Thomson Learning.

MILLS, S., & HANNA, J. (2002, April 29). Timing is right for torture probe. *Chicago Tribune Internet Edition*. Retrieved May 3, 2002 from the World Wide Web: http://www.chicagotribune.com/news/local/chicago/chi-0204290183apr29.story.

PEAK, K.J. (2001). *Justice administration: Police, courts, and corrections management* (3rd ed.). Upper Saddle River, NJ: Prentice-Hall.

PHILLIPS, P.W. (1987). *Task analysis report on the job of adult probation officer*. Huntsville, TX: Sam Houston State University, Criminal Justice Center.

———. (1999). De facto police consolidation: The multi-jurisdictional task force. *Police Forum*, *9*(3), 1–8.

REAVES, B.A., & GOLDBERG, A.L. (2000, February). *Local police departments 1997*. Washington, DC: Bureau of Justice Statistics [Monograph: NCJ 173429].

SPARROW, M.K., MOORE, M.H., & KENNEDY, D.M. (1990). *Beyond 911: A new era for policing*. New York: Basic Books.

STARLING, G. (1998). *Managing the public sector* (5th ed.). Fort Worth, TX: Harcourt Brace.

STRECHER, V.G. (1997). *Planning community policing: Goal specific cases and exercises*. Prospect Heights, IL: Waveland.

SWANSON, C.R., TERRITO, L., & TAYLOR, R.W. (2001). *Police administration: Structures, processes, and behavior* (5th ed.). Upper Saddle River, NJ: Prentice-Hall.

TERRITO, L., HALSTED, J.B., & BROMLEY, M.L. (1998). *Crime and Justice in America* (5th ed.). Boston: Butterworth-Heinemann.

WEBER, M. (1987). Legitimate authority and bureaucracy. In L.E. BOONE, & D.D. BOWEN (Eds.), *The great writings in management and organizational behavior*. New York: McGraw-Hill. (Original work published 1947.)

WELSH, W.N., & HARRIS, P.W. (1999). *Criminal justice policy & planning*. Cincinnati, OH: Anderson.

2

Special Units in Policing

Functionality v. Legitimacy Maintenance

Lorie L. Rubenser

INTRODUCTION

In 1995, the Omaha Police Department (OPD) received 7775 calls concerning dead storage, abandoned or illegally parked vehicles to their 911 system. This represented 3.31% of the total calls to the 911 system (Omaha Police Department, 1998). When calls to the Mayor's Hotline are considered, the problem increases in importance as it is estimated that the hotline received approximately 1000 calls per month for these problems.

This study examines how the police unit assigned to handle these complaints, the Nuisance Task Force (NTF), actually performs its task. The questions investigated are threefold: (1) whether this specialized unit engages in activities related to its official mandate; (2) whether officers assigned to the unit believe they are engaging in a legitimate police function, and (3) whether community residents, as consumers of the unit's services, are aware of and have positive attitudes about the unit's activities. The study explores Crank and Langworthy's (1992, pp. 343–344) argument that many specialized units are essentially "ceremonial", and simply create the appearance of accomplishment for the sake of preserving the legitimacy of the police department rather than contributing to the effectiveness or efficiency of a police department.

A recent study by Katz (1997) of a different specialized unit within the same police department also examines the activities of a specialized unit (the gang unit). Katz found that the majority of the officers' time was spent on behaviors not related to the functioning of the unit. It also found that the unit did not enjoy full legitimacy in the eyes of officers assigned to it. Finally, public controversy over the issue of gangs indicated that the unit had not achieved legitimacy in the eyes of public consumers of the unit's activities. These findings suggest the need to carefully assess the behavior of specialized units in order to determine if they serve an actual functional purpose or exist for the purpose of maintaining organizational legitimacy as argued by Crank and Langworthy (1992, pp. 343–344).

Organizational legitimacy is an important concept to understand as it underlies many of the behaviors in which organizations engage. Organizational legitimacy is the process of creating public support for the organization by making the organization's behaviors seem appropriate or right. In a police organization, reliance upon the rule of law is a major source of this legitimacy. It gives the public a much needed reason to accept police authority as appropriate (Kelling & Moore, 1995; Reed, 1999; Walker, 1984).

This study examines the extent to which the NTF has achieved legitimacy as measured by (1) the observed activities of the unit, (2) the attitudes of the officers assigned to the unit, and (3) the perceptions of the consumers of the unit's services.

THE HISTORY OF THE NUISANCE TASK FORCE

Originally, when the problem of abandoned vehicles became an issue for the city of Omaha, two workers from the city impound lot were assigned to handle all of the vehicle complaints in the city. All vehicle complaints that came to the police department and all such calls that came in to the Mayor's Hotline were referred to these two workers. Upon the discovery that these two employees were not effective in addressing this problem, the impound supervisor wrote their jobs out of the annual budget and their positions were thus eliminated.

When these positions were eliminated, the police were forced to take responsibility for handling all vehicle complaints. The calls were divided up into the four precinct divisions and forwarded to the captains of these precincts. The captains would then distribute the calls to officers in the patrol division to handle. These calls were handled on top of all the other calls that the officers were supposed to respond to in their daily work. It quickly became apparent that the officers did not have adequate time to respond to all of the complaints about vehicles due to the volume of 911 calls to which they were dispatched.

In January of 1997, after two years of researching the issue, the NTF was created by OPD. Each of the four precinct captains donated two officers to this new unit. In addition, Mayor's Hotline personnel, the City Planning Department, the Municipal Board of Appeals, and the city tow lot personnel were brought on board to engage in a cooperative effort.

The task force was originally designed as a six-month project. The idea was to have a unit of officers devoted to towing cars, thus freeing regular patrol officers from having to wait for the tow truck. The unit would conduct a total door-to-door sweep of the city, handling all violations. Once this was accomplished, it was thought that the unit would no longer be needed. However, the sheer volume of complaints that continue to occur (approximately 1000 per month) has led to the unit becoming permanent.

The Officers

The unit is currently composed of seven sworn line officers, one sergeant, one lieutenant, and one secretary. Additionally, officers from other units, especially those assigned to work in local schools, occasionally work with the NTF officers. The unit is housed in the basement of the community store-front station and therefore is not in immediate contact with the majority of other officers and units in OPD.

The officers of the NTF do reflect some of the diversity of the communities they serve. There are currently three women officers and four men. One of the women is African American and one of the men is of Korean descent. The rest of the officers and both the sergeant and lieutenant are white males. The unit secretary is a white female.

Four of the seven officers have been with the unit from the beginning and only two have been with the unit for less than one year. The average length of time with OPD is nine years, with a low of six years and a high of 13 years. The current sergeant for the unit has been with OPD for 30 years. All of the officers except one were assigned to the unit voluntarily, with one essentially forced to take this assignment.

Each officer in OPD must spend a minimum of three years in the Patrol Bureau before they are eligible to be assigned to a specialized unit. Three of the officers currently assigned to the NTF have been assigned to other specialized units before this one. Two of the officers came from the unit assigned to the public schools. The other officer had spent time with both the bicycle patrol and the Weed and Seed units.

SPECIALIZED UNITS

Many policing strategies have not changed significantly since the time of Robert Peel: The overall directive of the police still is crime control. Their second task is apprehending offenders. The traditional strategy of policing, commonly referred to as the professional model, emphasizes such tactics as preventive patrol, rapid response to calls, investigations and identification of offenders, and evidence development (Alpert & Moore, 1997, pp. 265–266; Kessler & Duncan, 1996). The search for effectiveness in their response to crime has caused the police to consciously narrow their focus and refine their skills to cover these areas (Alpert & Moore, 1997, p. 267). This search has also caused many departments to create special units whose mandate is to respond to a myriad of problems, which can plague a city or neighborhood.

Specialized units in American policing emerged around the turn of the century. New specializations included criminal investigations and juvenile units among others. These units were cited by mid-century reformers as a sign of professionalism in the police department (Walker, 1999; Woods, 1991). These units also represent a fast and easy way for a police department to implement new ideas without having to totally restructure their organization (McGarrell, Langston, and Richardson, 1997).

The significance of these units goes beyond professionalism. They provide the department with important sources of legitimacy in the communities they serve. This legitimacy is accomplished through providing the appearance of, if not a substantive response to, the special or intense problems of the community (Brooks, 1997; Crank & Langworthy, 1992, pp. 343–344; Langworthy, 1986; Lipsky, 1998: 36; Trojanowicz & Bucqueroux, 1990).

Adaptations and Crises

OPD is a large public organization. As such, it is subject to pressure from the public concerning how it does its work. OPD is also like other public organizations in that it is forced to respond to low-level crisis situations when they arise. Unlike private organizations, the police cannot move to another city or county to conduct their business, and because they are an institutionalized organization, society cannot simply abolish them. While this may insulate the police from some pressure, in order to achieve support for their actions some response to public pressure must be made (Crank & Langworthy, 1992). These responses promote support by making the police appear legitimate in the eyes of the public.

For OPD, the volume of complaints about vehicles can be conceived of as a low-level crisis situation. There is great pressure on the department from both the public and the mayor's office, which forces them to implement a response to this issue. In order to explore the implementation process OPD undertook in setting up the NTF, it is necessary to employ aspects of institutional theory.

Institutional theory deals with the ways that the external environment of an organization affect its internal structure. The forces of the external environment which contribute to the shape of the organization include rules and regulations, which may take the form of laws, and public opinion, which often emerges as communication from interest groups (DiMaggio & Powell, 1983; Scott, 1995; Zhao, 1999).

There are two core components to institutional theory; the normative, and the cognitive. Normative components refer to the values of an organization and the rules, or norms, which are in place to ensure that the actions of the organization conform to these values. Rules and regulations may also be imposed upon the organization from external sources through passage of laws (Meyer & Rowan, 1977; Scott, 1995; Zhao, 1999).

The cognitive components of the theory deal with the process of developing shared meanings, or definitions, during interactions. As the organization interacts with its environment, it can create meanings or can have meanings thrust upon it. It is through the creation of meanings that an organization acquires or maintains its legitimacy (Scott, 1995; Zhao, 1999).

The organization experiences pressure for action through its interaction with the external environment. Organizations in a democratic society such as the United States should respond to these pressures in order to accurately represent the changing interests of the public. Changes in the external environment can create situations where an organization must either adapt to the new situation or risk becoming obsolete. In the case of a public organization like the police, extinction is not an issue but a loss of legitimacy can be. It will also be easier for the organization to maintain efficiency and high levels of performance if the organization can find a good fit between itself and its environment (Meyer & Rowan, 1977; Wamsley & Zald, 1973; Zhao, 1996, 1999).

Political issues also force an organization to adapt to changing situations in the external environment. Budgets, personnel and equipment allocations, and support are all affected by changes in political power. Legitimacy in the eyes of the politically powerful can protect the organization from many changes such as budget cuts (Wamsley & Zald, 1973). Police Chiefs can also be replaced (within civil service limits), thus adding some pressure from the top of the organization to conform to environmental expectations.

Continuous changes in the external environment can create uncertainty for an organization. This uncertainty arises from a lack of information and/or a lack of predictive ability about the nature of environmental changes. Pressures to adapt to new situations come from many sources in the external environment. Some of these pressures will be more relevant than others and therefore are more important for the organization to heed. Pressure is also often in conflicting directions, and therefore the organization suffers from uncertainty as to which direction to take (Stojkovic, Kalinich, and Klofas, 1998). In addition to the pressures discussed above, there are often problematic situations that emerge either as a new phenomenon or as the result of a shift in priorities. The ability to address these problems effectively is an important source of legitimacy.

A police department which faces this situation may find itself failing to meet community demands for its services. This failure occurs due to the inability of the organization to predict or plan for changes in demand. Often the organization will suffer from inadequate resource allocation and/or support to adjust to these changes and will thus be incapable of adapting to the new situation that it faces (Stojkovic et al., 1998).

Where issues are complicated, an organization will often undertake programs designed to influence the public to support it. These programs are not necessarily reflective of either the real work of the organization or the true needs of the public being served by the organization. They do, however, give the organization a public image portraying the organization as modern or even "cutting edge" (Meyer & Rowan, 1977).

The organization will also employ special rituals or visual symbols designed to reinforce the value of its activities (Stojkovic et al., 1998). Many of these rituals hold a legitimacy status in the external environment even if they do not actually contribute to the effectiveness or efficiency of the organization. Rather, these rituals demonstrate to the external environment that the organization is acting on the collective values of the environment and further, they provide a way for the organization to account for its activities (Meyer & Rowan, 1977). In other words, the organization can say that it is in touch with the values of the environment and it is doing "x" in support of these values. The creation of specialized units in policing also fits here.

Police departments create many special units such as the Gang Task Force, Drug Abuse Resistance Education, and the NTF. These units create a mechanism through which the department can visibly show that they are focusing some part of their attention on a particular problem. As demonstrated by Katz's (1997) study of the Gang Task Force, however, these units do not necessarily engage in any substantive work related to the mission of the organization, but exist only for symbolic purposes. (It may be true, also, that some police agencies form special units solely because federal or state funding, directly or through grants, is available to maintain them.)

Specialized units may serve only to protect the core activities of the police department from scrutiny by the actors in the external environment. In other words, they serve as window dressing, which deflects attention from the "real" work of the department. In this fashion, the organization can attempt to exert control over the environment and therefore become less dependent on that environment (Stojkovic et al., 1998). By thus decoupling these special units from the core work of the department, the police department is able to reach a broader section of the public they serve while still maintaining their image as crime fighters (Crank & Langworthy, 1992; Meyer & Rowan, 1997).

This problem also occurs because legitimacy and efficiency may require different approaches. The most efficient means of accomplishing the work of the organization is not always the best means for creating legitimacy (Meyer & Rowan, 1977). The reverse is also true. A police department may create a program such as DARE in order to appear legitimate. This program does not seem to have any real connection with the efficiency of the department in accomplishing their crime control mission.

The solution to the problem of legitimacy versus effectiveness is reflected in the loose-coupling approach presented by institutional theory. Loose-coupling involves the deliberate restructuring of an organization in order to gain legitimacy while at the same time preserving the integrity of the daily work activities. For an example, consider the foot patrol program of the Newark Police Department. The program was immensely popular with the public, but had little or nothing to do with the crime control mission of the department. Because the program was operated through a special unit, it also provided a minimal amount of interference with the daily activities of the regular patrol officers (Police Foundation, 1981; Zhao, 1999).

Another problem that may confront an organization is a lack of clear goals. The ambiguity present in this situation leaves the organization open to pressure from many outside forces, some of which may be conflicting. Along with this problem is the issue of effectiveness in reaching goals. Theoretically at least, the more effective the organization is, the less pressure. A crisis situation can create a sudden amount of pressure, thus forcing the organization to react to the crisis effectively in order to reduce this pressure (Wamsley & Zald, 1973). This situation also presents an opportunity for the organization to engage in the loose-coupling approach.

Not all changes in the external environment will lead to adaptations, however. Changes that result in adaptations do so for several reasons, most of which relate ultimately to organizational legitimacy. These reasons include the provision of future resources and support from powerful actors outside of the organization. Public agencies are particularly vulnerable to this problem as they are often overseen by another agency which may control their future budget and personnel allocations (Wamsley & Zald, 1973). Essentially, the public organization is looking to acquire permission (or an "approval license") to operate within the external environment (Zhao, 1999).

The first thing an organization must do when confronted with a problem is to determine what, if anything, is to be done about it. If the organization is going to do something, the next step is to develop and implement a response strategy. Implementation was the focus of Pressman and Wildavsky's (1973) book on the Economic Development Administration's (EDA) efforts to address minority employment in Oakland, California. The book describes a three-year attempt to implement a variety of programs to promote minority employment and the problems that plagued this attempt. Of particular relevance to the current study is the discussion of the problems stemming from the attempts to coordinate multiple actors.

EDA can be conceived of as facing two crises at the same time. The first is a budgetary problem. In order to meet their budgetary requirements for the next fiscal year, EDA had to spend $23 million in a very short time period. Oakland became the perfect project as there were already plans and contacts with relevant personnel in place. This would save EDA a great deal of time in the planning phases. The second crisis EDA was facing was the possibility that riots would occur in Oakland. This possibility applied pressure to EDA to ensure that whatever programs they implemented would be effective, not just for the sake of success, but for the sake of the city itself (Pressman & Wildavsky, 1973).

In accomplishing the implementation of their programs, EDA was attempting to deal with businesses, city and state agencies in California, and other federal agencies. The first problem was in getting all of the actors to agree on the goals of the overall program and the responsibilities each would have. This became an ongoing problem as leadership changes in the relevant agencies introduced new priorities and new issues. For example, original EDA leadership gave the Oakland project top priority, but after a change in leadership the project was relegated to a more back burner status (Pressman & Wildavsky, 1973). In part due to these coordination problems, the project was forced to undergo changes in its goals and plans. This in turn forced changes in implementation strategy. This shows that implementation is not always a smooth process. Further, it shows how implementation often occurs in stages, particularly as circumstances change.

Another good example of the implementation process can be found in Selznick's (1966) analysis of the Tennessee Valley Authority (TVA). In this analysis he shows that TVA, like EDA, was forced to modify many of its original agricultural program plans in the face of pressure from outside groups. By adjusting these plans, TVA was able to gain support from the external environment for its other programs.

The fact that TVA adapted to the preferences of the local community is not so unique as the strategies it was using to develop and implement its plans. TVA operated with a grass roots ideology, meaning that it was organized around the goal of promoting the interests of the local community rather than state or federal interests. This strategy in fact may have placed more pressure on TVA to adapt to changing circumstances (Selznick, 1966).

Both of these examples of implementation demonstrate the convoluted nature of the process. They provide a guide for studying other implementation efforts by showing the necessity of defining the steps of the process. Also, the actors in the external environment who create the pressure that causes the changes in the implementation process and the means by which they create this pressure must be identified.

As mentioned in the sections above, organizations use symbols and rituals as ways of achieving and maintaining their legitimacy. These symbols and rituals give the appearance of organizational concern for the demands and requests of actors in the organization's external environment (Scott, 1995; Zhao, 1999). They may not, however, be truly reflective of the work of the organization.

As this section demonstrates, institutional theory provides an explanation for the presence of special units within a police department. This explanation centers on the idea that these units may only provide legitimacy in the eyes of the external environment, as opposed to performing a real activity related to the core mission of the organization. Police departments are similar to other public agencies in their need for support from clientele. They therefore engage in a variety of activities designed to achieve this support. Among these activities one finds the implementation of special units.

METHODOLOGY

Research Questions

The present research has been undertaken to explore several research questions concerning the work of specialized units. The theoretical perspective is provided by Crank's and Langworthy's (1992) institutional perspective on law enforcement agencies, and in particular their argument that specialized units are largely ceremonial, do not contribute to the effectiveness or efficiency of the organization, and exist largely to create the appearance of legitimacy in the context of the organization's external institutional environment.

The first question involves the day-to-day work of the NTF, and whether those activities are related to the unit's official mandate. This question includes issues related to the process by which OPD created the NTF in response to the problem of junk vehicles in the city.

The second question involves the attitudes of officers assigned to the Task Force regarding the function of the unit and their own activities. Do these officers regard the unit as having legitimacy in terms of the role of the police department? Or, as Crank and Langworthy (1992) suggest, do they perceive the unit to be largely ceremonial?

The third question involves the perceived legitimacy of the unit in the eyes of relevant external institutions. Do these external institutions perceive the unit to have legitimacy? Most important, are the principal consumers of the unit's activities aware of its activities and do they have a positive perception of its effectiveness?

Data Collection

In order to address the questions listed above several types of data were collected. As Babbie (1992) has indicated, use of several different methods allows the researcher to take advantage of each method's strengths while overcoming the weaknesses. This provides the research with increased validity by providing the ability to double check the results acquired through each measure against those of other measures.

This is particularly important where qualitative measures are used, as they are here. The validity of observations like those used here is often questioned due to the potential for researcher bias to affect these results (Babbie, 1992). This problem is overcome here by combining these observations with quantitative measures.

Three different sets of data are utilized in this study. The first deals with the activity of the unit, as measured through both observations, interviews, and official records. The second data set addresses the perceived legitimacy of the unit through interviews with the officers assigned to the unit itself. The third and final data set looks at legitimacy as seen through the eyes of external agencies and consumers. This data set is made up of interviews with these agencies and with the community groups that the unit serves. These data sets are described in detail in the sections that follow.

UNIT ACTIVITIES

A description of the unit itself and its internal workings was the first step in the research process. This was accomplished through an examination of official documents related to the unit, the collection of quantitative data concerning unit's workload, and through interviews and direct observations of the unit in operation.

Unit Work Load   Data were collected on the number of complaints received about vehicles and the number of tows completed by the unit officers during the two-month period of January and February, 1999. This information has been disaggregated by precinct areas so as to determine patterns of complaints throughout the city.

The data were collected from the unit officers' daily reports, as this is the only place where the work is broken down into small enough categories to be useful for the current purposes. There is no other record source that provides the ability to break down the work into sections of the city. The reports contain information on addresses, complaint type, mayor's complaint number, and disposition of the complaint.

One of the problems with using this source of data is that the officers may either lie or file incomplete reports, which would cause the data gathered to misrepresent the reality of the situation. This problem is minimized in several ways.

The first way in which the problem of inaccuracy is minimized is through review of records. If the report is incomplete, the sergeant may return it to the officer for completion. The sergeant also could make field visits to observe the officers at work, thus giving him a method of ensuring that the reports were accurate.

The other officers also provide an accuracy check as they often work together and occasionally follow-up on each other's initial work. If the reports are inaccurate or incomplete, it is impossible for another officer to effectively complete the task. While this check is not explicitly discussed, it operates informally to promote accurate and complete reporting. Often this check occurs at role call where one officer may ask another about a case. If an officer did not file accurate reports, he/she must rely on memory to discuss the case. Lies would become complicated, and incompleteness would make an officer look sloppy. Thus, it is in the officer's best interest to file accurate and complete reports.

Also, the unit secretary may question an officer if the reports are incomplete. The secretary is responsible for entering completed complaint forms into the computer system and also for generating complaint forms for cases that are officer initiated. If she cannot understand the report or cannot find all the information she needs to do this job, she will speak to the officer, thus enforcing the need to be complete and accurate the first time.

The nature of the work provides further pressure for accuracy. The work of the unit is the result of a complaint. If this complaint is ignored, it may be repeated. Also, the owner of the vehicle that is complained about may call in to protest the actions of the officer. In most cases, these types of complaints result from a misunderstanding of the law rather than the actions of an officer. Accurate reports strengthen the officer's case if questions over conduct do arise.

Quantitative data were also collected from the tow lot facility. These data revealed the percentage of vehicles towed to the facility by the police which were the result of action by the unit's officers. It also allowed for a check on the reliability of the officer's daily reports concerning their towing activity.

Observations   At least one ride-along was conducted with each officer in the unit. Additional ride-alongs were conducted so that a sample could be collected for each day of the week. Approximately 200 hours of observations were undertaken. The purpose of the observations was to determine whether or not the officers were engaging in activities related to their official mandate. That is, were they spending all, some, or only a little of their time dealing with problem vehicles. If the officers spend all or most of their time dealing with problem vehicles, this would provide some evidence that the unit was both effective and efficient. If, on the other hand, officers spend little or no time dealing with problem vehicles, this would provide some evidence that the unit is ineffective.

UNIT OFFICERS

Interviews were conducted with each of the seven officers in the unit as well as with the sergeant and lieutenant. The original interviews with each officer were conducted at the same time as the first ride-along with that officer. Follow-up questions were handled during later rides.

These interviews followed a semi-structured format, with each officer being asked the same questions, but allowing for individualized questions as well. Interview questions dealt with issues such as officer perceptions of the job, the other officers in the unit, and the response from the civilian community. These questions were designed to uncover both the officer's feelings and the underlying reasons for those feelings; i.e. Do you like working for the unit? Why? Do you consider the work the unit does important? How so?

Incorporating the interviews with the ride-alongs served two purposes. The first was to save time. The officers did not have to take up either work time or after hours time for the interviews but instead could get on with their daily routine with a minimum of disruption. The second purpose was to allow the interviewer/observer the opportunity to tailor the interview to the work experiences of the officer. Questions could be added to the interview in order to capture nuances of the job that became apparent during the course of the observation period. This process followed the suggestion of Mastrofski and Parks (1990), that researchers conducting observational studies ask questions immediately after the actions that are witnessed rather than trying to recall the incident later for interview purposes. This keeps the situation fresh in the minds of both the observer and the observed.

EXTERNAL AGENCIES AND CONSUMERS

Other Government Agencies   The research also focused on actors who are external to the unit. Each of these actors either works with the unit or is serviced by the unit. The goal of these interviews was to determine perceptions of the usefulness/effectiveness of the unit, satisfaction with the work done by the officers, and the extent to which various outside persons have involvement with the unit.

The City Zoning Inspector, the Mayor's Hotline supervisor, and the leaders of organized community groups were interviewed. Telephone interviews were used in each case with the exception of the City Zoning Inspector. This procedure was employed due to the difficulty in scheduling appointments and also due to the brief amount of time needed for each interview.

The City Zoning Inspector was interviewed in person as per his request. This enabled him to use visual aids in demonstrating how he functions in relation to the NTF. Additionally, the City Zoning Inspector was subjected to observations during the course of the study as he often works in the field with the officers of the NTF.

Community Groups   Organized community groups were also interviewed during this research project. A list of all of the organized neighborhood associations in the city limits and their presidents was acquired from the police department. This list contained 164 community associations and was used as the population pool for the interviews. Approximately one-third of the list was used as the sample. This sample was selected by taking every third entry from the list. Where that entry could not be used, the next one down on the list was to be substituted.

ANALYSIS

As mentioned, several types of data were collected for this study. In part this was due to the variety of research questions this study is designed to answer and in part it was due to the desire to paint the most accurate picture of the NTF as possible.

Each type of data provides a piece of the puzzle that this research is designed to solve. The observations revealed what the officers do and how they do it. By combining these observations with the interviews, common meanings were developed concerning the work, and therefore the observations gained additional validity.

The quantitative data supplements the observational data and reveals just how much work the officers do. These data also make it possible to determine if the work being done by the officers conforms to their official mandate. This was made possible by breaking down the categories of work the officers are to perform and then measuring the percent or number of cases the officers handle that fall into each of these categories.

The interviews with actors outside of the task force serve to reinforce the findings of the other methods by showing whether these outsiders believe the officers are accomplishing the work they are mandated to undertake. These interviews also measure the level of satisfaction the outside actors have with the unit. This gives a sense of the value the unit holds for the community and the other city agencies.

Finally, the interviews with the officers and the outsiders reveal the history of the unit. This history demonstrates the implementation process and the legitimacy that the officers feel the unit commands.

FINDINGS

Workload Data

Data of two months were collected concerning the activity of the NTF. These data were collected from the daily reports that officers are required to fill out for every shift they work. These forms list the addresses of the vehicles that were the subjects of the complaints and the action taken by the officers in reference to them.

One of the things that can be determined from these data is the geographical distribution of the calls handled by the task force officers. Due to the small number of calls occurring in some areas, the data have been disaggregated into the four primary precincts of the city rather than by beat. The four precincts are labeled Northwest, Northeast, Southwest, and Southeast.

As Figure 2-1 indicates, all cases handled by the NTF officers during the month of January, 1999, occurred in two precincts; Southeast and Northeast. This finding does not indicate that there were no cases in the other two precincts. In examining the daily report sheets filed during this month it becomes apparent that the officers who generally work in the two other precincts did not work during this month.

The February data that appears in Figure 2-2 reveals that cases were handled in all four precincts of the city. The majority of cases continued to occur in the Southeast and Northeast sections however. This can be interpreted in several ways. As the officers have freedom in deciding where they are to work on a given day, it might at first appear that the officers simply ignored the missing two precincts in January. However, when the data are combined with the observations of the officers it becomes apparent that this pattern does reflect the distribution of calls for service.

Additional evidence supporting this pattern of usage can be seen when looking at the pattern of calls to 911 for other police business. When 911 calls are broken down into precincts as has been done to the task force cases, the Southeast and Northeast precincts emerge as the heaviest users of police services (Omaha Police Department, 1998).

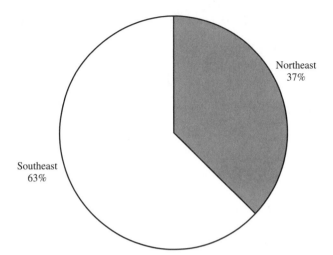

FIGURE 2-1 Nuisance Task Force activity by area, January, 1999.

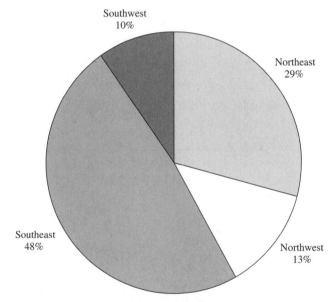

FIGURE 2-2 Nuisance Task Force activity by area, February, 1999.

Finally, evidence of the accuracy of this pattern can be found in the survey answers from the organized community association presidents. While the surveys were not labeled with identifiers, the researcher did notice that the presidents of associations located in the Northwest and Southwest areas were more likely to reply that they had never had occasion to use the services of the unit, nor did they anticipate having such occasion. They felt that they did not have vehicular problems that the task force could address.

Figures 2-3 to 2-6 indicate the type of activities that the task force officers engaged in during the months of January and February, 1999, and the number of incidents of each type. Figures 2-3 and 2-4 indicate the breakdown of the cases that the task force is officially mandated to handle plus a miscellaneous category which covers the work the officers engaged in that was not part of their official unit mandate (e.g., answering radio calls and

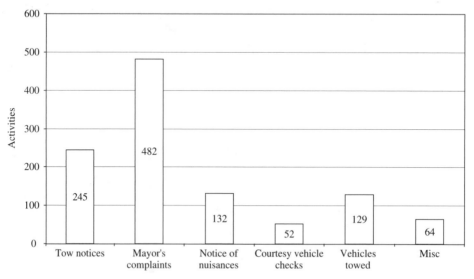

FIGURE 2-3   Nuisance Task Force activity, January, 1999.

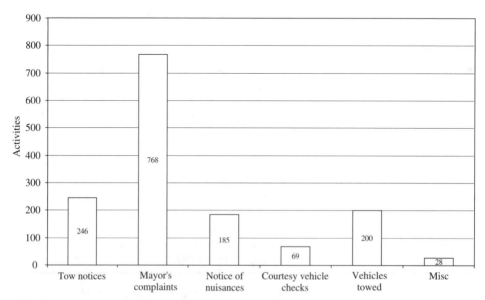

FIGURE 2-4   Nuisance Task Force, February, 1999.

writing parking tickets). The Mayor's complaint category in these figures deals with cases the officers handled because there was a call to the Mayor's Hotline. The other categories may or may not have been the result of a call to this hotline as the officers are free to initiate activity where they see fit.

One important thing to note from these figures is that the task force does appear to be doing the work that it is officially mandated to do. This does not imply, however, that the officers handle all of this work. If the Mayor's Hotline staff forwards roughly 1000 calls to the unit each month, the unit does not appear to handle their workload. In January they appear to have handled almost half, and in February almost two-thirds of these calls.

Tow Lot Data   An additional source of quantitative data that is used here to address the issue of workload comes from the tow lot. Personnel from this agency keep records on the total number of vehicles towed to their facility by officers from OPD. This data can be broken down into categories that indicate the percent of this total that can be attributed to the NTF officers. Data for January and February, 1999, are summarized in Figures 2-5 to 2-8.

As Figures 2-5 and 2-6 indicate, the entire police department towed 305 vehicles to the tow lot during the month of January, and 371 during February, for a total of 676. Officers from the NTF were responsible for 42% of the January tows and 54% of the February tows (n = 128 and 200, respectively). As can be seen, the NTF is responsible for a significant portion of the total towing activity engaged in by OPD.

Vehicles towed by the NTF would fall under the categories of unregistered, motor vehicle litter, dead storage, nuisance, and abandoned. It is possible that officers who were not members of the NTF completed these tows; however, the general practice is for patrol officers to refer these problems to the task force officers. The percentage breakdown of NTF tows in these categories is presented in Figures 2-7 and 2-8. As the figures show, the largest category during both months was unregistered vehicles.

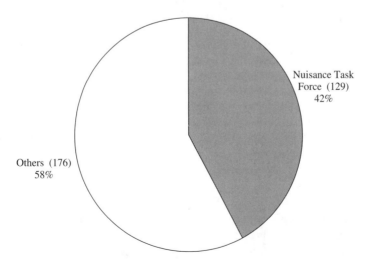

FIGURE 2-5   Nuisance Task Force, total vehicles impounded, January, 1999.

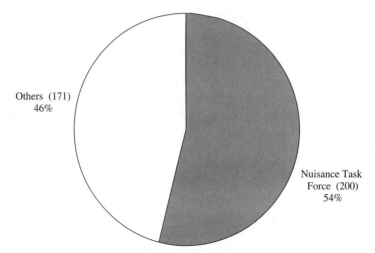

FIGURE 2-6   Nuisance Task Force, total vehicles impounded, February, 1999.

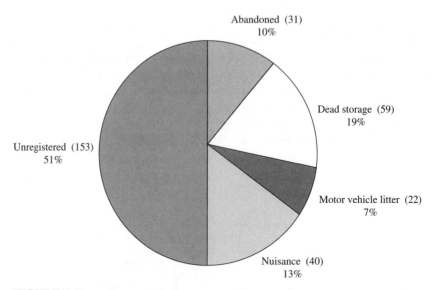

FIGURE 2-7   Nuisance Task Force, types of impound, January, 1999.

Observational Data

The section that follows will summarize the observations into a description of the typical behaviors that occur during an NTF workday. These observations will include the other city agencies which the task force normally interacts with on any given day. This will be followed by some specific examples, which demonstrate the variety of cases the officers will be called upon to handle.

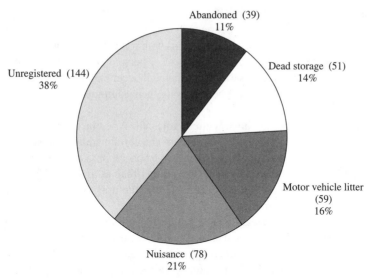

FIGURE 2-8 Nuisance Task Force, types of impound,
February, 1999.

The Mayor's Hotline

The Mayor's Hotline was established in January 1995, when the new mayor took office. It serves as a resource for people throughout the city of Omaha to call with complaints about city services and other problems they may be having. A great deal of the work done by the hotline staff is referring people to the appropriate city agency for their specific problem.

The hotline is staffed by two full-time and two part-time employees who answer phones, take complaints, and make referrals. Approximately 1000 calls are received each week. Of these calls, around 250 deal with nuisance vehicle complaints. This is the entry point for most of the complaints that are handled by the NTF.

Each vehicular complaint that comes in is checked in a database maintained by the hotline staff. If the complaint has already been recorded, the caller is informed and no duplicate paperwork is produced. If the complaint is new, the address, description of the vehicle, and the reason for the complaint are entered into the database. Approximately 75 such reports are bundled up and sent through inter-office mail to the task force each day.

The Nuisance Response Process

The officers of the NTF work seven days a week on the day shift. They use a rotating shift schedule so that at least one officer is on duty every day. When the officers arrive at work each day they select the area of the city they wish to work in and take the appropriate section of complaints. They then organize the complaints according to the route they wish to take for the day. Many of the officers have a specific section of the city where they prefer to work. The unit is fortunate in that there are enough officers whose preferences differ that the entire city is covered without assigning areas. Due to this naturally occurring breakdown, the sergeant never indicates that an officer should work in a certain section of the city for that day (although such supervisory action well may be necessary in the future).

Monday through Friday, the day begins with roll call at 7:00 a.m. For the NTF, roll call is an informal event. There are no equipment or personnel inspections. The officers simply gather around a table to discuss the events that will occur that day. Roll call is so informal that many times the officers will still be gathering their equipment and sorting their paperwork while it is occurring.

The sergeant does not always attend role call. When he is present he will often ask an officer what he/she knows about the problem at a certain address or ask an officer to check into a problem. The requests by the sergeant to check on a problem are phrased more as requests that could theoretically be refused rather than as an order. This is in keeping with the informal nature of roll call. The sergeant also attempts to give complaints to the officer who prefers to work in that section of the city.

Additionally, the sergeant makes no demands of the officers concerning which area of the city needs to be focused on or what they will do with their time while out on the street. Only when there is a serious backlog of complaints or when someone such as a city council person has complained will he indicate that more work needs to be done in a certain area or that they are behind on handling complaints. These comments are directed at the entire unit rather than one person, and are general in nature rather than an order to handle something. The officers are trusted to use their discretion to get the job done in a timely fashion. Once roll call is completed the officers filter out of the office and onto the street and working by 8:00–8:30 a.m.

Each officer checks in with dispatch once on the street. Calling in to dispatch is meant to insure that dispatch is aware of which officers are on the street at any given time. If an officer had an emergency and dispatch was not aware that this officer was on the street, valuable time could be wasted searching. This also allows for dispatch to assign calls to the officers in this unit if it is necessary. Under normal circumstances this would not happen as the task force officers are freed from answering radio calls, but there are exceptions or emergencies that may arise and it is important that dispatch know these officers are on the street and available if needed.

Once on the street, the officers will proceed to the first case for the day. This address may be related to a new complaint from the mayor's hotline, or may be a follow-up from previous action taken by the officer. It is at this point that each day begins to take on its own character. Even though the officers theoretically do the same thing every day, each day presents its own unique pattern. The following sections will lay out the range of possible activities the officers are faced with in their work. It is important to recognize that not all scenarios present themselves to all of the officers every time they go onto the street. There will be some days where only one or two of the scenarios present themselves and other days where there will be greater variety.

The simplest or least time-consuming scenario is when the vehicle in question is no longer present. This can occur when the officer is handling a new complaint or when the officer has already been to the address and informed the owner of the vehicle of the problem. If the vehicle is gone, the officer simply makes an entry onto the complaint sheet and the daily report and moves on. Removing the vehicle from the premises is seen as a form of compliance and thus police involvement in the situation is terminated.

Another easy scenario to deal with occurs when the vehicle is present but is no longer in violation. Again an entry is made into the paperwork and police involvement in the situation is terminated.

A third scenario concerns the handling of new complaints. If the police arrive at the address of a new complaint and find the vehicle to be in violation, there are several options open to them depending on the nature of the violation. The most common action taken is to place a sticker on the vehicle, which indicates the violation, and the date that a follow-up check will be made by the officer. This sticker includes the name and serial number of the officer, so the citizen can call this particular officer if any questions arise—and the officer is required to return these calls.

If at the time of the original investigation, the officer finds an unlicensed vehicle parked on public or private property, the officer can tow that vehicle immediately. This includes vehicles with no license plates or with expired plates. There is no requirement that the officer issue a warning notice to the owner.

The officer can, however, choose to issue a warning indicating that the car must be moved within ten days or it will be towed. The officers are entrusted to use their discretion on this issue. There is no formal demand from the sergeant that the officers tow each vehicle they find suffering from this violation. In many cases the officer's actions at this point will depend on the circumstances of the situation. Some of the factors that influence the officer's decision include presence of the owner, demeanor of that person, condition of the vehicle (tires up, running), and location of the vehicle (blocking street or sidewalk access).

If the owner of the vehicle is present, most of the officers indicate a willingness to work with this person. This can include informing them of the law and giving them a warning rather than towing the vehicle immediately. However, consistent with the research on demeanor (Black, 1980; Klinger, 1994, 1996), the willingness of the officer to work with the vehicle owner depends in large part on the attitude displayed toward the officer. If the owner is respectful and shows a willingness to comply with the officer's request to license or move the vehicle, the officer is usually willing to give the warning. If, on the other hand, the owner is hostile or rude to the officer, the vehicle is likely to be towed immediately.

If the vehicle is in a condition that creates a potential hazard of some form, perhaps to children who may play on or around it, the vehicle is also more likely to be towed. The same is true of vehicles that are blocking access or vision for other drivers. On the other hand, if the vehicle is in proper working condition, the officer may allow the owner to either move it or promise to get proper license plates for it before the ten-day warning period expires.

If the vehicle lacks the proper license plates but is on private property, the officer will place a tow notice sticker and two copies of the Notice of Nuisance form on the vehicle. The sticker and the forms indicate to the owner the violation and the time frame in which the vehicle must be brought within code before being towed away. Additionally, the forms indicate whom the owner can contact if he/she has questions or wishes to appeal the citation.

The situation may also depend on the attitude of the officer. If the officer is having a bad day or has decided at the beginning of the shift that he/she is going to tow all improperly licensed vehicles, there is often nothing the owner can say that will stop the tow. Conversely, if the officer is having a good day or simply doesn't want to take the time to tow, the vehicle will be left with a warning sticker placed on it.

Two additional factors affect the decision to tow; when in the shift the officer encounters the violation and whether the officer has been at this address before. These factors affect the decisions made at all types of violations but are particularly salient where an immediate tow is a possibility. It takes on average 15–30 minutes for a tow truck to arrive on the scene. If the officer has plans for lunch, needs to use a restroom, or is close to ending his/her shift, he/she may not wish to wait this long to complete the job. In this case he/she will issue the warning sticker and return at a later time. When time is not a factor, the officer will use the time waiting for the tow truck to complete paperwork, conduct work related or personal business via the cellular phone, or perhaps just rest.

For addresses where the officer has responded to complaints before, officers are often less willing to give breaks. Repeated calls about the same addresses, and sometimes the same vehicle, indicate a pattern of refusal to comply. Excuses and promises of vehicle owners at these addresses are not believed and, other factors held constant, vehicles are more likely to be towed.

Another common violation encountered by task force officers is dead storage or abandoned vehicles. These are vehicles on public property that are parked for prolonged periods of time in the same spot and appear to never be moved. The vehicle may or may not be in running condition. As indicated in the city ordinances, vehicles cannot remain parked in the same spot on public property for more than 48 hours at a time. The vehicle must be moved at least one mile every two days. Officers place a tow notice sticker and copies of the Notice of Nuisance form on these vehicles, giving the owner ten days to move it. Additionally, officers will place chalk marks on the tires and take down the mileage of the vehicle. This enables them to determine if the vehicle has been moved at least one mile since they put the tag on it.

Motor vehicle litter cases and vehicles not licensed for Douglas County but which are owned by someone living in the county are treated in the same way. Vehicles which are wrecked, or not running are considered motor vehicle litter. License plates from the wrong county or state are subject to wheel tax and thus to tow if proper plates or a wheel tax sticker are not purchased for the vehicle.

In all the cases where a tow notice sticker is placed on the vehicle, officers can tow the vehicle after ten days if the violation is not corrected. While this is the ideal, it is more common for the officers to return to the address of the violation a week to a month after the ten days have expired. This lapse is the result of officer work schedules and the caseload the officers handle. There simply is not time to return to every vehicle within ten days. Additionally, officers attempt to conduct their own follow-ups, as they are the most familiar with the problem at hand. Thus, the rotating shift they work where days off are not on the same day each week will often mean that an officer is off duty on the day the follow-up should be conducted.

Officers are also free to initiate the issuance of notices for vehicles they encounter but for which they do not have an official complaint. Several officers have indicated that they will do this in situations where they are handling a complaint at one address and the next address is also in violation. This is an attempt to appear fair rather than appearing to pick on one person in a neighborhood.

The ability of the officer to tow any vehicle still in violation after the expiration of the ten-day waiting period does not automatically mean that the vehicle is in fact towed. Many of the factors affecting this decision have been discussed above in the context of unlicensed vehicles. An additional factor that affects decisions for dead storage and motor vehicle litter cases would be progress made toward compliance. If, for instance, the individual had several vehicles in violation when the original investigation was made, but now has only one or two vehicles left in violation, the officer may be inclined to give this person a few extra days to finish cleaning things up.

Additionally, after the original notice has been given, citizens can appeal to the city's Municipal Board of Appeals. This board is staffed by three citizens appointed by the mayor with the approval of the city council. These individuals function to ensure that city ordinances are being applied properly and fairly to the situation. This board can determine that the citizen is not in violation, that the citizen can have extra time (typically 90 days) to achieve compliance, or can deny an appeal. The officer's future involvement in the case is determined by the finding of the board.

One of the final violations that the task force officers encounter deals with vehicles abandoned on freeways or highways. These vehicles may have broken down or been involved in minor accidents. They must be moved within 24 hours. Often, the state patrol officers have placed a tow notice sticker on the vehicle but have failed to perform a follow-up investigation and tow the vehicle. Task force officers generally find these vehicles while traveling from one section of the city to another by way of the freeway system. Most of the officers do not actually search the freeways for these violations.

If the officer decides to tow the vehicle, he/she remains on the scene waiting for the tow truck to come and collect the vehicle. This serves two purposes. It enables the officer to ensure that the tow truck driver takes the right vehicle, and it ensures that the owner or some other person does not appear on the scene and create a confrontation with the tow truck driver.

Specific Cases   One specific case that stands out involves an interaction between one of the task force officers and the owner of a vehicle that was being towed away. This officer had previously placed a tow notice sticker on this vehicle because of a complaint that the vehicle did not run and that it had been parked on this street for several weeks. The officer returned to follow-up on the case and found that the vehicle was still in the same condition. As the tow truck was hooking up the vehicle, the owner emerged from the house and approached the officer. The owner explained to the officer that the vehicle was being repaired and that he would move it into the garage if the officer would not tow it away. The tow truck was already on the scene, however, and the officer explained that it was too late. The vehicle was towed and the owner returned to his home after thanking the officer for the attention.

This case is typical of those observed during the study. The officers frequently interact with citizens in this fashion. At only one case that occurred during the observation period was hostile behavior displayed by a vehicle owner. In general, the vehicle owners are respectful, even if the officer is not cooperating with their requests. Often the officers gave the vehicle owner extra time to take care of the problem even though they were not required to do so. Every officer indicated that the polite demeanor of the owner played a role in his/her decision to be lenient.

At the case where hostile behavior was observed, the officer was working in conjunction with the City Zoning Inspector. There were two cars and a pickup truck that were to be towed. The City Zoning Inspector had ordered these tows as part of his effort to bring the owner into compliance with a variety of zoning ordinances. The City Zoning Inspector was taking pictures of the vehicles and the yard so that he would have evidence if the owner appealed to the Municipal Board of Appeals. It appeared that the picture-taking behavior angered the owner of the vehicles because he began to scream obscenities and gesture toward the City Zoning Inspector.

At this point the City Zoning Inspector requested that the task force officer arrest this individual for violating city ordinances concerning litter and parking. A back-up officer was called to the scene in case the individual refused to sign the arrest form and had to be taken into custody. The Humane Society was also called to deal with the two large dogs who were keeping the tow truck operators from being able to reach the vehicles safely.

In the end, another individual arrived on the scene and was able to convince the first person to sign the arrest form. The second individual informed the officer that the first had a mental illness and had stopped taking medication. The vehicles were towed and everyone departed the scene in peace.

The two cases discussed above illustrate the types of treatment the officers may be forced to confront. In addition, several cases were observed where the apparent owner of the offending vehicle was unable to speak English. In all of these cases the officers were able to resolve the issue peacefully. Sometimes a mixture of languages and gestures helped and in some cases the officer was able to provide material concerning the problem that was written in Spanish.

Dogs and other animals also play a role in the work of the task force. One officer was observed drawing her weapon because of a large unsecured dog that suddenly appeared around a corner. This officer had been attacked by a pack of dogs previously and had ended up shooting at least one of them. Any time there is a question about handling an animal, the Humane Society is called.

The City Parks Division maintains a Weeds and Litter Unit that is often called to work in conjunction with the task force officers. Many of the places the City Zoning Inspector refers the officers to suffer from problems related to the general upkeep of their yards. In some cases the officers cannot tow a vehicle until some of the other junk in the yard is removed, and in some cases the Weeds and Litter Unit cannot finish cleaning up a yard until vehicles have been removed.

Some of the cases observed provided moments of entertainment for both the officer and the observer. One case involved two large work trucks parked on opposite sides of the street. The officer was checking these vehicles to determine if they exceeded the width limits for parking on a public street. When the officer approached the vehicles to check for the VIN numbers, it was discovered that both trucks had purple bowling balls on their front seats. As the officer said, despite the fact that the work seems repetitive, there is always something different to be seen in this job.

The City Zoning Inspector

The City Zoning Inspector of Omaha also plays a role in the work of the NTF. This role is particularly apparent in cases where the officers of the NTF have a question concerning what action should be taken for a complaint. In many cases, the complaint may or may not be something on which the officers can legally act. The City Zoning Inspector can clarify the rules, and invoke zoning ordinances in order to supply the officers with the proper actions to be taken.

The work of the NTF can also originate from the City Zoning Inspector. In the process of enforcing other zoning regulations, the City Zoning Inspector may find violations of the ordinances that deal with vehicles. He then asks the NTF officers to deal with this situation for him.

The relationship is thus reciprocal, with each side assisting the other in finding the most appropriate way to handle problems with vehicles. The work of both the NTF and the City Zoning Inspector is made easier through this relationship.

The Tow Lot

All tows are handled through one company in the city. This company takes the vehicle to the city impound lot. At this point, unless the owner of the vehicle takes legal action to protest the taking of their vehicle, or calls the unit for some reason, the task force officers are finished dealing with the case. They will not become involved with this person again unless future violations arise.

At the impound lot, the vehicle is run through the state computer to attempt to locate the last registered owner. This individual and any other individual listed on the form that was filled out by the task force officer when the vehicle was towed, will be sent a letter indicating that they have 30 days to retrieve their vehicle or the title will be forfeited to the city and the vehicle will be auctioned to the public. If there is no response within ten days a certified letter will be sent.

After thirty days the vehicle title is forfeited to the city and the vehicle is auctioned to the public. The impound facility holds a vehicle auction every Saturday and a private property auction the first Saturday of every month for any property that was left in the vehicles when they were brought in. If an individual wishes to claim property from the vehicle but not take the vehicle, they must prove ownership of said vehicle. In this case, the impound personnel will attempt to get the owner to sign over the title so the vehicle can be auctioned immediately.

In order to remove a vehicle from the impound facility it must be brought up to code which includes current license plates and being in running condition. The vehicle inspectors for the Department of Motor Vehicles (DMV) will come to the impound lot to conduct the required inspections for a small fee. If the vehicle does not run it can be towed to a licensed auto garage to be worked on, but must still be currently licensed. The owner must pay for this tow in addition to the fees from the impound lot. The vehicle cannot, however, be towed to the owner's residence to be worked on or to be stored if it is not running.

At any given time the impound facility contains around 850 vehicles. These vehicles come from all units of the police department, not just the task force. They include vehicles used in the commission of a crime, vehicles involved in accidents, and vehicles towed both by the unit and by other officers for dead storage or motor vehicle litter.

The auctions are a way of attaining much needed space for incoming vehicles. The problem is of such size that the facility is currently leasing an additional section of property, bringing its size up to 14 square acres. Additionally, the auctions brought in over $2.5 million in 1998. This money goes into the city's general fund, some of which is then redistributed back into the budgets of the agencies involved in dealing with the vehicular problems of the city.

The Municipal Board of Appeals

The Municipal Board of Appeals is another government agency that regularly interacts with the officers of the NTF. Any person may appeal to this board for exemption or extension of the time allotted them to remove problem vehicles. The board will look into the situation and grant relief where appropriate. According to the City Zoning Inspector, however, the board usually sides with the NTF officers or at the most will grant the citizen a longer time period in which to deal with the problem. Notice of these decisions is forwarded to the NTF so that they will be able to abide by them.

Neighborhood Associations

The city of Omaha has an extensive network of neighborhood associations. These associations are designed to represent the interests of the various areas of the city and to communicate to public officials about problems that need attention. For the purposes of this study, the neighborhood associations represent the principal consumers of the NTF Unit's work. Neighborhood association leaders were surveyed to determine whether they were aware of the unit and believed that it effectively responds to the problem of abandoned vehicles.

Commentary   Any time a researcher engages in observation, there is a chance that the party being observed will alter his/her behavior. This chance is particularly acute in a study such as the present one, where the subjects have been explicitly informed that their behavior is being observed.

There are several reasons to conclude that the behavior of the NTF officers did not change significantly when they were under observation. The first of these reasons deals with the amount of time spent in the observation phase of this study. Over 200 hours of observations were completed. Six of the seven officers were accompanied for at least two full shifts. The amount of time the researcher was able to spend with the officers allowed a comfort zone to develop. The officers understood the purpose of the research and the researcher understood the purpose of the officer's behaviors. Combining the interviews with the observation phase contributed significantly to the development of this comfort zone. Any questions on the part of any of the participants, be they researcher or subject, were answered during this interview. Observations thereafter were accompanied by informal, friendly conversation and there was never a feeling of suspicion or hostility. In fact, more than one officer expressed excitement at being the subjects of a research study.

The second and perhaps the best reason for concluding that the behaviors being observed did not change significantly due to the presence of the researcher is simply the range of behaviors that were observed. Beginning with roll calls in the morning, no effort was made to make these situations formal. There were jokes, stories, and other friendly exchanges. Roll call did not always begin on time, and some days lasted over an hour even though the actual substantive content would have taken only ten or 15 minutes. There was never an attitude of needing to hurry things along so the officer could get to the street. In fact, after roll call almost all of the officers would stop for a cup of coffee or soda before actually beginning their day. These stops usually lasted no more than ten minutes, but were indicative of the lack of rush the officers felt in getting to their work.

Other behaviors observed also gave the impression that the officers were not trying to hide anything about the real process. The officers who did not stop for coffee first thing would stop later in the morning for a bagel and soda. They would then attempt to find a place to park that was hidden from plain view so that they could eat in peace. These officers would often read the paper or a magazine while taking this break.

All officers would take a lunch break eventually. Some officers were more conscientious about how long this break took than others. Several officers would take at least the entire hour down to the minute while others would be more lax about timing this break. Some of the officers would eat in a restaurant while others would get the food to go and eat either in the car or at a police substation breakroom. On more than one occasion, the researcher was taken to a private home for lunch.

Most of the officers would either run personal errands while working or spend large amounts of time on cellular phones dealing with personal business. The researcher was taken on unexpected stops to personal homes, dentist offices, stores, and other errands. Officers who had phone calls to make or other personal business they did not wish to conduct in the presence of the researcher would return to the mall substation early in the shift. The shift officially ends at 5:00 p.m. and the officers all returned between 4:30 p.m. and 5:00 p.m. so as to complete paperwork anyway, but it was not abnormal for the officers to return to the station as early as 3:30 p.m., with no intention of returning to the streets for the remainder of the shift.

Officers would also try to time their work so that they would not be engaged in towing a vehicle or another time-consuming task close to the end of the shift. On more than one occasion the officers would be finished working for the day but would not return to the station because of the early hour. These officers would find a place to park and do paperwork, or would drive around the city aimlessly.

When asked about these seemingly time-wasting behaviors all of the officers expressed the feeling that they did not need to hide things from the researcher. Several even pointed out the need to have some flexibility with their day in order not to become burned out from running to and from cases all day. They noted that if they did spend every minute of their shift engaged in productive work, they would be busier than most patrol officers and would tire quickly.

UNIT OFFICER INTERVIEWS

All of the officers in the NTF were subject to loosely structured interviews during the first ride-along. Many of the officers were asked supplemental questions during subsequent ride-alongs. Several of the questions the officers were asked dealt with their satisfaction with their assignment to the task force.

All of the officers expressed satisfaction with their assignment to the NTF. While the specific reasons for this satisfaction varied, the most often mentioned reasons included working hours (day shift only for NTF officers) and freedom to move anywhere in the city rather than being bound to a single beat. The wider range of discretion in what to work on and how to handle the cases was also mentioned as a source of satisfaction.

Several officers expressed a preference for this assignment over their previous one. Two officers preferred this work to working with the unit in the public schools, and at least one indicated that the work was an improvement over patrol work on the night shift. Officers with children also expressed a preference for this assignment as it allowed them to be home with their children in the evenings.

These responses give an indication of the job satisfaction the officers found in this unit. A better indication perhaps is that none of the officers indicated a desire to leave the unit. One officer in fact indicated that reassignment would never occur voluntarily.

Other questions asked in the interviews dealt with the officer's perceptions of the work itself; i.e. could they make a difference, were they appreciated, etc. Overall the officers felt that they could make at least some difference. Many felt that they could at least make a difference for the person who had complained, while others felt this difference went further by impacting the entire area from which the offending vehicles were removed.

Most of the officers felt that the problem of junk vehicles in the city would never disappear. There are simply more junk vehicles than there are officers to attend to them. One officer even went so far as to speculate that there are three vehicles no one has complained about for every one that someone has called about. At the most the officers hope to cut the size of the problem and maybe get some people to stop storing junk vehicles on or about their property.

Several officers indicated that there were certain people who continued to re-offend even though the officer had been to the address more than once. Some of these people appeared to be doing auto rebuild work in their yards. The officers would come and remove multiple vehicles and some months later would have to return and remove more vehicles. These people are the ones the officers feel they cannot reach.

One officer said that the job was really about changing people's behavior rather than their attitudes since some people were never going to change. Others would change only after the officer had removed vehicles on several occasions. These people probably thought the officer was wrong to do this, but they would eventually comply to avoid the hassle.

Two officers indicated that they felt their work held meaning for the community because it dealt with the things the community thought were important. They noted that one junk vehicle in a neighborhood may not seem like a big problem to most people, but when it is in your neighborhood it is a big deal.

The officers all indicated that they felt the community residents appreciated their work. Even the people who had their vehicles taken away were at least polite in most cases. The sergeant also indicated that he did receive phone calls from community people who were satisfied and/or thankful that the officers had taken action in their neighborhoods.

There are of course some people who call and complain, but the officers all indicated an understanding of this occurrence. People are upset when their vehicles are taken away. Only when the individual who called to complain was disrespectful or in some other way "stupid" in the eyes of the officer that the officers indicate irritation.

The officers also indicated that they believe the other city agencies such as the City Zoning Inspector, and the Weeds and Litter Unit, as well as the regular patrol officers have a positive view of their work. The officers feel they have a good working relationship with these individuals. This relationship is seen as important to their work and the officers see the relationship as a two-way street. They do things for the other agencies, and the other agencies do things for them.

EXTERNAL AGENCIES AND AUDIENCES

City Agency Interviews

Representatives of other city agencies were also interviewed for this study. All of these individuals indicated that they held the work of the unit, and in particular the officers themselves, in high regard. The Mayor's Hotline personnel indicated that this work was important to the mayor, as well as being important to the communities being served.

These individuals felt that the unit provided another way for the city to provide people with the services they felt they needed.

The head of the Tow Lot Facility felt that the work was important, but also that the unit itself was effective. Given that the persons who used to be in charge of dealing with junk vehicles had been her employees, she is in a good position to estimate whether the unit is more or less effective than previous strategies.

The City Zoning Inspector also spends a significant amount of time working directly with the task force officers. In his official capacity, he can order vehicles towed for violating city ordinances. He can also order persons arrested for continuous violations. By working closely with the unit, he has an enforcement mechanism for these activities. The unit in turn is able to use the City Zoning Inspector's orders to justify actions in cases where the appropriate action is not clear. The relationship between the City Zoning Inspector and the NTF is mutually beneficial. Additionally, the City Zoning Inspector feels strongly that the unit can and does make a difference through its work.

Neighborhood Association Leaders

The original plan for conducting the interviews with the presidents of the organized community associations called for every third entry on the list to be interviewed. Once this process was underway, however, it became apparent that this plan would not work. The list of presidents acquired from the police department had several flaws that made the original research plan impossible. To begin with, not all of the entries on the list were eligible for the survey. The list contained eight listings for phone numbers that had been disconnected, and one listing was for an organization that had not existed for over three years.

Another problem that was encountered was that the list contained twenty entries that belonged to organized business associations. These associations do not deal with issues such as abandoned or junk vehicles. Rather, they exist to coordinate activities between the local business people in their area. Further, these individuals refused to participate in the survey. Most cited the fact that they were at work when giving a reason for declining. It was necessary, therefore, to remove both these businesses and the disconnected or non-existent entries from the pool of possible respondents.

One additional problem occurred in selecting those entries that would be a part of the pool from which the sample would be selected. The same president ran two community associations. This person was allowed to respond, but was counted as being eligible for only one association.

Table 2-1 indicates the distribution of listed organizations that fell into the various categories of response possibilities. As can be seen, the original pool of 164 possible respondents from which the sample was to be taken had to be narrowed down to 134 entries due to the problems discussed above.

Simply reshaping the possible pool of applicants did not enable the sampling strategy of using every third entry on the list to be reinstated. The timing of the calls became an issue. The surveys were undertaken during a six-week period from mid-June to the end of July. This time period proved to be problematic as many persons on the list appeared to be on vacation and were unreachable during this time frame.

After the first week of encountering this problem a decision was made to abandon the sampling strategy and instead attempt to call every member in the eligible pool. This new strategy was more successful; however, a one-third-response rate proved difficult even at this

TABLE 2-1   **Organized Community
Association Surveys**

Disconnected	8
Doesn't exist	1
Businesses	20
Same president	2
Refused	14
Never reached	72
Answered	48

point. Each member of the eligible pool was called at least three times over the six-week time frame. Calls were made during the workday, in the evening, and over the weekend in order to maximize the chances of contact. Additional calls were made as time permitted, and resulted in only one extra individual being reached.

In the end, 62 persons were reached. As Table 2-1 shows, 14 persons refused to participate in the survey. The remaining 48 persons each answered the survey questions. The results of these interviews are discussed below.

The interviews with the organized community association presidents reveal that most (80%) of these individuals are aware of the existence of the NTF. Several persons mentioned that they had acquired this knowledge through a community group meeting put on by OPD. At this meeting the presidents were given information concerning which city agencies to contact for an assortment of problems including junk vehicles. They were also given a contact sheet listing the agencies and their phone numbers.

Fifty percent of the individuals who responded that they knew of the unit reported having used the unit's services at some time in the past. Of these persons, four reported that usage was not personal but that they had referred members of their organization to the unit and knew that these persons had used their services.

Of the persons who had used the services of the task force, almost all were satisfied with the response they received. Two individuals mentioned dissatisfaction with the amount of time it took for the response to occur, but were satisfied once there was a response.

Only one individual responded that the unit could not make a difference and that Omaha did not need this unit. All other respondents felt that the unit could and did make a difference. Four respondents indicated that the unit could only make a difference under certain circumstances; i.e. if people would faithfully call to complain, or depending on the state of the neighborhood. Two persons who felt the unit made a difference were still unsure of whether Omaha really needed the unit. One suggested that the money spent on the unit might be better spent on other things.

DISCUSSION AND CONCLUSIONS

The principal findings of this study are:

1. The NTF of OPD engages in activities closely related to its official mandate regarding the removal of abandoned or illegally parked vehicles;

2. The officers assigned to the NTF have a positive attitude toward their job and value the work of the unit;

3. The principal consumers of the unit's services are aware of the unit's activities and value its work.

These findings do not support Crank's, and Langworthy's (1992) argument that specialized units are largely "ceremonial," contribute little to the effectiveness and efficiency of law enforcement agencies, and exist largely for symbolic purposes. The NTF engages in meaningful work closely related to the mission of the police department, and is a more efficient means of handling abandoned or illegally parked vehicles than the old approach of having patrol officers handle them. The task force is valued by the officers assigned to it, and they do not regard their work as merely ceremonial. Finally, the principal consumers of the unit's activities, community leaders, are aware of the unit and value its activities.

The data indicate that the NTF does actually perform a functional purpose that relates to their official mandate. The analysis of the quantitative data from the officers' daily reports and from the city tow lot facility revealed that the officers assigned to the task force do engage in activity related to removing the problem of junk vehicles from the city. The fact that the officers could be more efficient in this work becomes apparent when examining the observation data; however, this fact must be balanced against the risk that the officers would become burned out and reluctant to engage in any type of proactive work if they were to use every minute of their shift in an efficient manner.

The popularity of the unit, as shown in the surveys of community association presidents, indicates that the unit also performs a legitimacy-related function for OPD. This unit is popular, well respected, and considered necessary by most of those who were surveyed. The surveys also indicated that the unit was seen as responsive to community needs.

IMPLICATIONS

Specialized Units

This research also has implications regarding the nature of the function that specialized units perform within a police department. Katz's (1997) research indicated that the OPD gang unit was not performing along functional lines, but rather it existed for the purpose of maintaining organizational legitimacy. The results of the current research indicate that these findings cannot be generalized to all specialized units within OPD. There is no doubt that the NTF serves some legitimacy-related function; however, it also accomplishes a significant amount of the work it is officially mandated to perform.

Institutional theory is not contradicted by a special unit that actually does work if it can be shown that the unit also assists the department in maintaining legitimacy. It may be that the success of the unit is a happy coincidence or a side effect. As demonstrated here, functional purposes and legitimacy maintenance are not mutually exclusive. This means that Katz's (1997) findings are also not generalizable to specialized policing units across the country.

Additional research will be needed to determine if the difference in findings between this study and Katz's (1997) study is based on the type of special unit studied. In other words, it may be that only certain types of specialized units are employed solely for legitimization purposes while others are functional. It may also be that certain tasks performed by specialized units are easier to frame in terms of functionality and success. Both other types of special units and the same units in other police departments will need to be studied to make these determinations. Thus, the conclusions drawn here should be interpreted as part of a beginning discussion on the legitimacy of special units rather than the end.

SUGGESTED READINGS

CRANK, J.P., & LANGWORTHY, R. (1992). An institutional perspective of policing. *The Journal of Criminal Law and Criminology, 83*(2).

MANNING, P.K. (1997). *Police Work: The Social Organization of Policing* (2nd ed.). Prospect Heights, IL: Waveland Press, Inc.

McGARRELL, E.F., LANGSTON, E., & RICHARDSON, D. (1997). Implementation issues: the specialized unit or department-wide community policing. In Q.C. THURMAN, & E. McGARRELL (Eds.), *Community policing in a rural setting* (pp. 59–64). Cincinnati, OH: Anderson Publishing Company.

MEYER, J.W., & ROWAN, B. (1977). Institutionalized organizations: formal structure as myth and ceremony. *American Journal of Sociology, 83*, 340.

WAMSLEY, G.L., & ZALD, M.N. (1973). *The political economy of public organizations: a critique and approach to the study of public administration.* Bloomington, IN: Indiana University Press.

REFERENCES

ALPERT, G., & MOORE, M.H. (1997). Measuring police performance in the new paradigm of policing. *Performance Measures for the Criminal Justice System.* Washington, DC: U.S. Department of Justice, Bureau of Justice Statistics.

BABBIE, E. (1992). *The practice of social research* (6th ed.). Belmont, CA: Wadsworth, Inc.

BLACK, D. (1980). *The manners and customs of the police.* San Diego, CA: Academic Press.

BROOKS, L.W. (1997). Police discretionary behavior: A study of style. In R.G. Dunham, & G.P. Alpert (Eds.), *Issues in Policing: Contemporary Readings* (3rd ed.). Prospect Heights, IL: Waveland Press, Inc.

CRANK, J.P., & LANGWORTHY, R. (1992). An institutional perspective of policing. *The Journal of Criminal Law and Criminology, 83*(2).

DiMAGGIO, P., & POWELL, W. (1983). The iron cage revisited: Institutional isomorphism and collective rationality in organizational fields. *American Sociological Review, 48*, 147.

KATZ, C.M. (1997). *Police and gangs: A study of a police gang unit.* Unpublished doctoral dissertation, University of Nebraska, Omaha.

KELLING, G.L., & MOORE, M.H. (1995). The evolving strategy of policing. In V.E. Kappeler (Ed.) *The police and society: Touchstone readings.* Prospect Heights, IL: Waveland Press.

KESSLER, D., & DUNCAN, S. (1996). The impact of community policing in four Houston neighborhoods. *Evaluation Review, 20*, 627.

KLINGER, D.A. (1994). Demeanor or crime? *Criminology, 32*, 475.

———. (1996). More on demeanor and arrest in Dade County. *Criminology, 34*, 61.

LANGWORTHY, R.H. (1986). *The structure of police organizations.* New York, NY: Praeger.

LIPSKY, M. (1998). Toward a theory of street-level bureaucracy. In G.F. COLE, & M.G. GERTZ (Eds.) *The criminal justice system: Politics and policies* (7th ed.). Belmont, CA: Wadsworth Publishing Company.

MASTROFSKI, S.D., & PARKS, R.B. (1990). Improving observational studies of police. *Criminology, 28*, 475.

MCGARRELL, E.F., LANGSTON, E., & RICHARDSON, D. (1997). Implementation issues: The specialized unit or department-wide community policing. In Q.C. THURMAN, & E. MCGARRELL (Eds.), *Community policing in a rural setting* (pp. 59–64). Cincinnati, OH: Anderson Publishing Company.

MEYER, J.W., & ROWAN, B. (1997). Institutionalized organizations: formal structure as myth and ceremony. *American Journal of Sociology, 83*, 340.

Omaha Police Department Crime Analysis Unit. (1998). *Calls for police service: Annual report – 1997.*

POLICE FOUNDATION (1981). *The Newark foot patrol experiment.* Washington DC: Police Foundation.

PRESSMAN, J.L., & WILDAVSKY, A.B. (1973). *Implementation: How great expectations in Washington are dashed in Oakland.* Berkeley, CA: University of California Press.

REED, W.E. (1999). *The politics of community policing: The case of Seattle.* New York: Garland Publishing, Inc.

SCOTT, W.R. (1995). *Institutions and organizations.* Thousand Oaks, CA: Sage Publications, Inc.

SELZNICK, P. (1966). *TVA and the grass roots: A study in the sociology of formal organization.* New York, NY: Harper & Row Publishers.

STOJKOVIC, S., KALINICH, D., & KLOFAS, J. (1998). *Criminal justice organizations: Administration and management.* (2nd ed.). Belmont, CA: Wadsworth Publishing Company.

TROJANOWICZ, R., & BUCQUEROUX, B. (1990). *Community policing: A contemporary perspective.* Cincinnati, OH: Anderson Publishing Co.

WALKER, S. (1984). "Broken Windows" and fractured history: The use and misuse of history in recent police patrol analysis. *Justice Quarterly, 1*(1).

——. (1999). *The police in America: An introduction* (3rd ed.). Boston: McGraw-Hill.

WAMSLEY, G.L., & ZALD, M.N. (1973). *The political economy of public organizations: A critique and approach to the study of public administration.* Bloomington, IN: Indiana University Press.

WOODS, D. (1991). *An Analysis of Peace Officer Licensing Revocations in Texas.* Unpublished doctoral dissertation, Sam Houston State University: Huntsville, Texas.

ZHAO, J. (1996). *Why police organizations change: A study of community-oriented policing.* Washington, DC: Police Executive Research Forum.

——. (1999). Personal Communications.

3

An Examination of the Compatibility Between COP and Special Units

Stan Shernock

❖

INTRODUCTION

At a time when American policing is supposedly undergoing a revolution in community-oriented policing (COP), there continues to be the development of special police units of various types. Moreover, a number of police departments deploying special units have at the same time claimed to have implemented COP. Despite the development of various types of special units within police departments, most of what has been written about COP and specialist units has focused on whether COP itself can be embodied by and organized into special COP units comprised of COP specialists. When COP is considered an overall philosophy and strategy to policing and not merely a program or special unit with specific activities, the question arises whether these other special units are antithetical to the basic principles of COP and thus undermine its more complete implementation.

In order to understand the relationship or compatibility between special units and community policing, it is first important to: (1) understand the meaning of a special unit as a distinctive type of unit within a police organization; (2) realize that there are three entities interacting—the special unit, the larger police organization, and the community; and that special units have effects on the role that officers in other units play regarding the community; (3) distinguish between two very different interpretations or models of community policing; and (4) consider whether special units can be incorporated into the roles of the generalist COP officer.

Special units should not be seen as merely traditional institutionalized units within police organizations performing routine specialized functions, but instead as units that have been developed as *ad hoc* or improvised units that address, often in an innovative or different fashion, a special need or purpose. Special units such as police reserves and SWAT (Special Weapons and Tactics) teams, for example, were established to be used in emergency situations.

A special unit must be analyzed in terms of its relation to the organization besides the community, insofar as it is not only a part of the organization, but an entity that tends to institutionalize or entrench itself in a way that affects the organization's structure in relation to the community or task environment. It is important to understand how special units engage in the processes of role expansion, routinization, normalization, institutionalization, and usurpation, as well as coordination, within the organization, while the organization engages in co-optation of, as well as cooperation with, the community. When examining special units in regard to their primary and expanding functions, it is not only important to understand the type of roles that they perform, but also how they affect the role of other parts of the police organization and other officers within the police organization. As special units move away from their original purpose and expand into routine roles and activities, they begin to threaten to usurp the roles of the generalist police officer, and thereby threaten conventional community policing. In the special case of the volunteer auxiliary, as with the Civil Guard in Israel (Yanay, 1993), the police organization engages in co-optation of the community through the creation of a special unit.

Many police scholars (Rosenbaum & Lurigio, 1994) have commented that the concept of community policing is itself ambiguous and means different things to different people. Nevertheless, DeMichelle and Kraska (2001) identify two basic versions or models of community policing: one they label the "peace corps" and the other the "zero-tolerance" model. The peace corps model, better characterized as a police–community partnership model, largely derived from the work of Robert Trojanowicz and Hermann Goldstein, emphasizes partnerships between the police and the community, community empowerment, community development, joint problem-solving, and the service function. According to this version of community policing, private citizens and private entities participate in helping police their own communities. The police institution and the criminal justice system in general are seen merely as components of a larger community complex of interconnections and overlapping functions with various private and public agencies.

The zero-tolerance model, derived from the "broken windows" concept of Wilson and Kelling (1982), emphasizes order maintenance and the re-establishment of social order in communities by enforcing all laws, even in regard to the less serious forms of deviance, and cleaning up signs of community decay, all of which are assumed to be correlated with more serious crime. The signs of disorder or of communal disregard and incivilities, coupled with the breakdown of informal social controls, presumably indicate that no one cares about the community. Greene (1999, p. 506) says that zero-tolerance policing in New York City put emphasis on "quality-of-life" issues. Consequently, "'squeegee men,' the petty drug dealers, the graffiti scribblers, and the prostitutes who ruled the sidewalks in certain high crime neighborhoods were targeted in candidate Guiliani's campaign promise to reclaim the streets of New York for law-abiding citizens." DeMichelle and Kraska (2001) note that this type of policing also involves little tolerance for disorderly behaviors and conditions such as loitering and vagrancy, public drunkenness, peddling, etc. What is identified as conditions creating criminal opportunities become as important a focus as serious crime itself. By attending to problems of decay and disorder, the police and community can work together to clean up the community through linking order maintenance and crime prevention. Police then attend to these crime-producing conditions by engaging in aggressive order maintenance policing. Thus, the "zero-tolerance model seeks to proactively 'clean up' a community, thereby reducing the potential for crime and diminishing citizens' fears, while the other model is concerned with building responsible and knowledgeable communities through police–citizen partnerships" (DeMichelle & Kraska, 2001, p. 9).

Because of the very interdependence of different police roles, it is analytically difficult to distinguish only one role that an officer engages in when involved in an incident. Numerous service or order maintenance activities require officers to exercise law enforcement skills; and order maintenance or law enforcement calls may require both types of skills on the part of the responding officer (Gaines, Kappeler, & Vaughn, 1994). Traffic control, likewise, while incorporating a law enforcement function in citing violators and arresting drunken drivers, can be viewed as an essentially protective service (Langworthy & Travis, 1994). The view that categorical distinctions of police roles are artificial is related to the generalized and interdependent role of the police according to the philosophy of community policing. When examining the compatibility between special units and community policing, the question arises whether the functions performed by special units can be assumed by COP generalists? This implies a more integrated police officer role identified with COP, where such activities as report taking, information collection, crime investigation, and apprehension are assumed by the same officer. At first glance, special police units would seem theoretically antithetical to COP insofar as their specialization *ipso facto* presumably undermines the people-oriented and generalist skills that police scholars have associated with COP (Trojanowicz & Bucqueroux, 1992; Peak & Glensor, 1996). Cordner (1995), for one, in his classic article on community policing, has stipulated that the number of specialized units and personnel can sometimes be reduced with more resources devoted to the direct delivery of police services to the general public. During the first part of the Guiliani administration in New York City, Police Commissioner Bratton argued that community policing in New York was hampered by an unwieldy and highly over-specialized police bureaucracy. He then integrated many of the police functions previously held by specialized units so as to empower patrol officers to move directly to address drug and gun crimes in the neighborhoods they served (Greene, 1999).

However, it has been found that certain special units perform necessary functions which cannot be performed by generalist police officers nor be incorporated into their role within a community-policing structure (Pate & Shtull, 1994). At the beginning of Lee Brown's administration as Commissioner of Police in New York City in 1990, he replaced the earlier Community Patrol Officer Program (CPOP) with the Special Operations Unit (SOU), and then replaced a number of specialist positions that were eliminated because their responsibilities were assumed by SOU members. These functions included those previously performed by anti-crime unit officers, street narcotics enforcement unit officers, fingerprint technicians, warrant officers, crime prevention officers, and highway safety officers in the 72nd Precinct (Pate & Shtull, 1994). Although some functions were smoothly absorbed into the new community policing structure, those functions requiring much expertise and training were not performed well. Consequently, the SOU relinquished the crime prevention and latent prints functions, the anti-crime function was carried out by a core of SOU officers, and the Street Narcotics Enforcement Unit (SNEU), warrants, and highway safety function remained under the purview of the entire SOU (Pate & Shtull, 1994). Thus, attempts to make the SOU total generalists did not work.

Given these four considerations about the compatibility between special units and community policing, this paper goes on to analyze this problematic compatibility in greater detail by examining two special units, the volunteer auxiliary and paramilitary police unit (PPU), which represent diametrically opposing but simultaneously occurring movements within policing.

CIVILIANIZATION THROUGH VOLUNTEER AUXILIARIES AND MILITARIZATION THROUGH PPUs

The volunteer auxiliary involves a form of civilianization of the police, which has been conventionally identified as facilitating and enhancing community policing, and the other, the PPU, involves a form of militarization of the police, which has been conventionally identified as undermining community policing. The two units, one emblematic of civilianization and the other of militarization of the police, nevertheless, can be seen as related to two different models of community policing. The volunteer civilian auxiliary, fits in with the community-policing model emphasizing police–community partnerships, while the paramilitary police unit is more compatible with the zero-tolerance order maintenance model of community policing. Auxiliary police units comprised of unsworn volunteers, by co-opting and directly involving citizens in policing, have been characterized as compatible with COP because they have been seen as providing a bridge between the public and the police and as a conduit for communication and community relations. PPUs, to the surprise of many, have been defined by some police departments as a form of community policing because they emphasize proactive policing (Kraska, 1996).

Technically, volunteer police officers, most recently estimated to number six hundred thousand persons (Doerner, 1998), are generally unpaid citizens who are recruited, trained, and supervised by law enforcement agencies (Cooper & Greenberg, 1997), even though some have their own command structure or own bureau (Dubin, 1973). Auxiliary police are comprised of part-time civilians that tend to provide general ongoing help to the police, whereas civilian reserve organizations originally have arisen during periods of national or localized emergency or unrest, and have been used in emergency situations, such as wartime, when there is a manpower shortage (Banning, 1986; Gill & Mawby, 1990). However, several of the organizations officially titled reserve police perform essentially the same functions as defined for auxiliary police and clearly have not been within the formal definition of reservists (Banning, 1986). Auxiliaries or reserves may or may not possess regular police powers, be sworn officers, or be compensated for their services (Greenberg, 1979). Since we are interested in civilian auxiliaries, we will be focusing on those who do not possess police powers. In New York City in 1973, auxiliaries did not possess powers of arrest beyond that of other citizens while in uniform and on duty (Dubin, 1973). However, there is a definite trend in the United States to provide these volunteers with peace officer status after they have passed a background investigation and completed a primary training period (Greenberg, 1979).

In terms of the relationship between the volunteer auxiliary unit and community policing, the defining characteristic of this unit is that it is composed of civilians. The volunteer auxiliary unit as a civilian special unit provides an opportunity to understand a type of civilianization of policing that goes beyond assuming the incorporation or co-optation of community members as merely personnel and requires greater focus on these civilians as an organized group or unit that performs certain functions that are similar or complementary to sworn officers. It thus goes beyond assuming civilians who enter a police organization have a particular set of values, attitudes, and norms that are inherently different from sworn officers, and analyzes how different they are once they assume roles within the police organization.

Doerner (1998) sees the greater reliance on volunteers to provide more services as one outgrowth of the civilianization movement; and Gaffney and Kelleher (1999, p. 5) state that the concept of civilianization provides a rich basis for analysis and further study in defining the dynamic relationship between policing and citizen volunteers. Greenberg (1984) suggests the importance of the civilian status of police auxiliaries for democracy, stating in another context (Greenberg, 1979, p. 272) that "volunteer police work provides a meaningful role for citizen's wishing to fulfill their civic obligations."

This type of civilianization of volunteer auxiliaries is different in important ways than full-time civilian positions within police departments. First, civilian volunteers take over functions rather than positions. Secondly, they have taken over the service functions that involve interaction with the public and community rather than routine activities removed from the public, which Guyot (1979) identifies with most regular civilianized positions. While full-time civilians have taken over office work and other non-operational positions in order to free sworn officers to go into the field, auxiliaries have further freed sworn officers in the field from performing service functions.

Moreover, because of the type of functions that they perform in interaction with the public, volunteer auxiliaries have a special effect on community policing. The relationship between the police and full-time civilians in police departments is an intraorganizational relationship involving both formal and interpersonal relations, usually between units performing different, but often complementary, functions. While the relationship between the full-time sworn officers and auxiliaries can also be intraorganizational and involving formal and interpersonal relations, it additionally can be either a relationship within units performing similar functions or support functions or a relationship between separate units performing similar or support functions.

Thurman, Zhao, and Giacomazzi (2001, p. 304) have observed that "during the last 10 years when students of American policing were focusing mainly on an expansion of community policing, there also has been a less noticeable trend toward the militarization of the American police." While the movement toward greater militarization of policing has also been analyzed in terms of recent circumventions of the *posse comitatus* law and the provision of military training, weapons, and other equipment to the police by the armed forces (Kraska, 1996), Peter Kraska, in a number of publications, has identified the paramilitary police unit as one of the major indications of an increased militarization of the police. Although, as will be described, it has taken on greater dimensions, the term "tactical unit" has traditionally been defined to mean a "formal police unit organized to respond to hostage taking, barricading, riot, bombings, or other terroristic incidents" (Stevens & MacKenna, 1989, p. 4). Kraska and Cubellis (1997) specify its meaning further by noting three basic characteristics that define it: (1) it must train and function as a military special operations team with a strict military command structure and discipline; (2) it must threaten or use force collectively, and not always as an option of last resort; and (3) it must operate under legitimate state authority, and its activities must be sanctioned by the state and coordinated by a government agency. The PPU or SWAT unit provides an opportunity to examine actual paramilitary units within police organizations rather than assuming that the entire traditional police organization takes on a paramilitary ethos and organization that presumably contradicts the basic tenets of community policing. Such units provide a contrast to routine policing, particularly as routine policing is characterized under the community-policing model.

One aspect of the growth of militarization of policing, according to Kraska and Cubellis (1997) has been an exponential growth of PPUs within a short period of time.

While only 20% of today's police paramilitary units were formed prior to 1980, between 1985 and 1995 paramilitary units in agencies serving small jurisdictions increased by 157%. Kraska and Kappeler (1997) found that in nearly 90% of the police departments surveyed in cities with populations over 50,000 had paramilitary units, as did 70% of the departments surveyed in communities with populations under 50,000.

TWO WAYS OF INTEGRATING SPECIAL UNITS AND COP: CO-OPTATION AND PROBLEM-SOLVING

Since the functions performed by special units cannot always be assumed by COP generalists, the question arises is how, if at all, can presumably indispensable special units be compatible with a COP department. The two special units examined, the civilian auxiliary and PPU, provide examples of two different ways special units can be integrated or made compatible with community policing. The civilian auxiliary unit represents the case of *co-optation*, where a presumably representative segment of the community is co-opted into the police organization; whereas, the PPU represents the case in which community-policing philosophies and strategies, such as *problem-solving*, can be integrated into special units as well as other units. In the first case, community policing is conceived as an organizational system, and in the second case as units imbued with a community-policing approach or a philosophy, such as problem-solving.

As early as 1967, the President's Commission on Crime and the Administration of Justice recommended both the civilianization of certain positions within police departments and the creation of community service officers (CSOs), which presupposed some form of nexus between civilianization and community policing (Tien & Larson, 1978). More recently, Skolnick and Bayley (1986) have suggested that civilianization of police departments be one of the main thrusts in the movement toward community policing. They believe that civilianization, where unsworn employees do jobs formerly done by police officers, leads almost directly to the creation of successful community organization and crime prevention programs. Likewise, Cox (1990, p. 172), in considering policing in the 21st century, states that "in keeping with community-oriented policing efforts, increasing numbers of civilians are likely to become employees of police agencies and as volunteers." And, in their community-policing textbook, Peak and Glensor (1996) also predict that in keeping with COP philosophy, increasing numbers of civilians will likely be hired to assist police agencies with tasks where sworn officers are not essential. It is no surprise, then, that Gardiner (1987) found that the driving force behind civilianization in Santa Ana, California was COP.

Civilianization might be seen as the last step in bringing the community into greater involvement in policing by bringing it into the police department, and thus inaugurating a new era of community policing. Presumably the more citizenry are involved in the police function, the better they understand it and cooperate with it. Viewed in this way, civilianization can be seen as a type of co-optation, especially when police departments oppose citizen-initiated programs, such as occurred in Israel with its Civil Guard (Yanay, 1993). Randall (1978) also describes how co-optation occurred in Boston, when in 1968 Mayor Kevin White attempted to civilianize traffic work within the Boston Police Department as a means of providing blacks with municipal public service patronage, and thus co-opting them, promoting their allegiance and muting conflict. In terms of community policing, the major positive effects of this type of co-optation are improved involvement and interaction between police organization and community, but not necessarily between sworn officers and community. Actually, the major reason for bringing civilians into police departments has not been to get civilians to do police work, but rather to extricate police from doing what has been defined as civilian work; nor has it meant making police more like civilians, as one might interpret community policing has done. Instead it has meant giving civilians so-called civilian functions in police organizations and police law enforcement functions.

Civilian auxiliaries have been envisioned as a potential bridge between the public and the police. Skolnick and Bayley (1986) recommended that citizens be hired to specialize in community liaison and sworn officers to concentrate on law enforcement. Pohl (1988, p. 76) adds "that civilian employees provide a valuable link between the police and the public by fostering a general awareness of the problems confronted by the police. Greene and Pelfrey (2001, p. 456) state that "the use of public safety or community service officers, who are essentially civilians performing many citizen contact functions, is expected to reduce the social isolation of the community from the police." Falcone and Wells (1995, p. 49) state that "unsworn volunteers constitute a conduit for communication and community relations," and that they are an indicator of openness and accessibility. Greenberg (1979) says that unsworn volunteers provide an avenue for establishing mutual respect between the people and the police through joint participation in crime prevention activities, and that improved police–community relations is a secondary advantage of using police auxiliaries and reserves. He goes on to state that the presence of citizen volunteers also fosters positive police performance and accountability.

The Civil Guard, a volunteer auxiliary in Israel (Yanay, 1993), as well as auxiliaries in Britain, China, Cuba, Singapore (Gill & Mawby, 1990), became a tool for local contacts between police and local government and a legitimate way for them to cooperate. This Civil Guard formed a new community-policing unit and became representative and liaison to the community, as well as possessing valuable information about the community. And, in the Rochester-Monroe Co. CJ Program (Rochester-Monroe Co. CJ Program, 1973, p. 5), some para-police were recruited from the area of the city that they were to be deployed to encourage rapport with beat residents and a sense of "proprietary" interest on the part of the team in the neighborhood itself. Not only were civilians teamed with police officers and recruited from the communities where they would help police, but they were also authorized to take the lead in relating the team to the community and in developing the circumstances in which the police partner may find it possible to take a more "client-centered" view of his services.

If there is a department-wide philosophical commitment to COP rather than the development of isolated COP specialist units in which COP specialists have been assigned, specialist unit may not contradict COP if each unit is infused with the community-policing philosophy and can incorporate basic COP strategies, such as problem-solving. There, in fact, have been specialist units that have been infused with a community-policing philosophical orientation when focusing on specific types of problems and victims. Butzer, Bronfman, and Spivak (1996) note that a specialized unit, a domestic violence reduction unit (DVRU) in the Portland (Oregon) Police Bureau, demonstrates how acting on the values of COP can result in the integration of fresh perspectives and new approaches into police work, and result in a closer alignment between the police agency and the community. Here, police made a conscious effort to consult with various groups within the community and to incorporate these groups into the process of creating the DVRU unit, its priorities, and strategies. Moreover, Jolin and Moose (1997), also writing on the very same DVRU unit, report that apparently because of this community involvement, community support for the unit appeared to exist independently of the program's documented effectiveness, and that DVRU officers derived status and satisfaction from the unit's community support.

Likewise, Decker, Bynum, Curry, and Swift (2000) note that the multi-jurisdictional task force (MJTF) in Illinois they studied adopted a proactive problem-solving approach, in part a consequence of folding in the problem-oriented policing (POP) unit and the county gang tactical unit. The approach adopted through the reformulated MJTF was upon solving problems, which led to a shift in emphasis toward street suppression and rapid response to neighborhood complaints. The reformulation of the MJTF unit marked the shift from an approach that was based principally upon enforcement to one that concentrated upon solving problems through a range of strategies including enforcement.

In their study on the assignment and coordination of tactical units, Stevens and MacKenna (1989) surveyed tactical teams for the methods that they used in information-gathering activities. In contrast to community involvement that was achieved by the Kankakee MJTF through joint community–police problem-solving, none of the methods Stevens and MacKenna list in Table 3 of their article (1989, p. 6) involved the community in anyway. Even if SWAT teams were used exclusively for emergency situations, it might be helpful and prudent to prepare for such contingencies by joint police–community problem-solving.

IDENTIFICATION WITH AND ALIENATION FROM THE COMMUNITY

The identity of the personnel within special units is very relevant to the way those units will function within a community-policing structure. The question arises whether co-opted civilian auxiliary volunteers, who are supposed to serve as liaisons between the police and the community, maintain their civilian community identity or whether they assume an identification with the police and become alienated from their communities. Skolnick and Bayley (1986), Pohl (1988), and Mapstone (1994) believe that civilians maintain a greater identity with the community than do sworn officers and the police organization, and that civilianization will improve involvement with and relations with the community. Skolnick and Bayley (1986) link civilianization with COP in terms of the assumption that civilian employees, especially if drawn from within the inner city communities that are being policed, will identify more with the community. They will presumably be more sensitive, receptive, and responsive to community needs and values than sworn police officers because they will possess special cultural understandings and special linguistic skills.

However, when considering volunteer auxiliary units in police departments, it is a mistake to consider the way in which the police organization has co-opted civilians, then merely to assume that given their presumed identity they will perform certain desirable roles. Instead it is paramount that their actual identity with the community be examined, besides considering how they usurp certain roles from police officers. Volunteer civilians may not contribute to community policing unless they see themselves as representatives of the community. Here, if, in their capacity as volunteer auxiliary civilians, they do not act as shepherds of community interests, they at least are supposed to act as liaisons or brokers between the community and police. Unlike most full-time civilian police employees, they are more likely given police functions that allow them greater direct exposure to and involvement with the public and community groups, and therefore the opportunity to act as liaisons to the community.

Gaffney and Kelleher (1999) state that the possibility exists that the result of the relationship between volunteers and the police is the inculcation of police culture in citizen volunteers instead of the civilianization of the force needs to be recognized and explored. If volunteers develop or have a greater identity with police than with their community, they may not fulfill their liaison function well, and may merely be co-opted. And, if their civilian identity becomes attenuated as they are co-opted into and perform functions for the police organization, volunteer auxiliaries themselves may become somewhat alienated from the community. In fact, according to Greenberg (1979), rather than reinforcing their civilian identity, there has been a trend toward giving volunteers sworn status, having them wear uniforms, and participate in all phases of agency activities.

Gaffney and Kelleher (1999) found that the police volunteers they studied rated police higher on six of seven criteria than did citizens in general: that by specifically, quality of service, fairness, honesty, courtesy, equality, and concern. Most significant to community policing, they found that auxiliaries had a higher attachment to their neighborhoods, and, most strongly, a higher level of fear of most types of crimes surveyed than did the general public. Richman (1973) notes that English civilian traffic wardens also share many values with the police. He found that almost three-quarters of civilian traffic wardens in England regard police work as the job which their work resembles most, and almost one quarter had an immediate family member in the police force. In Worcester, Massachusetts, Tien and Larson (1978) found that 88% of service aides aspired to become police officers. Because of their identity with police, as well as their experience in policing, they have been given preference in hiring for police positions. Only those who had been civilian police aides were hired as officers with the Scottsdale (Arizona) Police Department (Tien & Larson, 1978); and four (of ten) finalists selected for the position of full-time police officer in Paramus, New Jersey, were active Paramus reserve members, with two other reserves on the list selected by the police departments in New York City and Mesa, Arizona (Cooper & Greenberg, 1997). Conversely, Hill (1975) notes that all but two of the thirty civilian Executive Officers replacing Sergeants in the Sub-Divisional Administrative Units were former police officers in the London Metropolitan Police Department.

Civilians in police organizations, operating as liaisons or brokers, not only are supposed to break down barriers between police and other citizens, but are also supposed to reduce police alienation from the public by breaking down police subcultural insulation, which is a *sine qua non* for community policing. Pohl (1988, pp. 74–74iii) suggests that subcultural values that develop from the paramilitary organization and functions of the police agency and that conflict with democratic values can be countered by "mixing civilians with sworn officers." He goes on to state that this interaction presumably "brings new life and versatility to the police organization," and that "the practice of employing civilians within the police organization should tend to normalize the atmosphere within the agency." Here, presumably police officers learn how to interact with citizens in the community by first learning how to interact with civilians in the police department. However, Pohl (1988, p. 75) realizes that "civilians in lower level routine positions are more likely to adapt to the quasi-military police environment, even though not accepted into the quasi-military police subculture; and would have little impact on police subculture or would enhance to any significant extent the democratisation [sic] of North American police agencies. Only a much wider range of civilianized posts all throughout the police agency structure could very well carry much more impact in these areas by infusing civilians throughout the police agency, the very character of individual police employees, and the agency itself, changes."

Community police must see themselves as part of the community rather than as some form of occupying force. Because of their intensive special subculture, paramilitary police in SWAT teams are alienated from the public. The military mind-set and the accouterments of military dress and weaponry foster the development of an identity separate from the community, and thereby contribute to alienation from the community. Thurman et al. (2001) believe that rather than becoming more open and progressive, with the increasing establishment of PPUs, police culture may be returning to a more paramilitaristic orientation that was largely ineffective in dealing with the conditions and causes of crime in the 1970s. Moreover, Weber (1999) says that SWAT units share not only training and technology with the military, but most significantly, a common mind-set, and this mind-set is the warrior mentality of the military's special forces. Weber (1999, p. 7) quotes the former police chief of New Haven, saying that he "was offered tanks, bazookas, anything [he] wanted . . . [but] turned it down because it feeds a mind-set that you're not a police officer serving a community, you're a soldier at war." Weber (1999, p. 10) says that SWAT teams have developed an outlook in which "American streets are viewed as the 'front' and American citizens as the 'enemy.' " They are then likely to act like the military, which is trained to inflict maximum damage on enemy personnel rather than like civilian police, who are expected to use minimum force. Kraska (1996) even more specifically notes that many in these paramilitary units see themselves instead of the whole department or the military as the ones who will be the frontline defense against the inevitable emergence of civil disturbances in ghetto neighborhoods.

Kraska (1996) furthermore learned the officers enjoy the paramilitary activity to the boredom of their regular work, and apparently enjoy the greater technologically sophisticated equipment, weapons, and tactics they use to that of their normal work. This runs counter to those who feel it is a *sine qua non* to demilitarize the police in order to inaugurate the era of community policing (Falcone & Wells, 1995), and that demilitarization requires dispensing with the symbolic aspects of militarization. Fairchild (1988) recommends changing the military appearance and bearing of the police, as well as eliminating heavy armaments and military equipment, military style uniforms, and military terminology, in order to demilitarize the police successfully so that it will be less dysfunctional and alienating.

EXPANSION INTO AND USURPATION OF OTHER POLICE JURISDICTIONAL DOMAINS

When special units maintain their very limited missions that do not involve much interaction with other police units, they are less likely to interfere or undermine existing community-policing structures and practices. However, there is a tendency for special units to attempt to institutionalize, entrench, and normalize themselves by expanding their functions beyond their original purpose in order to justify their existence.

Often prompted by budget constraints and cost savings, civilianization customarily has been instituted to free sworn officers from non-critical areas so that they can perform the roles and responsibilities that require police power and that focus on skills for which they have been trained and are supposedly being paid (Cox, 1990). It has been assumed that the non-law enforcement police work that has been done by sworn officers—duties that do not relate to police officers' training, skills, expertise, or powers and that comprise 85% of a police department's calls for service—can be carried out just as effectively by civilians such as police service aides (Tien & Larson, 1978). Even when the service functions performed by civilians are viewed as important, they still have been characterized as detracting police officers from their law enforcement responsibilities (Leonard, 1980; Loveday, 1993). Skolnick and Bayley (1986), in discussing civilianization as one of the basic elements of community policing, also argue that civilianization functions to allow officers to concentrate on law enforcement. According to them (1986, p. 218), "with civilianization, police no longer need to concern themselves with traffic accident investigation, minor burglaries, victim assistance, victim follow-up, or teaching crime prevention to the public." They go on to point out that in Santa Ana, California, civilians wearing uniforms almost identical to sworn officers investigate traffic accidents, take crime reports, mediate neighborhood disputes, and organize communities for self-defense—which are largely community-policing tasks. It is quite likely, then, that the expansion of civilianization, especially in its volunteer auxiliary form, would have civilians replace sworn officers in much of what has become defined as community-policing tasks and functions.

Civilianization by use of auxiliaries goes further than traditional civilianization; volunteer auxiliaries no longer merely free police to do field activities that involve interaction with the public. Civilian auxiliaries take over service functions in the field, allowing sworn officers to devote more time and attention to fighting crime and maintaining order—or what has been considered "real police work" that they were trained to do. By freeing police from service functions, civilianization by use of auxiliaries may facilitate utilizing sworn officers almost exclusively for activities involving the use of force in general and possibly police paramilitary activities in particular.

Operational assignments or field duties that involve law enforcement or crime prevention become defined as the "real police work," while service activities become interpreted as peripheral to the "real business" of policing, and as "something to be done" on top of an officer's normal duties. In fact, the Worcester police were enticed to accept police service aides when they were told that they would have the opportunity to become more "professional" law enforcement specialists and would not have to spend their time on "garbage" calls unrelated to police work which they had complained about (Tien & Larson, 1978, p. 121). This orientation would certainly militate against the community-policing view that service activities provide a vital function and a great opportunity for police (Sparrow, 1988).

At the time that civilianization and police service aides were being introduced into policing, policing was moving away from the "general practitioner" model of the patrol officer (Tien & Larson, 1978, p. 119). By allowing sworn police officers to focus more on crime fighting and crime control, special units may essentially spawn greater specialization, and actually narrow the role of police officers. It is not involvement in the crime-fighting role that contradicts community policing, but instead only involvement in the crime-fighting role to the neglect of other roles that contradicts community policing.

While community policing is supposed to expand the role of police officers, giving them even greater and more general social control responsibilities and a generalist perspective, civilianization, as it has been conceived and implemented, has narrowed their role and fostered greater specialization; and therefore (like militarization) to primarily a focus on crime fighting, the role most compatible with the military model. It also may remove sworn officers from greater interaction with the community. This in turn, may lead to less identity with the civilian world, reinforcing a tendency for officers not to share civilian values and orientations. Ironically, it is possible, then, that civilianization, given the largely unanticipated effects of forcing sworn officers into law enforcement work only, can make the police even more remote from the community.

Consonant with the view of the police role as non-categorical and interdependent, the compartmentalization of police roles is unrealistic. When engaging in the service function, police familiarize and acquaint themselves with members of the community, which can possibly perform an important community-policing function: enhancing information-gathering activities in the future. Exclusive focus on the law enforcement role denies the police the opportunity to obtain valuable information from the public by developing more trust from the public through sworn officers' involvement in the service role. Also, as Gardiner (1987) points out, with civilianization there is a decrease in the flexibility of sworn officers to handle emergency situations as they focus on crime fighting. Therefore, when it is argued that civilians can do certain forms of non-police work as well as sworn officers (Hill, 1975), the availability and accessibility for performing other police functions is not being considered. Randall (1978), for example, notes that the Boston Police Patrolmen's Association argued that civilianizing traffic officers would deprive the downtown of much needed police protection since traffic officers made arrests.

There appears to be some indication that the civilianization of the service role in particular and of community policing in general has been occurring through use of volunteer auxiliaries. Tien and Larson (1978) state that unsworn Community Service Workers or Police Service Technicians in Worcester, Massachusetts had been performing what they refer to as para-police functions which have involved service-type functions to members of the public, as well as functions such as parking enforcement, traffic and accident investigation, and animal control. These service-type functions included taking written statements, non-hazardous calls for service, nuisance calls, locked vehicles, and response to public calls involving animals and pets. In Spokane, police volunteers in uniform, both senior citizens and youth, were directly involved in the provision of police services such as manning of information booths, various forms of public outreach, and crime target assessment, as well as the operation of a city-wide Block Watch program and the operation of the neighborhood COP Shops (Gaffney & Kelleher, 1999). In the Rochester-Monroe Co. CJ Program (1973), civilian training was supposed to prepare police service aides to undertake the short-circuiting of public discontent with police action and the securing of public cooperation by explaining to curious, uninformed, or hostile audiences at scenes of police action, the legal and practical constraints governing police behavior in the scene.

There also have been numerous recommendations to have civilian subprofessionals, often called community service officers, replace police in other significant types of work best identified with community policing that are non-routine in nature. In what certainly would also appear to be a considerable usurpation of many community-policing activities in which police officers themselves should be involved, Cooper and Greenberg (1997) suggest that volunteers could engage in numerous helpful delinquency prevention activities. In terms of primary delinquency prevention, they state that auxiliaries could be trained to conduct surveys to ensure that homes, commercial property, and streets possess adequate lighting, locks and appropriate levels of access control; could help recruit and train Neighborhood Watch groups, and increase the general level of crime deterrence by their own presence on routine patrols; could deliver presentations to community groups on how citizens can be safe from crime; could organize and staff after-school recreational centers and youth leagues; and could be used as security personnel at school functions, parking lots adjacent to convenience stores or shopping malls, and parks and community events. Alas, they could even track down "deadbeat" parents. In terms of secondary delinquency prevention, Cooper and Greenberg recommend that auxiliaries could be assigned to patrol areas where quality-of-life offenses (e.g., vandalism, graffiti) are most frequently committed by young people; could help locate runaways and work with social workers and police to ascertain whether runaway children can be safely returned to their homes or if they are in need of other types of assistance; and could staff hotlines and/or internet "chat lines" for aiding runaways, depressed young people, battered persons, or other individuals facing a crisis. They have even suggested that these volunteers mediate disputes, which they feel builds community cohesion. In terms of tertiary prevention, they suggest that these auxiliaries escort crime victims to and from court, serve as additional security personnel, and help investigate various family court cases (such as delinquency, status offenses, abuse, custody, visitation, etc.).

Also, in England, proposals were put forth for the development of specialized civilian units to deal with domestic disputes, and in the United States, Community Service Officers were to provide continuing assistance to families encountering domestic problems (Greenberg, 1979; Tien & Larson, 1978). As the circumscribed assistance role originally designed for volunteer auxiliaries changes and they assume more routine service and community-policing type activities by themselves, it is possible that in the future the civilians performing these expanded roles will be volunteer auxiliaries. Yanay (1993, p. 393) writes that the Civil Guard in Israel was used exclusively instead of regular police in community policing and crime prevention, and thus supposedly changed "the police" into one of community policing. All of these new responsibilities described that would be assumed by volunteer auxiliaries are opportunities to interact with the public that come under the rubric of community-policing functions. The usurpation of these functions and activities by civilian auxiliaries essentially becomes the civilianization of community policing rather than just the civilianization of policing contributing to the implementation of community policing.

While sworn officers' concern was originally with replacement of positions more than narrowing their jurisdictional domain, civilianization through the use of auxiliary police also has been perceived to encroach on and threaten to restrict or usurp the police officer's jurisdictional domain by replacing them in performing certain roles, rather than in supplementing them, as is assumed most special units do. Regular officers and their unions in the United States have been opposed to any form of civilian supplemental support such as community service officers, police service aides, or auxiliaries (Greenberg, 1979, 1984; Tien & Larson, 1978). Skepticism and hostility toward reserves by regulars can also be found in Britain, Holland, New Zealand, Canada, and Hong Kong (Gill & Mawby, 1990), as well as in the United States. Paradoxically, while the police want to do what they call "real police work," they have also opposed relinquishing functions not covered under that rubric. Even when civilians have moved into support positions which did not eliminate sworn positions, regular police officers have been suspicious. In fact, there has been a persistent apprehension by police in a number of societies that they would not only have some of their roles usurped, but that their numbers would be reduced or they would be replaced altogether; that there would be a wholesale transfer of the police role rather than sharing certain functions with auxiliaries and other civilians. For example, in his study in southeastern Michigan, Banning (1986) found that in 83% (20) of 24 police organizations he surveyed, full-time police officers and their unions interpreted the presence of the civilian volunteer police auxiliaries as a threat to the security of their positions. Interestingly, even after the Civil Guard in Israel was co-opted, police officers were opposed to them, feeling that they still infringed on the jurisdiction of the police. In terms of community policing, the positive effect of co-optation on improved involvement and relations between the community and the police organization does not necessarily translate to improved involvement and relations between the community and sworn officers.

Role expansion has also occurred with paramilitary police units. SWAT units were designed to react to specific situations and to perform a very specialized function (Gabor, 1993). However, Kraska and Cubellis (1997), and Twohey (2000) have pointed out that new paramilitary police units have played a much larger role than their originally designed functions involving response to hostage taking and barricaded subjects. Noting the problem with "mission creep," both Weber (1999) and Twohey (2000) state that inactive SWAT teams have a strong incentive to expand their original "emergency" mission into more routine policing activities to justify their existence. Consequently, there has been an expansion in terms of (1) more frequent use of SWAT units, (2) involvement in new crime-fighting and order maintenance activities, and (3) movement from reactive to proactive policing. The latter two developments coincide with the view of SWAT units as engaging in the zero-tolerance type of community policing.

Kraska and Cubellis (1997) found that there was a 238% increase of call-outs of paramilitary units between 1980 and 1995. The SWAT team of the St. Louis County Police Department became so institutionalized that it functioned as a group on a daily basis and never decentralized among various elements of the department (Wadsack, 1983). While Stevens and MacKenna (1989) state that routine incidents account for more gun battles and police officer injuries and deaths than do the more newsworthy conflicts between police and militant or terrorist groups, Gabor (1993) argues that SWAT call-outs are frequently unjustified and that roughly 99% of even barricade situations could be handled effectively by patrol units and negotiators alone. Twohey (2000) asserts that SWATs escalate rather than defuse the high-risk situations for which they were designed. It is not surprising, then, that according to Gabor, during a four-year period, negotiators for the Los Angeles County Sheriff's Office settled 99% of requests for SWAT assistance without shots being fired. Consonant with community policing, Gabor suggests that managers may find that uniform patrol units could actually handle the vast majority of situations that now result in SWAT call-outs.

Gabor (1993) believes that their overspecialization can create a sense within SWAT units that the specific abilities they possess represent the best response to almost every situation; and this runs counter to the evolving understanding that today's crime problems require a multi-faceted approach from law enforcement. In fact, SWAT teams have been used in many different types of law enforcement and order maintenance situations. They have been used to serve arrest and investigatory search warrants, conduct drug raids, to deal with crimes involving use of firearms, felony arrests, and mass arrests, and even to patrol high crime areas (DeMichelle & Kraska, 2001). Stevens and MacKenna (1989, p. 5) say that routine patrol activities were listed by 36% of respondents in their study as an activity in which tactical units are "often used" or "sometimes used," possibly indicating, according to them, the multiple assignment of tactical personnel in small departments. Furthermore, they have also been called out to deal with order maintenance problems, such as suicide calls, domestic disputes, and vicious dogs, besides used in training officers in special programs. The MJTF that Decker et al. (2000) evaluated became involved in activities such as nuisance abatement, responding to immediate neighborhood concerns over minor crimes, and reverse buy and bust activities that focus on short-term neighborhood problems rather than long-term undercover operations.

Not only has there been an increase in SWAT unit call-outs and a movement into new crime fighting and order maintenance functions and activities; SWAT units also have converted these functions from activities involving reactive policing to those involving proactive policing. DeMichelle and Kraska (2001, p. 10) found that 80% of PPU activity is now proactive in nature, and that "serving search and arrest warrants, most often as no-knock drug raids, account for more than three-quarters of all PPU activity." Kraska and Kappeler (1997) cite how SWAT teams are into saturation patrols in hot spots, and that they look for minor violations and do jump-outs, using an ostentatious display of weaponry. Expanding into new law enforcement and order maintenance functions, such as drug raids and street sweeps from previously dealing strictly with emergency situations such as hostage rescues, barricaded suspects, or civil disturbances has been interpreted by many officers, as Kraska (1996) has learned, as switching from a basically *reactive* function to emphasizing the *proactive* aspect of policing. They are now used as proactive forces, seeking evidence by carrying out searches based on no-knock search warrants. Even the community-oriented MJTF that Decker et al. (2000) evaluated engaged in aggressive preventive patrol, including such tactics as street stops, warrant checks, and neighborhood surveillance. According to them (2000, p. 7), "citizens wanted units that could respond quickly and resolve problems, and thereby gain their confidence."

Moreover, Kraska and Kappeler (1997) learned that officers narrowly interpret emphasizing proactive policing and the establishment of order, which involve aggressive paramilitary tactics and activities, as the logical extension of community policing. In fact, Kraska and Cubellis (1997, p. 623) found that 63% of the police agencies they surveyed that served 25,000 people or more agreed or strongly agreed that PPUs "play an important role in community policing strategies." It is, of course, the zero-tolerance version of community policing, which involves "the strict enforcement of all criminal and civil violations within certain geographical hot-spots . . . using an array of aggressive tactics such as street sweeps, proactive enforcement . . . and a proliferation of drug raids on private residences," as well as aggressive field interviews and Terry stops (DeMichelle & Kraska, 2001, pp. 7, 18). According to this model of community policing, police deal proactively with order maintenance problems and conditions assumed to give rise to crime as seriously as they do actual crime fighting, using tactics similar to aggressive policing. DeMichelle and Kraska (2001, p. 18), quote one PPU participant as noting that his unit "focus[es] on quality of life issues like illegal parking, loud music, bums, and neighbor troubles. We have the freedom to stay in a hot area and clean it up . . . Our tactical team works nicely with our department's emphasis on community policing." Similarly, Decker et al. (2000) concluded that the MJTF they studied in Kankakee contributed to the quality of life when it shifted to a police–community problem-solving structure and became involved in nuisance abatement, street suppression of minor crimes, and rapid response to other neighborhood complaints. DeMichelle and Kraska also quote (2001, p. 19) an official of a highly acclaimed community-policing department, who states that "we're into saturation patrols in hot spots . . . We look for minor violations and do jump-outs, either on people on the street or automobiles. After we jump out the second car provides periphery cover with an ostentatious display of weaponry." It is not surprising, then, that Kraska and Kappeler (1997), learned that officers have been hired for community policing and then shifted to staff new paramilitary police units.

If the community agrees in problem-solving meetings to the zero-tolerance strategies and tactics, the order maintenance patrolling and jump-out activities of PPUs that target problematic space (hot spots) with roving squads of tactical officers, as opposed to the cop-on-the-beat reacting to calls for service (DeMichelle & Kraska, 2001), might not contribute to alienation of the police from the community. However, the very appearance of PPUs, in battle garb with attack weapons, coupled with their mode of operation, would most likely contribute to their alienation from the community (Twohey, 2000). Also contributing to their alienation from the community is their tendency as a paramilitary force to view their task environment as involving interaction with masses in contrast to conventional police, whose tendency is to view their task environment as involving interpersonal interactions (Skolnick & Fyfe, 1993). And, because on their special proactive patrol, paramilitary officers have not been required to answer routine calls (Kraska & Cubellis, 1997), which would tend to isolate them even more from the public.

A SWAT team's expansion into other crime-fighting roles, such as serving warrants (and some order maintenance roles) rather than emergency situations would also tend to relegate regular police mainly to service roles and most order maintenance activities. When these PPUs have assumed patrolling responsibilities, they have usurped roles that regular patrol officers engage in as an important aspect of their involvement in community policing. Thus, the expansion of the roles of these special units can restrict the functions of other police units that have been engaged in more generalized roles, which, in turn, tends to undermine community policing.

Another criticism about PPUs is that their increased activity and expanding functions have not been linked to worsening conditions because there has been no increase in crime rates, drug use, fear of crime, or economic problems in the small cities in which they have been established which would even justify their use for non-emergency situations (Thurman et al., 2001). Thus the mission creep or expansion of these PPUs into routine police functions would appear to be more of a consequence of an attempt to institutionalize and entrench the unit within the police organization rather than being a genuine response to fulfilling a need for the greater specialized expertise of the paramilitary unit in dealing with increasingly complex functions.

THE EFFECTS OF THE TWO SPECIAL UNITS ON PROFESSIONALISM OF POLICE AND STRATIFICATION WITHIN POLICE DEPARTMENTS

The questions of jurisdictional domain and role incursion are directly related to the problem of police professionalism and community policing. The movement toward greater specialization and the use of unsworn police to perform non-police functions was traditionally seen as part of the professionalization process. It was felt that specialized work should be viewed as professional while social service roles should be civilianized. Professionalization has involved discarding trivial tasks. In fact, it may have the effect of leaving for police only professional activities involving the use of force. Essentially, the very maintenance of the larger role definition has been seen as deprofessionalizing the police, in part because they have not been trained for these ancillary roles (Leonard, 1980). In discussing the opportunity for police service aides to assume trivial tasks sworn officers previously performed, Tien and Larson (1978, p. 117) state that "the use of police service aides represents a major step . . . in the professionalization of police patrol." Even Skolnick and Bayley (1986) say there is a need for paraprofessionals so that police could be professionals.

The relationship between specialized work and professionalism has been extremely valued in policing, as well as the other professions, and directly related to status. For example, the Criminal Justice Commission in Queensland, Australia (1994) admits that the work of the general duties officer is still widely seen as involving low-level duties and as lacking prestige. Conversely, according to this perspective, it could logically be assumed that developing increased specialized skills for certain functions would tend to professionalize those functions. Thus, when certain law enforcement and order maintenance functions are assigned to PPUs because they become seen as requiring the special skills of PPUs, the type of work involved in performing those functions would be seen as becoming more professionalized.

Yet, in the community-policing model, the general duties police officer is supposed to take precedence over the specialist. Consonant with this community-policing perspective. Harring (1981) has described this attenuation of the police role by stripping it through civilianization of so-called ancillary functions as a form of deprofessionalization rather than professionalization. Likewise, Gardiner (1987, p. 50) has characterized the narrowing of the police officer's role to one of "enforcer" and away from the general practitioner role as the "death knell to the concept of professionalism." In somewhat of a different vein, Greenberg (1984) says that police officers feel that auxiliaries dilute their police authority and prestige, and thus also contribute to deprofessionalization. Police officers feel that when part-time job that doesn't require much. In any event, when police auxiliaries are assigned work previously undertaken by police regulars and subsequently redefined as paraprofessional, it certainly would seem to indicate that that type of work has been deprofessionalized.

Rather than community policing, a type of bifurcated (or even trifurcated) police organization could develop that is more likely to permit the simultaneous developments of civilianization and militarization and which, as it is implemented today, could involve the increasing use of paramilitary strike units in police departments. Skolnick and Bayley (1986) recognized that civilianization suggested a task-based system of stratification within policing alongside of, and in addition to, the ranking structure in all police departments. In view of the lower status of full-time civilians and part-time auxiliaries and reserves in police departments, it would be logical to assume the tasks they have been given in community policing and crime prevention have low status within the department. Furthermore, the traditional description of these tasks as paraprofessional has led to viewing roles incorporating them as inferior within the police organization. It is thus possible that a two-tiered system will arise with the first tier professional work, such as law enforcement, investigation, and even some forms of order maintenance, assigned to sworn officers (and technical specialist work to some full-time civilians), and a second tier community work and service-related responsibilities reassigned to paraprofessional civilians (both part-time auxiliary and some full-time civilian). It should be emphasized that in the field volunteer auxiliaries or civilians within police departments would work with the community, whereas sworn officers would focus exclusively on crime and disorder. As work is divided between sworn and unsworn personnel, specialization is increased within police departments. Insofar as paraprofessionals, for whatever reason (usually economic savings), assume work previously done by professionals, deprofessionalization could be said to have occurred.

The tendency toward a two-tiered system is especially true if community-policing tasks are perceived as secondary rather than as tasks involving police expertise, and those who engage in these community-policing tasks as having inferior status within police departments. In fact, a new position within the patrol division, the Community Service Officer, has been referred to as a paraprofessional position *vis-à-vis* the regular patrol officer.

Stratification has also occurred in police departments with the introduction of SWAT teams. The special status that SWAT units have been given has led them to view themselves, and has led other officers to view them, as an elite within the department (Kraska, 1996), which has not only created antagonisms, but also threatened cooperation. Gabor (1993) sees this elitism as divisive within police departments, and says that officers and supervisors in specialized units may lose a sense of identity with personnel in other sections of the department. If these views are allowed to persist, specialists begin to view patrol officers—the generalists—as second-class personnel. Even a perceived arrogance on the part of specialized units can lead to serious and organizationally devastating conflicts within agencies. Hence, according to Stevens and MacKenna (1989), those in their survey of tactical units attributed jealousy on the part of patrol officers and other supervisors as a source of friction between patrol units and tactical unit personnel.

COOPERATION AND COORDINATION BETWEEN SPECIAL UNITS AND THE COMMUNITY AND OTHER UNITS IN POLICE DEPARTMENT

In order for special units to be compatible with the team policing organizational aspects of community policing, those units must work together with other police units as well as with the community. Civilian auxiliaries need to team up with and accompany regular officers as they engage in community-policing activities. Greenberg (1979) believes that where auxiliary police and regular police have been kept separate and isolated from one another, unsatisfactory results have occurred. Already in 1973 the Rochester-Monroe Co. CJ Pilot Program appreciated this, having police–civilian teams ally themselves to other community agencies in behalf of the required community services in order to respond to the estrangement of police from the community.

Kaiser (1990) believes that tactical units also cannot operate within a vacuum and depend on the support of other units; that all police units must work as one, not as individual entities. Increased specialization requires increased cooperation between different units. According to Kaiser, there are specific roles for patrol units, investigators, K-9 teams, and tactical and negotiation personnel, and that establishing a team concept helps to minimize jealousy on the part of non-tactical officers.

However, there are problems of cooperation and coordination when integrating special units within a community-policing structure that must be overcome. In the model precinct in Brooklyn that Pate and Shtull described, not only did the attempt to absorb certain specialists in the generalist role of the SOU not work, but even an attempt to coordinate special units with SOU did not succeed. Although the original plan for the community-policing scheme in the model Brooklyn precinct was for members of the detectives, organized crime, narcotics, public morals, and mounted special units to become involved with SOU in problem-solving activities, their participation was minimal (Pate & Shtull, 1994). Pate and Shtull (1994) attributed this lack of involvement to the fact that the modes of operation and goals of these special units had not yet been fully integrated into the community-policing model, as well as to the fact that those units had a different reporting structure. Also, cooperation and coordination may not develop on a foundation where there is a division of labor and a compartmentalization of roles that eventuates in the development of elitism and socially distant strata in police departments, which is the case with the two special units examined and with special units in general.

CONCLUSION

Very little has been written theoretically or empirically about COP and specialist units other than whether COP itself should be established as an isolated specialist unit or developed as a department-wide philosophical commitment or strategic approach. It has been found that while certain specialist functions can be performed by community-policing generalists, certain other specialist functions that require significant training and expertise cannot be effectively absorbed in the larger role of COP generalists.

This chapter has examined two special units, the volunteer auxiliary and the paramilitary police unit, which can be analyzed from two frameworks representing diametrically opposing movements within policing: the first, civilianization of the police, and the second, militarization of the police. In turn, these two types of special units have been characterized as related to two different models of community policing: the volunteer civilian auxiliary fitting with the partnership model of community policing that emphasizes police–community cooperative endeavors and the service function; and the paramilitary police unit being more compatible with the zero-tolerance model that emphasizes order maintenance.

Community policing is supposed to be directed toward reducing social distance and boundaries between the police and community, which depends, in part, on the identity of those in special units. The civilian auxiliary unit represents the case of bringing the community into community policing through co-optation. Here, presumably representative segments of the community, maintaining an identification with the community, are brought into the police organization to provide a valuable liaison function in the development of a police–community partnership. If civilians in police organizations don't operate as liaisons or brokers, they will merely be co-opted by the police organization, and will fail to break-down police subcultural insulation, which is a *sine qua non* for community policing. Current evidence would seem to indicate that volunteer auxiliaries are demographically, but not necessarily attitudinally, representative of the community, particularly in regard to their stronger identification with the police. Therefore, volunteer auxiliaries themselves may become somewhat alienated from the community when their civilian identity is attenuated as they are co-opted into the police organization, and thus be incapable of performing the role of honest broker or liaison.

Paramilitary police have been alienated from the public because of their strong identification with their special intensive subculture and special tactics. Given their military organization and accouterments of military dress and weaponry, they develop a military mind-set or warrior mentality in which they see themselves as an impersonal force on a mission against enemies in neighborhoods treated as fronts. And, although civilian police are expected to use minimum force, SWAT units are notorious for using an overwhelming amount of force Their paramilitary appearance and mode of operation would very likely contribute to this police unit's alienation and social distance from the community in which they are supposed to be a part according to community policing.

Because they are not traditional units within police departments, there is a tendency for special units to attempt to institutionalize, entrench, and normalize themselves by expanding their functions beyond their original specialized purpose in order to justify their existence. When examining special units in regard to their primary and expanding functions, it is important to understand not only the type of roles that they perform, but also how they affect and even possibly usurp the roles of other units and officers in the police organization. It is probably better for the continued sustenance of community policing in police departments for special units to perform their own special and limited functions than becoming involved in routine functions performed better by police generalists. The expansion into and usurpation of routine police functions by these two units has basically meant the civilianization of the service function and the militarization of the crime control and, to a lesser extent, the order maintenance functions.

While full-time civilians took over non-operational positions that freed sworn officers to return to the field, civilian auxiliaries have assumed by themselves functions in the field that involve critical interaction with the community and closely resemble community-policing activities. Thus, the use of volunteer auxiliaries can and has led to civilians usurping rather than "partnering" with sworn officers in community-policing functions. This form of civilianization as it has been practiced has consequently had four potential major effects: (1) it shifts sworn officers more into law enforcement and order maintenance work only; (2) it deprofessionalizes and reduces in status the service function and many community-policing activities; (3) it creates specialized civilian community-policing units; and (4) it makes police officers even more remote from the community. By allowing sworn police officers to focus more on crime control and certain forms of order maintenance, it narrows the role of police officers, and thus spawns greater specialization. Given the interdependence of roles in policing, when volunteer auxiliaries replace police in duties involving the community (e.g., service functions) and assume sole responsibility for community-policing activities and roles, sworn officers interact less with and become less accessible and available to the public, acquire less valuable information from the public, and eventually become more alienated and socially distant.

Paramilitary police units have played a much larger role than their originally designed functions of response to hostage taking or barricaded subjects. There has been an expansion of these units in terms of (1) more frequent use, (2) involvement in new crime-fighting and order maintenance activities, and (3) movement from reactive to proactive policing. The movement into proactive order maintenance policing coincides with what is commonly characterized as the zero-tolerance version of community policing. The mission creep or expansion into routine police functions has been the consequence of attempts to institutionalize and entrench the paramilitary unit within the police organization rather than being a genuine response to fulfilling the need for the greater specialized expertise of the paramilitary unit in dealing with increasingly complex functions. Community policing in its zero-tolerance version, coupled with the perceived need, status, and professional skills of these units, has, in turn, provided these PPUs with the legitimacy to undertake order maintenance policing and routine crime fighting.

Whether talking about community policing in terms of the co-optation of the community or co-production of order with the community, what is fundamental is that real cooperation between police and community occur. If they are genuinely to engage in community policing, volunteer auxiliaries can only perform the liaison function between police and community adequately by maintaining identification with the communities they are supposed to represent and by working together with police officers in community-policing initiatives similar to what occurred in the Rochester-Monroe Co. CJ Pilot Program. In the case of PPUs, when even implementing a zero-tolerance version of community policing, they must incorporate into their units a joint police-community, as well as proactive, problem-solving approach, as was done in the Kankakee MJTF. Given the experience of the community-policing experiment in the model precinct in Brooklyn that Pate and Shtull described, it is not sufficient for special units merely to participate in problem-solving activities of other units in a community-policing structure, particularly specialist COP units. Instead it is necessary that these special units themselves inculcate problem-solving principles into the very modes of operation and goals of their own units even before coordinating with other units in a community-policing structure. As a priority, this means a community-policing philosophy or strategy be incorporated into special units rather than special units being incorporated into a community-policing organizational structure.

RECOMMENDED READINGS

CORDNER, G. (1995). Community policing: Elements and effects. *Police Forum, 5*(1), 1–8. Reprinted in R.G. DUNHAM, & G.P. Alpert (Eds.), *Critical issues in policing* (3rd ed.), pp. 451–468. Prospect Heights, IL: Waveland.

GREENE, J.R. (1999). Zero tolerance: A case study of police policies and practices in New York City. *Crime and Delinquency, 45*(2), 171–187. Reprinted in G.P. ALPERT, & A. Piquero (Eds.), *Community policing: Contemporary readings* (2nd ed.), pp. 505–519. Prospect Heights, IL: Waveland.

GREENE, J.R., & PELFREY, W.V. (2001). Shifting the balance of power between police and community. In R.G. DUNHAM, & G.P. Alpert (Eds.), *Critical issues in policing: Contemporary readings* (4th ed.), pp. 435–465. Prospect Heights, IL: Waveland.

JOLIN, A., & MOOSE, C.A. (1997). Evaluating a domestic violence program in a community policing environment: Research implementation issues. *Crime and Delinquency, 43*(3), 279–297.

MAGUIRE, E.R. (2003). *Organizational structure in American police agencies: Context, Complexity, and Control*. Albany, NY: State University of New York Press.

PATE, A.M., & SHTULL, P. (1994). Community policing grows in Brooklyn: An inside view of the New York City Police Department's model precinct. *Crime and Delinquency, 40*, 384–410.

PEAK, K.J., & GLENSOR, R.W. (1996). *Community policing and problem solving: Strategies and practices*. Englewood Cliffs, NJ: Prentice-Hall.

REFERENCES

BANNING, B.W. (1986). *The utilization of civilian personnel in the uniformed police function.* Unpublished masters thesis, Eastern Michigan University.

BUTZER, D., BRONFMAN, L.M., & SPIVAK, B. (1996). Role of police in combating domestic violence in the United States: A case study of the domestic violence reduction unit, Portand Police Bureau. In M. PAGON (Ed.), *Policing in Central and Eastern Europe: Comparing firsthand knowledge with experience from the west,* pp. 161–172.

COOPER, K., & GREENBERG, M.A. (1997). Auxiliary police help prevent delinquency. *The Police Chief, 64*(10), 116–118.

CORDNER, G. (1995). Community policing: Elements and effects. *Police Forum, 5*(1), 1–8. Reprinted in R.G. Dunham, & G.P. Alpert (Eds.), *Critical issues in policing* (3rd ed.), pp. 451–468. Prospect Heights, IL: Waveland.

COX, S.M. (1990). Policing into the 21st century. *Police Studies, 13*(4), 168–177.

CRIMINAL JUSTICE COMMISSION (1994, August). *Implementation of reform within the Queensland Police Service: The response of the Queensland Police Service to the Fitzgerald Inquiry Recommendations.* Unpublished report prepared by the Criminal Justice Commission.

DECKER, S.H., BYNUM, T.S., CURRY, G.C., & SWIFT, D. (2000, September). *Evaluation of the Kankakee Metropolitan Enforcement Group.* Unpublished report prepared for the Illinois Criminal Justice Information Authority. Justice Research Associates.

DOERNER, W. G. (1998). *Introduction to law enforcement: An insider's view.* Boston: Butterworth-Heinemann.

DEMICHELLE, M.T., & KRASKA, P.B. (2001, April). Community policing in battle garb: A paradox or coherent strategy? Paper presented at the annual meeting of the Academy of Criminal Justice Sciences, Washington, D.C.

DUBIN, M. (1973). The New York City auxiliary police. *Law and Order, 21*(8), 95–97.

FAIRCHILD, E.S. (1988). *German police: Ideals and reality in the post-war years.* Springfield, IL: C.C. Thomas.

FALCONE, D.N. & WELLS, L.E. (1995). The county sheriff as a distinctive policing modality. *American Journal of Police, 14*(3/4), 123–149. Reprinted in L.K. GAINES, & G.W. CORDNER (Eds.), 1999 *Policing perspectives: An anthology,* pp. 41–56. Los Angeles: Roxbury.

GABOR, T. (1993). Rethinking SWAT. *FBI Law Enforcement Bulletin, 62*(April), 22–25.

GAFFNEY, M.J., & KELLEHER, A.M. (1999, March). Paper presented at the annual meeting of the Academy of Criminal Justice Sciences, Orlando, FL.

GAINES, L.K., KAPPELER, V.E., & VAUGHN, J.B. (1994). *Policing in America.* Cincinnati, OH: Anderson.

GARDINER, J.M. (1987). *The impact of civilianization on law enforcement training: Year 2000.* Unpublished report for the Commission on Peace Officer Standards and Training Command College Class 111.

GILL, M.L., & MAWBY, R.I. (1990). *Special constables: A study of the police reserve.* Aldershot, England: Avebury.

GREENBERG, M.A. (1979). Police volunteers: Are they really necessary? In R.G. Iacovetta, & D.H. Chang (Eds.), *Critical issues in criminal justice,* pp. 266–275. Durham, NC: Carolina Academic Press.

——. (1984). *Auxiliary police: The citizen's approach to public safety.* Westport, CT: Greenwood.

GREENE, J. A. (1999). Zero tolerance: A case study of police policies and practices in New York City. *Crime and Delinquency, 45*(2), 171–187. Reprinted in G.P. ALPERT, & A. PIQUERO (Eds.), *Community policing: Contemporary readings* (2nd ed.), pp. 505–519. Prospect Heights, IL: Waveland.

GREENE, J.R., & PELFREY, W.V. (2001). Shifting the balance of power between police and community. In R.G. DUNHAM, & G.P. ALPERT (Eds.), *Critical issues in policing: Contemporary readings* (4th ed.), pp. 435–465. Prospect Heights, IL: Waveland.

GUYOT, D. (1979). Bending granite: Attempts to change the rank structure of American police departments. *Journal of Police Science and Administration, 7*, 253–284.

HARRING, S. (1981). Taylorization of police work: Prospects for the 1980s. *Insurgent Sociologist, 10*(4) & *11*(1), 25–32.

HILL, M.A. (1975). Civilianization in the police service. *Criminologist, 10*(35–36), 22–38.

JOLIN, A., & MOOSE, C.A. (1997). Evaluating a domestic violence program in a community policing environment: Research implementation issues. *Crime and Delinquency, 43*(3), 279–297.

KAISER, N.F. (1990). The tactical incident: A total police response. *FBI Law Enforcement Bulletin, 59*(August), 14–18.

KRASKA, P.B. (1996). Enjoying militarism: Political/personal dilemmas in studying U.S. police paramilitary units. *Justice Quarterly, 13*(3), 405–429.

KRASKA, P.B., & CUBELLIS, L.J. (1997). Militarizing Mayberry and beyond: Making sense of American paramilitary policing. *Justice Quarterly, 14*(4), 607–629.

KRASKA, P.B., & KAPPELER, V.E. (1997). Militarizing American police: the rise and normalization of paramilitary units. *Social Problems, 44*(1), 1–18.

LANGWORTHY, R.H., & TRAVIS, L.F. (1994). *Policing in America: A balance of forces.* Cincinnati, OH: Anderson.

LEONARD, V.A. (1980). *Fundamentals of law enforcement: Problems and issues.* St. Paul, MN: West.

LOVEDAY, B. (1993). *Civilian staff in the police force: Competencies and conflict in the police force.* Leicester, England: Scarman Centre for the Study of Public Order, University of Leicester.

MAPSTONE, R. (1994). *Policing a divided society: A study of part time policing in Northern Ireland.* Aldershot, England: Avebury.

PATE, A.M., & SHTULL, P. (1994). Community policing grows in Brooklyn: An inside view of the New York City Police Department's model precinct. *Crime and Delinquency, 40*, 384–410.

PEAK, K.J., & GLENSOR, R.W. (1996). *Community policing and problem solving: Strategies and practices.* Englewood Cliffs, NJ: Prentice-Hall.

POHL, B. (1988). *The impact of civilianization on police agencies in Canada.* Unpublished masters thesis, University of Ottawa.

RANDALL, F. (1978). Holding on: Union resistance to civilian employees in the Boston Police Department. In R.C. LARSON (Ed.), *Police accountability: Performance measures and unionism,* pp. 167–184. Lexington, MA: Lexington Books.

RICHMAN, J. (1973). Police auxiliaries: Traffic wardens. *Police Journal, 45*(2), 135–149.

ROCHESTER-MONROE COUNTY CRIMINAL JUSTICE PILOT CITY Program (1973). *"Pac-Tac": Police and citizens together against crime experimental action program.* Unpublished report prepared for the City of Rochester and the Rochester Police Department.

ROSENBAUM, D.P., & LURIGIO, A.J. (1994). An inside look at community policing reform: Definitions, organizational changes, and evaluation findings. *Crime and Delinquency, 40*(3), 299–314.

SKOLNICK, J.H., & BAYLEY, D.H. (1986). *The new blue line: Police innovation in six American cities.* New York: The Free Press.

SKOLNICK, J.H., & FYFE, J.J. (1993). *Above the law: Police and the excessive use of force.* New York: The Free Press.

SPARROW, M.K. (1988). Implementing community policing. Report in *Perspectives on policing* series. Washington, DC: National Institute of Justice.

STEVENS, J.W., & MACKENNA, D.W. (1989). Assignment and coordination of tactical units. *FBI Law Enforcement Bulletin*, *58*(March), 2–9.

THURMAN, Q., ZHAO, J., & GIACOMAZZI, A.L. (2001). *Community policing in a community era.* Los Angeles: Roxbury.

TIEN, J.M., & LARSON, R.C. (1978). Police service aides: Paraprofessionals for police. *Journal of Criminal Justice*, *6*, 117–131.

TROJANOWICZ, R.C., & BUCQUEROUX, B. (1992). *Toward development of meaningful and effective performance evaluations.* East Lansing, MI: National Center for Community Policing, Michigan State University.

TWOHEY, M. (2000). SWATs under fire. *National Journal*, *32*, 37–44.

WADSACK, L. (1983). High-risk warrant executions—a systematic approach. *FBI Law Enforcement Bulletin*, *52*(December), 13–16.

WEBER, D.C. (1999). Warrior cops: The ominous growth of paramilitarism in American police departments. *Cato Institute Briefing Papers*, No. 50, 1–13.

WILSON, J.Q., & KELLING, G.L. (1982). The police and neighborhood safety: Broken windows. *Atlantic Monthly*, March, 29–38.

YANAY, U. (1993). Co-opting vigilantism: Government response to community action for personal safety. *Journal of Public Policy*, *13*, 381–396.

4

Elder Abuse Units

Allen Sapp and Carla Mahaffey-Sapp

To date, elder abuse has been treated by most police agencies as crimes against persons (where the cases get little priority when mixed with homicides, aggravated assaults, and other more common violent crimes) or as an extension of the child abuse problem (where again they have relatively low priority). Financial abuse of the elderly gets little attention since there seems to be a dollar landmark value for priority cases. A few law enforcement departments are now organizing elder abuse units to focus on the specialized needs and problems of elderly persons. Given the rapid aging of America, the future needs in this area will increase dramatically and it is likely that most larger law enforcement agencies/departments will be called upon eventually to establish elder abuse units. Therefore, this chapter examines the concept of a specialized unit to address abuse of the elderly.

THE PROBLEM

In the United States, life expectancy has risen dramatically to 76 years in the past century. Anderson and Thobaben (1984) state that in 1900, one of eleven Americans was 65 or over; today it is one in eight (U.S. Census Bureau, 1996). By the year 2020, nearly one out of every five Americans will be 65 years of age or older (U.S. Census Bureau, 1996). Rapid growth of this population segment is accompanied by problems that are of concern to all society. The victimization of elderly persons leads the list of problems associated with the increase in the numbers of elderly.

As the aging population continues to increase from 3% of the population in 1900 to 19% today, family services and long-term care systems are identifying a syndrome of elder abuse and neglect. Elder abuse and neglect are forms of family violence, and a growing societal concern (Fulmer & Cahill, 1984).

Already, law enforcement agencies are being required to provide investigative and prosecutorial support in cases of elder abuse. The police administrator will be particularly concerned with unit and/or case management issues related to allegations of elder abuse. The agency creating a special unit for elder abuse will have to answer a number of questions and address a variety of issues that are discussed in this chapter.

Unlike other forms of violence, elder abuse cuts across socioeconomic and racial lines. "Professional families did it, working-class families did it, black families did it, and white families did it" (Yin, 1985, p. 11). Overt episodes of assault and battery against elders have been documented, as well as acts of omission, as in cases where elders have been left to deteriorate without adequate clothing, food, shelter, or health care intervention. Since the elderly are known to have substandard levels of income and an average of 3.5 chronic diseases per person, it is highly likely that their basic needs may not be met and they may suffer from the results of deprivation (Fulmer & Cahill, 1984).

Shapiro (1992) notes that the concept of elder abuse did not even exist until 1979 when initial studies confirmed that there was a form of abuse of the elderly similar to the abuse of children (Block & Sinnot, 1979; Gelles & Strauss, 1979). Shapiro (1992) also notes that following those studies an extensive network of legal reporting, investigation, and involuntary commitment of elderly for their own protection developed, modeled after the child abuse legal system. Shapiro (1992) argues that the elderly are not children and that treating them as such degrades and fails to provide the necessary services and support systems needed.

Researchers have confirmed that there are large numbers of victims of elder abuse in the United States (Block & Sinnot, 1979; Elder Abuse, 1990; FHP Foundation, 1995; Fulmer & Wetle, 1986; Gelles & Strauss, 1979; Pillemor & Finkelhor, 1986, 1988, among others). In 1999, the rate of violent crime victimization of persons aged 65 or more was 4 per 1000 (Rennison, 2000). Overall, elderly persons are the least likely of all age groups to experience crime of all types, including violent crime, personal theft, and household crime (Bureau of Justice Statistics, 1994). These estimates are based on interviews conducted as part of the National Crime Victimization Survey. The survey includes the violent crimes of rape, robbery and assault, personal theft, and the household crimes of burglary, larceny and motor vehicle theft (Bureau of Justice Statistics, 1994).

According to the Federal Bureau of Investigation (1997), nearly 5% of all murder victims in 1996 were aged 65 or more. Robberies make up about 38% of crimes against the elderly and the elderly victims are far more likely to receive serious injuries that require medical care or hospitalization than are younger victims (Bureau of Justice Statistics, 1994). The official data do not include many of the crimes committed against the elderly (Reid, 1990). These figures fail to reflect perhaps the most prevalent and damaging form of crime against the elderly, elder abuse.

The first National Elder Abuse Incidence Study estimates that a total of 551,011 elderly persons, aged 60 and over, experienced abuse, neglect, and/or self-neglect in domestic settings in 1996. Of this total, 115,110 (21%) were reported to and substantiated by adult protective service agencies and 435,901 (79%) not reported. Thus, new incidents of elder abuse, neglect, and/or self-neglect were nearly four times as likely to go unreported than those reported in 1996 (National Center, 1998).

Neglect of the elderly was the most frequent type of elder maltreatment (48.7%); emotional/psychological abuse was the second (35.5%); physical abuse was the third (25.6%); financial/material exploitation was the fourth (30.2%); and abandonment was the least common (3.6%) (National Center, 1998). Adult children comprised the largest category of abusers (47.3%) of substantiated incidents of elder abuse with spouses accounting for 19.3%. Other relatives were the offenders in 8.8% of the abuse cases and grandchildren were the abusers in 8.6% (National Center, 1998).

Three-fourths of the elder abuse and neglect victims were physically frail and 47.9% of substantiated incidents of abuse and neglect involved elderly persons who were unable to physically care for themselves. Another 28.7% of the victims were classified as marginally able to care for themselves (National Center, 1998).

Some experts estimate that as few as one out of fourteen domestic elder abuse incidents ever come to the attention of authorities. Based on these estimates, somewhere between 820,000 and 1,860,000 elders were victims of abuse in 1996 (Tatara, 1997). Over the decade from 1986 to 1996, a steady increase in the reporting of domestic elder abuse nationwide was noted, from 117,000 reported cases in 1986 to 293,000 reported cases in 1996—a 150.4% increase (Tatara, 1997).

According to the National Center on Elder Abuse (1998), 66.4% of victims of domestic elder abuse were white, 18.7% were black, 10.4% were Hispanic, and 1% each were Native Americans and Asian Americans/Pacific Islanders for the reporting year 1996. The National Center (1998) found that 22.5% of domestic elder abuse reports came from physicians and other health care professional. Family members reported the abuse in 16.3% of the cases and 15.1% were reported by other care service providers. Other reporting sources, including law enforcement, friends, neighbors, clergy, banks, and business institutions among others accounted for the remaining 46.1% of reports to authorities (National Center, 1998).

A 1990 survey of all of the state human service departments indicated that elder abuse was up 50% from 1980 and that over 1.5 million elderly citizens are abused by their loved ones annually (U.S. Congress, House Select Committee, 1990). According to the chairman of a subcommittee of the House Select Committee on Aging, "domestic violence against the elderly is a burgeoning national scandal" (1990, vi). Witnesses before that committee testified that abuse includes not only violent attacks upon the person of the elderly, but also such acts as withholding food, stealing their savings and social security checks, verbal abuse, and threats of sending the elderly family member to a nursing home (Reid, 1990).

The National Elder Abuse Incidence Study (1998) confirmed that reported elder abuse cases make up only the "tip of the iceberg." The study estimates that at least half a million older persons in domestic settings were newly abused, neglected, and/or exploited, or experienced self-neglect, in 1996. The study also found that for every reported incident of elder abuse, neglect, exploitation, or self-neglect, approximately five go unreported. The results of these studies suggest that 4–5% of all older persons suffer elder abuse in one form or another.

A hearing on June 14, 2001 before the Senate Special Committee on Aging heard testimony from the Director of the National Center on Elder Abuse that nationwide reports of abuse of senior citizens had increased from 117,000 cases in 1986 to 470,000 cases in 2000 (Avanis, 2001). These figures reflect an increase of 301% in reported cases. Paul Greenwood, Deputy District Attorney and head of San Diego's Elder Abuse Prosecution Unit, testified that elder abuse is "one of the most serious issues facing law enforcement and prosecutors within the next five years" (Craig, 2001).

Senator Larry Craig of Idaho called for coordination of existing resources to combat this growing problem. He stated, "What we need to do now is make sure that law enforcement, prosecutors and others involved in the system are aware of the laws, have the necessary specialized training, and are working together to bring these cases to trial" (Craig, 2001).

BACKGROUND OF THE PROBLEM

The research on elder abuse does not provide sufficient data to draw definitive profiles of victims and perpetrators. However, some general patterns are evident in the literature. That literature, however, is based almost totally on official reports and represents only those cases brought to official attention. It is important to realize that, despite the general profiles of victims, elder abuse can happen to anyone of any class, ethnic group, or gender (Police Executive Research Forum, 1993).

Steinmetz (1988) conducted the most extensive study on elder abuse and family care to date. The study was conducted with funding provided by the Division of Aging and took several years from 1980 to 1986 to complete the data-gathering process. There were several unanswered research questions that guided the study. First, what was the demographic profile of families that were caring for an elder? Second, what was the nature of tasks that these families were performing for the elderly? Third, was the elder's dependency, as measured by task performance, perceived as stressful or burdensome by the caregiver? Fourth, how did these caregivers and elders resolve conflicts or attempt to gain or maintain control? Finally, was there evidence of intergenerational transmission of patterns of interaction by adult children and elderly parents?

Steinmetz (1988) examined the impact that caring for an elderly parent had on the lives of the middle-aged and older care-giving offspring. He drew a sample of 104 caregivers with a total of 199 elderly relatives under their direct care. The daily tasks that were provided for the elderly by these caregivers and the resulting stress, conflict, and abuse were detailed. It was noted that in most instances, these caregivers probably represented model caregivers, yet the psychological, verbal, and physical abuse perpetrated on the elders was astonishing. The information was obtained through semi-structured in-depth interviews, which yielded the data that form the basis of this book. A second sample, 350 questionnaires mailed to service providers, produced 153 usable questionnaires that enabled comparisons of the general characteristics of abused elders in the study locale with those of other studies using third-party reports.

At the time of this study, there was no mechanism for systematically identifying elders under direct family care. Even using census data for developing a sampling frame was impractical since it is virtually impossible to ascertain whether the older person was the head or the dependent individual in these multigenerational families. A major goal of the study was to obtain reliable data that would provide information not only on abusive and neglectful interactions between caregivers and their elderly dependent parents, but also to illuminate the relationship between increased levels of dependency, stress, and feelings of burden and elder abuse (Steinmetz, 1988).

Although differences were observed in demographic profiles, relatively few of the findings reached statistical significance. The lack of systematic patterns or trends provides support that the sample was not skewed for the major demographic variables examined. Out of 182 possible relationships based on family characteristic variables, only 10, just over 5%, were significant. This suggests that variables such as race, residence, the length of time one has been providing care, the caregiver–elder relationship, and caring for more than one elder have little, if any, direct effect on the provision of dependency tasks, the stress produced by performing these tasks, family stress, caregivers' and elders' use of control maintenance techniques, and overall feelings of burden (Steinmetz, 1988). Steinmetz found that higher levels of verbal abuse, medical abuse, psychological abuse, physical, and total abuse characterized stressed caregivers. Caregivers who perceive that care-giving is stressful and a burden are more likely to use negative forms of control maintenance techniques (Steinmetz, 1988; Kilburn, 1996).

This study showed clearly that consistent patterns emerge in the methods used by elders and caregivers in their attempts to grasp the reins of control or to hold on to them. Families in which more negative forms of interactions occur tend also to be experiencing somewhat more stress and performing more dependency tasks. Longer periods of care-giving also tend to be associated with many abusive adult child–elderly parent interactions. Elders' use of physical violence increased the total effect of both dependency and dependency stress to elder abuse. Overall, 23% of the caregivers admitted to using physically abusive methods to control the elders (Steinmetz, 1988). Mistreatment seems to occur more frequently to elderly females (Scofield et al., 1999) who are frail, physically or mentally impaired, over 65 years of age, highly dependent on others for their daily needs, and living with relatives (Pillemer & Finkelhor, 1988; Soeda & Araki, 1999). Karmen (1990) indicates that most abused elders are living in relative social isolation from neighbors, friends, and other relatives who could interfere with the abuse.

The over-representation of females among the victims of elder abuse is a function of demographics as well as a matter of gender. In 1996, there were 20 million women 65 years or older, 145 women for every 100 men. The sex ratio increased with age, ranging from 120 women for every 100 men for 65–69-year-olds, to 257 women for every 100 men among persons 85 years and older. In 1996, women reaching age 65 had an average life expectancy of an additional 19.2 years while the expectancy was 15.5 years for men (U.S. Census Bureau, 1996).

Kosberg (1998) notes the high risk of abuse of elderly men when they are lonely, living in inner cities, homosexual, or incarcerated. Abuse of elderly men occurs both in domestic settings by spouses, children, relatives, and friends, and in institutional settings by paid employees. Additionally, Kosberg (1998) indicates that the rates of self-abuse and self-neglect among elderly men are very high. Landau (1998) points out that there are cases where the victim of abuse is also an abuser.

An abuser is likely to be a relative of the elderly victim (Block & Sinnot, 1979; Steinmetz, 1988), most commonly a sibling, spouse, or child (Karmen, 1990), and to have a dependency relationship with the victim (Wolf & Pillemer, 1989; Kilburn, 1996). The abuser tends to be overburdened, depressed, and hostile as he or she faces the prospect of long-term care for a dependent person who may have physical, emotional, and mental impairments (Karmen, 1990). Sons are the most frequent physical abusers when the victim is homebound while daughters are more likely to neglect or emotionally abuse the victim (Karmen, 1990).

The thought of an exhausted caregiver striking out at an impaired elder after years of frustration is probably the most understandable of the theories as to why elder abuse and neglect occur. It is important to realize, however, that although this theory of causation is the easiest with which to identify, it is not necessarily the right one (Stanhope & Lancaster, 1988). As an example, Soeda and Araki (1999) found that neglect by daughters-in-law to be the most prevalent type of elder abuse occurring in Japanese homes. The authors attributed this finding to "poor relationships" between daughters-in-law and their in-laws. This suggests there may be a strong cultural aspect to elder abuse in addition to the other factors involved.

Galbraith (1986) reviewed the literature on theories and causes of elder abuse and found that no one theory explains why the violence occurs. Instead, he used five theoretical frameworks to explain elder abuse. First, Galbraith identified the psychopathological theory, based on the premise that the offender has something psychologically wrong with him or her. Substance abuse by the perpetrator was included in Galbraith's first group (1986). Galbraith's second theoretical framework involved developmental dysfunctions based on the lack of interaction in the relationship between the parent and the child. This also includes the concept of the cycle of violence. The third grouping of theories includes mental and physical impairments of the victim which increase the dependency on the caretaker and increase the victim's vulnerability (Galbraith, 1986).

The fourth grouping of theories were concerned with stress, including physical, emotional, social, financial, and environmental stress, all of which may lead to violence or neglect of the elderly victim (Galbraith, 1986). There are undeniable pressures on middle-aged children to care for their frail elders at home, although the burdens of doing so can be enormous (Schiff, 1996).

Dementia, which is the most common cause of mental impairment among older people, has a long course, sometimes 7–10 years, and the period of total physical care may extend over several years (Insel & Roth, 1985). Dyer, Pavlik, Murphy, and Hyman (2000) found that depression and dementia are significant risk factors for abuse or neglect. Reed, Stone, and Neale (1990) examined sources of distress for a person caring for a relative suffering from dementia. Among those sources was the cost associated with such care. Weinberger, Gold, and Devine (1993) found that home care of a patient with dementia exceeded the cost of placement in a nursing facility and that the average time spent in direct care of the demented patient exceeded 100 hours a week.

Coyne, Reichman, and Berbig (1993) studied the relationship between dementia and abusive behavior in a sample of the elderly suffering from dementia and family members providing care. Findings of the study indicate that the physical and psychological demands of caring for a relative with dementia contribute to elder abuse. Coyne et al. (1993) found that abusive caretakers spent more time in care, cared for more seriously impaired relatives, and had been in a caretaker role for more years. The authors also found a "cycle of violence," where caregiver abuse occurred most frequently in retaliation to violence by the demented relative. Other stressors have to do with the relative maturity of the caregiver and the elder (Bland, 1998). Old battles can come back to life when the elder and adult child are thrown together. Quayhagen et al. (1997) support the need for investigation of abuse and possible early intervention to decrease the impact of progressive deterioration of family care over time.

Finally, Galbraith (1986) identified ageism as the fifth of the theoretical frameworks. Ageism is a way of viewing elderly people with stereotypical and prejudicial appraisals that result in degrading views of elderly persons.

Generally, research has tended to focus on five correlates of elder abuse, although there are probably as many suggested causes for elder abuse as there are studies of the problem. The major correlates noted in the literature include: (1) the personality attributes and characteristics of the abusing adult(s); (2) the transmission of violence across generations of families; (3) the web of dependency; (4) social isolation factors related to elder abuse; and (5) stress factors that contribute to the abusive situation.

Johnson (1992) did a comprehensive review of studies that focused on the personality attributes of the abuser. Those studies indicated that many of the abusive caretakers suffered from psychological or emotional problems and substance abuse, involving either drugs or alcohol. One surprising finding is that many abusers of the elderly are themselves elderly or near-elderly individuals (Johnson, 1992).

Spouse abuse that extends across many years may end up as elder abuse if continued into the late years of life (Floyd, 1984). Many of the risk factors present in spouse abuse cases are found in elderly couples and spouse abuse is relatively common in elder abuse. A study conducted by Harris (1996) indicates that the incidence of spouse abuse in older couples is significantly less than that of younger couples. Gesino, Smith and Keekich (1982) found that the reasons elderly women stay in abusive relationships included low self-esteem, limited social skills, social isolation, passive personality, the stigma associated with divorce, and a strong sense of loyalty. Older abusers commonly suffered from some form of stress due to the demanding needs of the elderly person (Steinmetz, 1988).

The transmission of violence across generations has been frequently discussed in child abuse literature (O'Malley, Daniels, O'Malley & Campion, 1983), but has not yet been adequately documented in elder abuse. Floyd (1984) suggests that members of multigenerational families may mutually reinforce the violent behavior of others in the family, especially in those families with extensive histories of violent behavior over several generations. The cycle-of-violence theory gained additional support from Steinmetz (1988) through an examination of the impact of interaction between elder adult–child and, thus extending the cycle into an additional generation.

However, Korbin, Anetzberger and Austin (1995) reviewed existing research on the intergenerational transmission of both child and elder abuse in families. The authors' findings suggest that child-abusing parents were more likely than elder-abusing children to have experienced severe domestic violence as children. Intergenerational elder abuse apparently is not as common as the case in child abuse. If the cycle-of-violence theory holds to be correct, it would contain elements of retaliation for violence experienced by the child at the hands of the parent (Korbin et al., 1995). Now the grown-up child can retaliate against the aged parent without fear (Pillemer, 1985).

The correlate concerning the web of dependency has been studied extensively (Kilburn, 1996; Pillemer, 1985; Teitelbaum, 1992). In his study, Pillemer found that the relationship between the dependent elders and caretakers was one that was stressful and contributed to the incidence of physical abuse. He noted that it was the dependency of the caretaker on the elder that was most stressful. The caretaker was dependent on the elder in the sense that he or she was powerless to lead an independent life because of the demands of the caretaker role and relationship. This web of dependency then becomes more and more complex as the elderly person requires more attention and resources from the caretaker (Kilburn, 1996; Pillemer, 1985).

Anderson and Theiss (1987) point out the degree of social isolation that is often a feature of the lives of the elderly, particularly when ill health is involved. Isolation may be related to unintentional neglect but it also provides cover for intentional confinement, exploitation, and violence against the elderly person (Teitelbaum, 1992). Stressors, either internal or external, or both, may contribute to violence and neglect of the elderly. The continual requirement for attention to the needs of the elderly person may be the most virulent of the stressors (Steinmetz, 1988). Caretakers may be taking care of their children at the same time they are caring for aged parents, creating even more stressful conditions (Hamilton, 1989; Police Executive Research Forum, 1993).

Indicators of abuse or neglect from the family or caregiver may include situations where the elder is not given the opportunity to speak for himself/herself or to see others without the caregiver being present. The family member or caregiver may have attitudes of indifference or anger toward the elder. Other indicators include too many "explained" injuries, or explanations inconsistent over a period of time, or a physical examination reveals that the client has injuries that the caregiver has not divulged (Quinn & Tomita, 1986).

There are some situations that are highly dangerous for elders such as when a caregiver has unrealistic expectations of an older person or an ill person and consequently feels that he/she should be punished for not following instructions. Some caregivers are incapable of taking care of dependent elders because of their own problems (Police Executive Research Forum, 1993). Some are young and immature and have dependency needs of their own. Other high-risk situations include those where the caregiver is forced by circumstances to take care of the elder or where the care needs of the elder exceed or soon will exceed the ability of the caregiver to meet those needs (Kapp, 1992; Quinn & Tomita, 1986).

Most cases of elder abuse occur in the home or domestic setting. However, as Beck and Phillips (1984) point out, some cases of elder abuse involve long-term nursing care facilities where most of the care is provided by nursing aides who typically are poorly trained and underpaid. In a study to assess interactions between staff members and nursing home residents, Meddaugh (1993) found a number of subtle forms of psychological abuse of nursing home residents by staff members. Included were issues of personal choice, resident dignity, isolation, and labeling. The study indicated that the most likely victims of abuse were the residents least able to influence their own care. Included in the residents who are least able to manage or influence their own care are those suffering from various degrees of mental disorders.

Mental disorders are common among nursing home residents with more than two-thirds having one or more psychiatric symptoms. Nursing homes have replaced state mental hospitals as the center of institutional care for older mentally ill and demented people (Jakubiak & Callahan, 1996). In the United States, 94% of all institutionalized elderly people with a mental disorder receive their care in a nursing facility, though very little treatment is being provided (Jakubiak & Callahan, 1996).

Yin (1985) notes that nursing homes have never been viewed favorably by most Americans and recurring anecdotes about the quality of care and the prevalence of abusive treatment have circulated throughout the society. Yin (1985) states that no systematic research has been conducted to determine the prevalence and incidence of elder abuse in care facilities. Mantese and Mantese (1995) note that nursing homes may be held civilly or criminally liable for neglect of residents.

TYPES OF ELDER ABUSE

Insel and Roth (1985) refer to the most prevalent form of elder abuse as "parent abuse" (p. 484). Other terms used include the "King Lear Syndrome," "granny bashing, granslamming, parental abuse and elder abuse" (Reid, 1990, p. 113). A major problem in the study of elder abuse is the lack of a uniform definition (Anderson & Theiss, 1987). In an effort to define more clearly the term "elder abuse," Block and Sinnot (1979) specified four types of maltreatment that they felt constituted the Battered Elder Syndrome:

1. Physical abuse—malnutrition and injuries such as bruises, welts, sprains, dislocations, or lacerations;
2. Psychological abuse—verbal assault, threat, fear, or isolation;
3. Material abuse—theft or misuse of money or property; and
4. Medical abuse—withholding or failing to provide medications or aids required such as false teeth, glasses, hearing aids, or prostheses.

The elder person could be a victim of any combination of the four types of abuse (Fulmer & Cahill, 1984). Elder abuse is also a recurrent phenomenon. One incident seems to lead to another and one form seems to stimulate another (Galbraith & Zdorkowski, 1985). Other researchers add to this listing such other forms of abuse as sexual abuse and active and passive neglect of care taking obligations (Fulmer & Wetle, 1986; Wolf & Pillemer, 1989).

Physical abuse consists of an intentional infliction of physical harm of an older person. The abuse can range from slapping an older adult to beatings to excessive forms of physical restraint. Physical abuse can be noted during routine physical assessments (Stanhope & Lancaster, 1988). Medical personnel must be alert for clinical symptoms that are inconsistent with the information collected from the patient's history (Insel & Roth, 1985). This becomes difficult if the elder individual has cognitive deficits and exhibits confusion and disorientation (Phillips & Rempusheski, 1985). It is necessary to locate the primary care providers and pursue the nature of the injuries seen in detail (Stanhope & Lancaster, 1988).

Medical abuse, as defined by Block and Sinnot (1979), can be detected by medical personnel who monitor an elderly person's response to various medications, therapies, and required prostheses. It is important for all medical personnel to explore circumstances and report cases of medical mistreatment (Fulmer & Cahill, 1984; Fulmer & O'Malley, 1987). Frost and Willette (1994) have introduced the concept of documentation of factors that place an elderly person at risk of neglect or abuse. By charting vulnerabilities and risks, the medical practitioner can more easily identify patterns that indicate abuse or neglect.

In the case of elders whose mental status is functional, careful questioning may indicate fear or resignation on their part regarding the nature and quality of their care. The fear of being institutionalized may outweigh their fear of abusive care providers and they may prefer to remain in a known abusive situation as opposed to risking a strange environment (Fulmer & Cahill, 1984).

In cases of passive and active neglect, the caregiver fails to meet the physical, social, and/or emotional needs of the older person. The difference between active and passive neglect lies in the intent of the caregiver. With active neglect, the caregiver intentionally fails to meet his/her obligations toward the older person. With passive neglect, the failure is unintentional; often the result of care-giver overload or lack of information concerning appropriate care-giving strategies.

Assessment of evidence of psychological or emotional abuse has many judgmental aspects (Floyd, 1984; Insel & Roth, 1985). However, Comijs, Pennin, Knispscheer, and van

Tilberg (1999) found that elder mistreatment had a negative effect on the psychological health of the victims. It is important to validate suspicions of psychological and emotional abuse with those who have personal knowledge of the individual as well as with the suspected elderly victim.

Material and financial abuse consists of the misuse, misappropriation, and/or exploitation of an older adult's material (e.g., possessions, property) and/or monetary assets (Frolik, 1991). Americans over 65 make up 12.5% of the nation's population, over 31 million people. But this same population makes up over 30% of the nation's fraud victims (Lester, 1981). Senior citizens are an accessible group because they are retired and are at home more often than younger generations. Older individuals who live alone are particularly popular victims because they have no one to discuss questionable deals with and may be more easily persuaded. This is especially true when the older individual is socially isolated and happy to have someone to talk with.

Blunt (1993) notes that financial exploitation is one of the most permanently devastating and overlooked forms of elder abuse. He further notes that a behavioral approach to the symptoms and behavioral patterns of the elderly will best aid in the identification of financial exploitation. Follow-up use of financial documents can aid in bringing the case to the appropriate judicial or non-judicial conclusion.

Material abuse may involve theft or misuse of money or property (Nerenberg, 1999). In the case where the elder is coerced to give belongings away in return for adequate care, the elder is clearly the victim of material abuse and in need of support to resolve the situation and regain the property (Beck & Phillips, 1984). Here again, there may be hesitancy to report theft, especially if the perpetrator involved is a relative or long-standing care provider (Fulmer & Cahill, 1984). Starnes (1996) states that elderly individuals are common victims of telemarketing fraud, health care fraud, and home repair fraud.

STATUTORY RESPONSES

Federal and state legislative actions have recently been targeted at various issues related to the problem of elder abuse. Federal laws on child abuse and domestic violence fund services and shelters for victims, but there is no comparable federal law on elder abuse (American Bar Association, 1998). The federal Older Americans Act (42 U.S.C. 3001 et seq., as amended) defines elder abuse and authorizes the use of federal funds for elder abuse awareness, training, and coordination activities. The Act also provides funds for the National Center on Elder Abuse but does not fund adult protective services or shelters for abused older persons.

However, all 50 states have passed legislation providing for Adult Protective Services to deal with elder abuse. The American Bar Association (1998) notes that Adult Protective Services laws include protection for disabled persons as well as for the elderly. In most states, Adult Protective Services have a dual mandate of providing social services for victims and for investigating reports of elder abuse (Krauskopf, Brown, Tokarz, & Bogutz, 1993).

The various state statutes use different ages for adult services, different definitions of abuse and the types of abuse that are included, and the types of protective services that are required (Byers & Hendricks, 1993). Other differences include variations in reporting requiremems, responsibility for investigation, and classification of the abuse as criminal or civil (American Bar Association, 1998).

In some states, institutional abuse is not included in the mandate of Adult Protective Services and a separate law has been passed (Goodrich, 1997). Those statutes dealing with institutional abuse statutes usually address reporting and investigating incidents of elder abuse that occur in long-term care facilities or other facilities (American Bar Association, 1998).

Politsky (1995) believes tort law and criminal law fail to provide adequate remedies to elder abuse victims. He states that states should criminalize physical and emotional abuse, and adds that such actions are the key to protecting the elderly from abuse.

A number of states have passed laws that provide criminal penalties for various forms of elder abuse (Krauskopf et al., 1993). Some Adult Protective Services laws include provisions for the criminal prosecution of cases of elder abuse. Many of the specific acts involved in elder abuse fit the definitions used in criminal laws. Thus, in those states or jurisdictions lacking a specific statute or provision for criminal prosecution of cases of elder abuse, the basic penal code can be used for prosecution when the acts against an elderly person fall within the boundaries of the code (Politsky, 1995). State statutes concerned with guardianship, conservatorship, durable powers of attorney, and fiduciary responsibilities also may apply to various types of elder abuse (Keith & Wacke, 1994; American Bar Association, 1998).

POLICE ELDER ABUSE UNITS

The incidence of elder abuse will continue to increase as the population ages. Despite the best efforts of social agencies, the abuse continues. What is needed is a coordinated effort involving tile social services agencies, law enforcement, and prosecutors. Law enforcement agencies serving large populations of the elderly need to devote resources to combat the problem of abuse of the elderly. An elder abuse unit is needed in such departments.

An elder abuse unit would be staffed with experienced investigators who receive specialized training in the issues related to elder abuse. In some states, law enforcement personnel already are learning about elder abuse as part of their basic curriculum at the police academy. Some state legislatures have mandated training in elder abuse for law enforcement personnel. Instructors at the Criminal Justice Academy in South Carolina train recruits in indicators of abuse and legal issues related to the problem of elder abuse.

At a conference on elder abuse, Randolph Thomas, a law enforcement instructor, noted that pictures of the physical impact of neglect on the body are particularly useful. He commented: "Most of my police officers know what a gunshot wound is. They do not know what a pressure sore looks like, and when you start showing them pictures, they start to have a frame of reference. That may be the most important thing, just that cognitive framework that says, 'Okay, this may or may not be abuse' " (Elder Justice, 2001). Police officers want training and can "tolerate a high level of sophistication, as long as it is done in laymen's terms," Thomas added. Since most police academy students are young they may have difficulty understanding and developing empathy with elderly adults (Elder Justice, 2001).

California's training for police officers teaches basic awareness of aging issues, plus information on effectively reporting elder abuse, including scene preservation and evidence collection. In California, advanced training on these topics is especially important because State law allows the State to become a conservator and take over the care of an older victim of abuse or neglect (Elder Justice, 2001). The California Commission on Police Officers

Standards and Training maintains a depository of training films on aging that are used to provide training throughout the state. California also is a leader in law enforcement distant learning with telecourses on issues related to the elderly (Elder Justice, 2001).

In Florida, a multimedia training course, presented by an interdisciplinary team including a psychologist and geriatric nursing specialist provides training for police departments across the state. The National College of District Attorneys has added elder domestic violence and elder abuse to its roster of training courses (Elder Justice, 2001).

It is important that training be an ongoing activity. Every member of the law enforcement agency should receive basic training and periodic updated training on issues related to the elderly population served by the department. Members of the elder abuse unit should receive more intensive training involving personnel from the long-term care industry, medical personnel, geriatric specialists, and others knowledgeable about the issues of aging.

What seems to be the best method to combat the problem abuse of the elderly involves a team approach. By combining efforts it is possible to form a cohesive group with common goals related to the protection of the elderly. The team approach insures that all officials have the same information and are working to meet the same goals.

When law enforcement and appropriate social service agencies jointly respond to reports of abuse, the presence of a law enforcement officer demonstrates the seriousness of the situation. That presence also provides safety for the elderly victim and the social service worker(s). Law enforcement officers are experienced in evaluating elements of crimes and assessing situations. When that experience is combined with the skills of the social service worker, the result is a more accurate and valid determination of the situation and possible approaches to resolution.

There are a number of examples of teamwork involving law enforcement and various other agencies. In San Luis Obispo County, California, law enforcement officials have joined with the District Attorney's office, adult protective services, the elder ombudsman, and mental health officials to investigate and assess reported cases of elder abuse (Senior Information Guide, 2001).

In several Wisconsin counties, teams of law enforcement officers and human service workers are investigating elder abuse. The Minneapolis Police Department is part of a multi-agency, multi-disciplinary approach to elder abuse. The Police Department joins Adult Protective Services, Department of Health, and Department of Human Services to work together as a team.

In six Northern California counties, a program called Operation Guardian is in operation. The program consists of medical providers, social services, and law enforcement officials who conduct spot checks on Nursing Homes and Licensed Care Facilities to identify any abuse or mistreatment within the facilities (Attorney General Press Release, 2001).

Ocean County, New Jersey has nearly 110,00 citizens aged 65 years and more living mostly in retirement communities and most have no family nearby. Ocean County has developed a program, Safe Operation Outreach for Seniors, to assist police officers in identifying signs and symptoms of dementia and possible elder abuse. Safe Outreach for Seniors brings law enforcement and Kimball Medical Center's Geriatric Evaluation and Management Service together to identify elders who are at risk for abuse or self-neglect (Kimball Medical Center, 2001).

A second approach to issues of elder abuse involves a reporting relationship between appropriate adult protective service agencies and law enforcement. Texas requires timely notification to law enforcement officials by Adult Protective Services of all completed investigations validated as an offense under any law. In Arizona, Adult Protective Services refers all substantiated cases of abuse, neglect, and exploitation to law enforcement officials.

The Eugene, Oregon Police Department has another approach to dealing with the problem of elder abuse prevention. The Department has developed a Personal Crime Prevention website that educates the public about possible indicators of physical, psychological, emotional, and financial abuse of the elderly. The website also provides a discussion of the Oregon statutes, criminal negligence, the Elder Abuse Prevention Act, and the availability of restraining orders (Eugene Police Department, 2001).

Not everyone believes that law enforcement has a role in protecting elderly residents in nursing homes and long-term care facilities. The long-term care industry is up-in-arms due to the increased criminal jurisprudence used to enforce standards that the industry believes are regulatory and not designed as criminal laws (Burguess, 2000). Those in the industry believe that law enforcement officials and prosecutors need to become educated in how the long-term care industry is regulated.

Burguess (2000) points out that criminal law requires stricter burdens of proof than does the regulatory law that controls the nursing home industry. If criminal sanctions were the only ones used in the industry, many cases that are sanctioned under regulatory law would not be prosecuted and would go unpunished. The argument is that the regulatory agencies already provide severe sanctions for abuse within the long-term facilities, including monetary fines and loss of operating licenses and require much less stringent proof (Burguess, 2000).

OPERATIONAL CONSIDERATIONS

Intervention in cases of elder abuse requires a knowledge of abuse-reporting requirements and legal constraints of professional action (Anderson & Thobaben, 1984). A major issue is the elderly person's degree of competency. If, for example, the elder is fully competent and prefers to stay in a relative's home where abuse and neglect may be occurring, that is the decision of a capable, competent person and it must be honored. Professionals have been threatened with bodily harm by caregivers who resent the attempt to break up what may be a lucrative financial exploitation situation for the caregiver (Anderson & Thobaben, 1984).

The person intervening must also be aware of mandatory reporting and mandatory arrest laws in the various states (Nerenberg, 1993; Police Executive Research Forum, 1993). In most states, elder abuse falls within the definitions of domestic abuse and may require mandatory arrests. Additionally, Blair (1997) discusses the legal issues surrounding elder abuse and the duty of care and notes that most states have statutes that deal with the duty of care for dependent persons. Duty of care includes omission of needed care as well as provision of such care.

In many, if not most, cases of elder neglect or abuse, the needed intervention does not involve removing the victim. Instead, the caregivers are assisted with respite and other home services (Anderson & Thobaben, 1984). Choosing an intervention strategy is not easy and there are many possibilities (Phillips & Rempusheski, 1985; Nerenberg, 1993).

Some interventions are not related to the diagnosis of a given situation. As an example, a case may be closed even when there is clear evidence of abuse or neglect or a decision may be made to continue education efforts within the family even when the life of the elderly person is clearly at risk. Some decision-makers will decide not to decide because any decision has risks and others will regret whatever decision is made (Phillips & Rempusheski, 1985). Landau (1998) notes that there may be ethical dilemmas facing professionals dealing with cases of elder abuse. Those cases involving family members of the elderly victim present a number of issues that professionals must answer (Baumhover & Beall, 1996).

Politsky (1995) discusses the problem of abuse of elders in vulnerable situations and concludes that neither tort law nor criminal law provides adequate remedies to elder abuse victims. Politsky (1995) concludes that state statutes criminalizing physical and emotional abuse are essential if elderly persons are to be protected from abuse. Schiamberg and Gans (1999) recommend development of community prevention and intervention strategies to deal with intergenerational aspects of elder abuse within families.

ADMINISTRATIVE CONSIDERATIONS

Management and supervision of the elder abuse unit would not be significantly different from that required by other investigation units within the agency. Selection and training of officers assigned to the unit would be more critical than most other units in the department. Officers selected for service in an elder abuse unit should have several attributes that are specific to the requirements of working with elderly people. They need to communicate well with older citizens, be patient in dealing with the elderly, and have excellent listening skills. The assigned officers should have extensive training in all aspects of aging, elderly life styles, and all of the various forms of elder abuse.

Since the unit will probably soon be dubbed by some departmental wit as "the granny squad" or some similar term, a sense of humor and a sense of dedication to serving the elderly are essential. Individuals who are selected for the unit should work well with representatives from other agencies. Law enforcement has a long history of secretiveness and a reluctance to share information outside the department. To be effective, an elder abuse unit must share information with a variety of other agencies. Nothing kills a working relationship quicker than the revelation that one or the other is withholding information.

FUTURE DIRECTIONS FOR ELDER ABUSE AND ELDER ABUSE UNITS

Stresses of modern living have been known to promote violent behavior in individuals. Current abuse statistics support the unfortunate fact that ours is a violent society. Few public-counseling, financial-aid, or daycare programs are available to ease the pressure on children, and few laws protect their parents (Insel & Roth, 1985). Those families who act aggressively toward children and mates do not exclude their elders from the same acts. The needs in this area are many.

There is a need to develop programs which help to reduce the victimization of the elderly, both in traditional crimes and in the care environment of homes and nursing facilities.

Traditional crime prevention techniques should be augmented by programs designed to fit the peculiar needs of the elderly. The public needs to understand the many reasons for concern about elder abuse and victimization.

Many of the elderly live below the poverty level and any financial or material loss has great impact (Malinchak, 1980). Elderly people who live alone and lack good health and stamina are prone to repeated victimization. If the older persons try to defend themselves, then they are more likely to be injured and to suffer more serious injuries (Malinchak, 1980). Finally, the vulnerability of the elderly who rely upon their children or other relatives for care is similar to the vulnerability of young children. They are subject to every form and type of abuse and neglect.

Even without much help from the government, the elderly are starting to fight back. Groups like the American Association of Retired Persons are lobbying for such services as senior-citizen daycare, and they expect to gain increased political power as the aged population continues to grow (Insel & Roth, 1985).

Educational programs are being developed in a number of jurisdictions. As an example, California is trying to fight elder abuse with the introduction of a video. This video is designed to alert Californians to the growing problem of crimes and other forms of abuse directed at senior citizens. The instructional video will be distributed to approximately 5000 hospitals, law enforcement agencies, social service providers, and other professionals to alert them to the various warning signs and forms of elder abuse (California Attorney General, 1993).

Mandatory reporting laws are now in place in almost all of the states, Mandatory reporting is one of the steps necessary to collect adequate data that can help in establishing criteria for identification of the various forms of elder abuse. The need for additional public policy actions is not evident until such time as documentation and research reflects such a need (Floyd, 1984). Such public policy actions as mandatory arrests may hinder the victim more than help in some cases. If the victim must be removed from the home because the offender is in jail, then the victim also suffers. In many cases education and training in care techniques may be all that is needed (Floyd, 1984; Yin, 1985).

There is a great need for additional research in the incidence of elder abuse. Current research and generalizations that can be made from the findings are limited. The lack of reliable and valid research continues to be an increasing concern. Although recent research studies have examined the extent and incidence of elder abuse and presented characteristic profiles of the abused and the abuser, the generalizations that can be made from their findings are limited (Galbraith & Zdorkowski, 1985). Most of the research was conducted in the 1980s and there has been a noted decrease in further research in the 1990s.

The elder abuse surveys have many methodological, definitional, and categorical limitations, although their similar findings suggest a broadly accurate image of elder abuse. The surveys have been hampered by non-random sampling techniques, small samples, and narrow research objectives (Galbraith & Zdorkowski, 1985). History has clearly illustrated that any group that is perceived to be politically and economically weak is unlikely to receive adequate attention. The phenomenon of elder abuse has suffered from selective inattention as have other forms of violent family interaction. After reviewing historical issues of elder abuse, one can only come to the conclusion that finding answers to the many unanswered questions about the topic is imperative.

RECOMMENDED READINGS

AMERICAN BAR ASSOCIATION COMMISSION ON LEGAL PROBLEMS OF THE ELDERLY (1998). *Information about laws related to elder abuse*. Available on the Internet at: http://www.elderabusecenter.org/laws.

ELDER JUSTICE: MEDICAL FORENSIC ISSUES CONCERNING ABUSE AND NEGLECT (2001). (Draft Report) National Institute of Justice Conference. Washington, DC: U.S. Department of Justice, National Institute of Justice.

NATIONAL CENTER ON ELDER ABUSE (1998). *National Elder Abuse Incidence Study: Final Report*. Washington, DC: U.S. Department of Health and Human Services, Administration for Children and Families and Administration on Aging.

POLICE EXECUTIVE RESEARCH FORUM (1993). *Understanding domestic elder abuse*. Washington, DC: Police Executive Research Forum.

NERENBERG, L. (1993). *Improving the Police Response to Domestic Elder Abuse: Instructor Training Manual and Participant Training Manual*. Washington, DC: Police Executive Research Forum.

REFERENCES

AMERICAN BAR ASSOCIATION COMMISSION ON LEGAL PROBLEMS OF THE ELDERLY (1998). *Information about laws related to elder abuse*. Washington, DC: American Bar Association. Available at: www.elderabusecenter.org/laws/index.html.

ANDERSON, J.M., & THEISS, J.T. (1987). *The etiology of elder abuse by adult offspring*. Springfield, IL: C.C. Thomas, Publisher.

ANDERSON, L., & THOBABEN, M. (1984). Clients in crisis: When should the nurse step in? *Journal of Gerontological Nursing*, *10*(12), 6–10.

ATTORNEY GENERAL PRESS RELEASE (2001). *Attorney General Lockyer issues report on first year of surprise nursing home inspections in Northern California*, Press release 01-036. Sacramento, CA: Office of the Attorney General. Available at: http://caag.state.ca.us/press/2001/01-036/htm.

AVANIS, S. (2001). *Reports of abuse of senior citizens up 300 percent*. Testimony before the United States Senate Special Committee on Aging hearing on abuse of senior citizens. June 14. Washington, DC.

BAUMHOVER, L.A., & BEALL, S.C. (Eds.) (1996). *Abuse, neglect, and exploitation of older persons: Strategies for assessment and intervention*. Baltimore, MD: Health Professions Press.

BECK, C.M., & PHILLIPS, L.R. (1984). The unseen abuse: Why financial maltreatment of the elderly goes unrecognized. *Journal of Gerontological Nursing*, *10*(12), 26–32.

BLAIR, J. (Fall 1996–1997). Honor thy father and thy mother—But for how long?—Adult children's duty to care for and protect elderly parents. University of Louisville, *Journal of Family Law*, *35*, 765.

BLAND, W.D. (1998). *Aging parents: Continuity and change in adult life*. East Rockaway, NY: Cummings & Hathaway Publishers.

BLOCK, M.R., & SINNOT, J.D. (1979). *The battered elder syndrome: An exploratory study*. College Park, MD: Center on Aging, University of Maryland.

BLUNT, A.P. (1993). Financial exploitation of the incapacitated: Investigation and remedies. *Journal of Elder Abuse and Neglect*, *5*, 19–21.

BUREAU OF JUSTICE STATISTICS (1994). *Elderly crime victims*. Washington, DC: Department of Justice, Office of Justice Programs, Bureau of Justice Statistics.

BURGUESS, K. (2000). Too many cooks cause SNFs indigestion. *Provider*, *26*(6), 47–48.

BYERS, B., & HENDRICKS, J.E. (1993). *Adult protective services: Research and practice*. Springfield, IL: C.C. Thomas.

CALIFORNIA ATTORNEY GENERAL INTRODUCES VIDEO TO FIGHT ELDER ABUSE (1993). *Crime Control Digest*, *27*(1), 8–9.

COMIJS, H.C., PENNINX, B.W.J.H., KNIPSCHEER, K.P.M., & VAN TILBURG, W. (1999). Psychological distress in victims of elder mistreatment: The effects of social support and coping. *Journal of Gerontology*, *54B*(4), 240–245.

COYNE, A., REICHMAN, W., & BERBIG, L. (1993). The relationship between dementia and elder abuse. *American Journal of Psychiatry*, *150*(4), 643–646.

CRAIG, L. (2001). *Craig will call on U.S. Justice Department to focus more effort on elder abuse.* Press release on testimony before the United States Senate Special Committee on Aging hearing on abuse of senior citizens. June 14. Washington, DC.

DYER, C., PAVLIK, V., MURPHY, K., & HYMAN, D. (2000). The high prevalence of depression and dementia in elder abuse or neglect. *Journal of the American Geriatrics Society*, *48*(2), 205–208.

ELDER ABUSE: A DECADE OF SHAME AND INACTION (1990). Hearing Before the Subcommittee on Health and Long-Term Care of the House Select Committee on Aging, 101st Cong., 2nd Session 3.

ELDER JUSTICE: MEDICAL FORENSIC ISSUES CONCERNING ABUSE AND NEGLECT (2001). (Draft Report) National Institute of Justice Conference. Washington, DC: U.S. Department of Justice, National Institute of Justice.

EUGENE POLICE DEPARTMENT (2001). *Elder Abuse.* Eugene, OR: Eugene Police Department. Available at: http//:www.ci.eugene.or.us/DPS/police/Crime-prevention/elder.htm.

FEDERAL BUREAU OF INVESTIGATION (1997). *Crime in the United States 1996*. Washington, DC: United States Government Printing Office.

FHP FOUNDATION (1995). *Silent suffering: Elder abuse in America: Elder abuse in rural and urban settings: Report on focus group activities and recommendations for the 1995 White House Conference on Aging.* Long Beach, CA: FHP Foundation.

FLOYD, J. (1984). Collecting data on abuse of the elderly. *Journal of Gerontological Nursing*, *10*(12), 11–15.

FROLIK, L.A. (1991). Abusive guardians and the need for judicial supervision. *Trusts & Estates*, *130*, 41.

FROST, M.H., & WILLETTE, K. (1994). Risk for abuse/neglect: Documentation of assessment data and diagnoses. *Journal of Gerontological Nursing*, *20*(8), 37–45.

FULMER, T.T., & O'MALLEY, T.A. (1987). *Inadequate care of the elderly: A health care perspective on abuse and neglect.* New York: Springer Publishing, Inc.

FULMER, T., & WETLE, T. (1986). Elder abuse screening and intervention. *Nurse Practitioner*, *11*(5), 33–38.

FULMER, T., & CAHILL, V.M. (1984). Assessing elder abuse: A study *Journal of Gerontological Nursing*, *10*(12), 16–20.

GALBRAITH, M.W. (1986). Elder abuse: *Perspectives on an emerging crisis*. St. Louis, MO: Mid-America Congress on Aging.

GALBRAITH, M.W., & ZDORKOWSKI, R.T. (1985). Teaching the investigation of elder abuse. *Journal of Gerontological Nursing*, *10*(12), 21–25.

GELLES, R.J. & STRAUSS, M.A. (1979). Determinants of violence in the family: Toward a theoretical integration. In BURR et al. (Eds.), *Contemporary theories about the family*. New York: Free Press.

GESINO, J.P., SMITH, H.H., & KEEKICH, W.A. (1982). The battered woman grown old. *Clinical Gerontologist*, *1*(1), 59–67.

GOODRICH, C.S. (1997). Results of a national survey of state protective services programs: Assessing risk and defining victim outcomes. *Journal of Elder Abuse and Neglect*, *9*(1), 69–86.

HAMILTON, G.P. (1989). Using a family systems approach: Prevent elder abuse. *Journal of Gerontological Nursing*, *15*(3), 21–26.

HARRIS, S. (1996). For better or for worse: Spouse abuse grown old. *Journal of Elder Abuse and Neglect*, *8*(1), 1–34.

INSEL, P.M., & ROTH, W.T. (1985). *Core concepts in health*. Palo Alto, CA: May field Publishing Company.

JAKUBIAK, C., & CALLAHAN, J. (1996). Treatment of mental disorders among nursing home residents: Will the market provide? *Generations, Winter*, 1995–1996.

JOHNSON, T.F. (1992). *Elder mistreatment: Deciding who is at risk*. New York: Greenwood Publishing Company.

KAPP, M.B. (1992). Who's the parent here? The family's impact on the autonomy of older persons. *Emory Law Journal, 41*, 773.

KARMEN, A. (1990). *Crime victims: An introduction to victimology* (2nd ed.). Pacific Grove, CA: Brooks/Cole Publishing Company.

KEITH, P.M., & WACKE, R.R. (1994). *Older wards and their guardians*. New York: Praeger Publishers.

KILBURN, J. (1996). Network effects in caregiver to care-recipient violence: A study of caregivers to those diagnosed with Alzheimer's disease. *Journal of Elder Abuse and Neglect, 8*(1), 69–80.

KIMBALL MEDICAL CENTER (2001). *Safe Outreach for Seniors Program*. Ocean City, MD: Kimball Medical Center. Available at: http://www.sbhcs.com/hospitals/kimbal_medical/mservices/.

KORBIN, J., ANETZBERGER, G., & AUSTIN, C. (1995). The intergenerational cycle of violence in child and elder abuse. *Journal of Elder Abuse and Neglect, 7*(1), 1–15.

KOSBERG, J. (1998). The abuse of elderly men. *Journal of Elder Abuse and Neglect, 9*(3), 69–88.

KRAUSKOPF, J.M., BROWN, R.N., TOKARZ, K.L., & BOGUTZ, A.D. (1993). *Elderlaw: Advocacy for the aging*. West Publishing, St. Paul Minn. [1137 pages, 2 volumes].

LANDAU, R. (1998). Ethical dilemmas in treating cases of abuse of older people in the family. *International Journal of Law, Policy and the Family, 12*(3), 345–355.

LESTER, D. (1981) *The elderly victim of crime*. Springfield, IL: C.C. Thomas, Publisher.

MALINCHAK, A.A. (1980). *Crime and gerontology*. Englewood Cliffs, NJ: Prentice-Hall, Inc.

MANTESE, T.M., & MANTESE, G. (1995). Nursing homes and the care of the elderly. *Journal of Missouri Business, 5*, 1155–1163.

MEDDAUGH, D.I. (1993). Covert elder abuse in the nursing home. *Journal of Elder Abuse and Neglect, 5*, 21–37.

NATIONAL CENTER ON ELDER ABUSE (1998). *National Elder Abuse Incidence Study: Final Report*. Washington, DC: U.S. Department of Health and Human Services, Administration for Children and Families and Administration on Aging.

NERENBERG, L. (1993). *Improving the Police Response to Domestic Elder Abuse: Instructor Training Manual and Participant Training Manual*. Washington, DC: Police Executive Research Forum.

NERENBERG, L. (1999). *Forgotten victims of elder financial crime and abuse: A report and recommendations*. Washington, DC: National Center on Elder Abuse and San Francisco, CA: Goldman Institute on Aging.

O'MALLEY, H., DANIELS, E., O'MALLEY, H., & CAMPION, W. (1983). Identifying and preventing family mediated abuse and neglect of elderly persons. *Annals of Internal Medicine, 98*(6), 98–105.

PHILLIPS, L.R., & REMPUSHESKI, V.F. (1985). A decision-making model for diagnosing and intervening in elder abuse and neglect. *Nursing Research, 34*(2), 134–139.

PILLEMER, K.A. (1985). The dangers of dependency: New findings on domestic violence against the elderly. *Social Problems, 28*(2), 227–238.

PILLEMER, K.A., & FINKELHOR, D. (1986). The prevalence of elder abuse: A random sample survey. *American Journal of Orthopsychiatry, 59*, 179–187.

PILLEMER, K.A., & FINKELHOR, D. (1988). Elder abuse. In V. Van Hasselt (Ed.), *Handbook of family violence*. New York: Plenum Press.

POLICE EXECUTIVE RESEARCH FORUM (1993). *Understanding domestic elder abuse*. Washington, DC: Police Executive Research Forum.

POLITSKY, R.A. (1995). Criminalizing physical and emotional elder abuse. *Elder Law Journal, 3*, 377–411.

QUAYHAGEN, M., QUAYHAGEN, M., PATTERSON, T., IRWIN, M., HAUGER, R., & GRANT, I. (1997). Coping with dementia: Family caregiver burnout and abuse. *Journal of Mental Health and Aging, 3*(3), 357–364.

QUINN, M.J., & TOMITA, S.K. (1986). *Elder abuse and neglect: Causes, diagnosis and intervention strategies.* New York: Springer Publishing Company.

REED, B., STONE, A., & NEALE, J. (1990). Effects of caring for a demented relative on elder's life events and appraisals. *The Gerontologist, 30,* 200–205.

REID, S.T. (1990). *Criminal justice.* New York: Macmillan Publishing Company.

RENNISON, C. (2000). *Criminal Victimization 1999, Changes 1998–99 with Trends 1993–99, NCJ 182734.* Washington, DC: U.S. Department of Justice, Bureau of Justice Statistics.

SCHIAMBERG, L., & GANS, D. (1999). An ecological framework for contextual risk factors in elder abuse by adult children. *Journal of Elder Abuse and Neglect, 11*(1), 79–103.

SCHIFF, H.S. (1996). *How did I become my parent's parent?* New York: Viking Books.

SCOFIELD, M., REYNOLDS, R., MISHRA, G., POWERS, P., & DOBSON, A. (1999). *Vulnerability to abuse, powerlessness and psychological stress among older women.* Callaghan, NSW: University of Newcastle.

SENIOR INFORMATION GUIDE FOR SAN LUIS OBISPO COUNTY (2001). *Investigating elder abuse.* San Luis Obispo County, CA: Area Agency on Aging. Available at: http//www.slonet.org/~seniors/sloguide/abuse2.htm.

SHAPIRO, J. (1992). The elderly are not children. *U.S. News & World Report,* January, 26–27.

SOEDA, A., & ARAKI, C. (1999). Elder abuse by daughters-in-law in Japan. *Journal of Elder Abuse and Neglect, 11*(1), 47–58.

STANHOPE, M. & LANCASTER, J. (1988). *Community health nursing.* St. Louis: C.V. Mosby Company.

STARNES, R.A. (1996). Consumer fraud and the elderly: The need for a uniform system of enforcement and increased civil and criminal penalties. *Elder Law Journal, 4*(1), 201–224.

STEINMETZ, S.K. (1988). *Duty bound: Elder abuse and family care.* Beverly Hills, CA: Sage Publications.

TATARA, R. (November, 1997). *Reporting Requirements and Characteristics of Victims. Domestic Elder Abuse Information Series #3.* Washington, DC: National Center on Elder Abuse.

TEITELBAUM, L.E. (1992). Intergenerational responsibility and family obligation. *Utah Law Review, 765,* 77–99.

U.S. CENSUS BUREAU (1996). *Population report: 65+ in the United States.* Statistical brief from the Economics and Statistics Administration of the U.S. Department of Commerce. Washington, DC: U.S. Department of Commerce.

U.S. CONGRESS. A REPORT BY THE CHAIRMAN OF THE HOUSE SUBCOMMITTEE ON HEALTH AND LONG-TERM CARE, HOUSE SELECT SUBCOMMITTEE ON AGING (1990). *Elder abuse: A decade of shame and inaction.* 101st Congress, 2nd session. Washington, DC: U.S. Government Printing Office.

WEINBERGER, M., GOLD, D., & DEVINE, G. (1993). Expenditures in caring for patients with dementia who live at home. *American Journal of Public Health, 83*(3), 338–341.

WOLF, R. & PILLEMER, K. (1989). *Helping elderly victims: The reality of elder abuse.* New York: Columbia University Press.

YIN, P. (1985). *Victimization and the aged.* Springfield, IL: C.C. Thomas, Publisher.

5

Gang Units

Jeffrey P. Rush and Gregory P. Orvis

INTRODUCTION

It cannot be contested that gang problems are intensifying in the United States. Surveys over the past 15 years have reported gangs in more and more cities, with gangs committing more violent crimes, using more lethal weapons, and creating more serious injuries (Howell, 1994). In a recent survey of all police departments serving cities with a population of 25,000 or more, all suburban county and sheriff's departments, a random sample of police departments serving cities with populations between 2500 and 24,999, and a random sample of rural county police and sheriff's departments, 48% of the respondents said they had active youth gangs in their jurisdictions, with more than two-thirds of those respondents stating that the gang problems were either getting worse or staying about the same. In fact, the term "youth" gang may be a misnomer. Respondents reported that 60% of their "gangsters" were over the age of eighteen, representing a 10% increase from what was reported two years previously (Wilson, 2000).

Gang problems are no longer just big city problems. Counter to a nationwide trend from 1996 to 1998, gang membership has increased 43% in rural counties and 3% in small cities, and gangs there are extensively involved in property crimes and drug sales. In fact, total respondents reported that one-third of all youth gangs are organized specifically to traffic in drugs (Wilson, 2000). It appears, therefore, there is a need for society to organize against these gangs relying on local policing as its main tool.

In the 1930s, early public policy aimed at gang activities emphasized community prevention programs designed to prevent youth involvement in gangs, but had only moderate success. Law enforcement began programs aimed at gang suppression when youth and adult gang problems grew significantly in the southwest in the early 1980s. During that time, gang units were specifically created as solutions to the communities' gang problem by police departments and were expected to investigate, suppress, and prevent gang's criminal activities. Modern gang unit tactics such as gang sweeps and hot spot targeting can be traced to "experimental" programs in the eighties such as the "Community Resources Against Street Hoodlums" (CRASH) implemented by the Los Angeles Police Department (Howell, 2000).

DOES A GANG PROBLEM EXIST OR DO YOU HAVE A GANG PROBLEM?

One of the more important questions for a police agency to answer is whether they should specialize in a particular area or not. This is an especially critical question regarding specialized gang units. It is generally recognized that America has a drug problem and that this problem exists in many if not most or all areas of our country. As a result, police administrators and politicians are not opposed to specialized narcotics units (often associated with consensual crimes, i.e., vice, as well). Neither are they opposed to tactical units nor, at least until the shooting of Amadou Diallo, street crimes units. The development of gang units, on the other hand, is not as readily acceptable. For many politicians and community leaders as well as many police administrators to have a gang unit means there is a gang problem, and this is all too often not something that wants to be admitted. An example of this is the following GANGINFO discussion list question that was recently posted:

> I work in several rural communities, with one particular one home to a regional university. All of these communities are on major interstates and close to the second largest urban area in the state.
> We have the graffiti, the colors, the signs, the drugs in the schools, the increase in particular types of juvenile criminal activity – *But no one will officially confirm the reality of the gang activity*. I need suggestions of data to collect or ways to approach the official community to get them *to open up and accept the reality of the presence so we can systemically begin intervention and prevention*. The gang activity is not entrenched yet, but we are the fastest growing area of the state. Thanks for any help. (Personal communication; Emphasis ours)

This community is clearly involved in the **DID** syndrome: **D**eny, **I**gnore and **D**elay (Sliwa, 1987). Communities involved in this syndrome are simply not willing to accept the fact that a gang problem exists. Political considerations may be the basis for a community's leaders denying that a gang problem exists (Curry, 1995). All to often denial is based on ignorance. They feel like since whatever they are seeing is not "like" Los Angeles or Chicago or maybe even New York, then it is not a gang(s) problem. In point of fact, many cities' entrenched gang problem is because they, too, operated by the **DID** syndrome and as such their problem may never go away. Says McBride (1990), "once gangs have arrived (and become entrenched), it is next to impossible to get rid of them" (emphasis in the original). This is not to say that a community and its agencies (including the police) should necessarily assume that gangs are on the horizon. It is to say that when the outward signs begin to appear (graffiti, colors, etc.), the existence of a gang problem should be assumed, and then, like every other criminal problem, a plan needs to be developed to address it. An increase in certain crime rates may also forewarn a community that it is time to plan for and create a gang unit:

> Certain types of crimes are especially likely to involve gang members, and a sudden increase in those crimes may be viewed as a potential "distant early warning signal" that

crime in the community may be increasingly gang-related. Crimes that may be especially worth monitoring closely are assaults involving rival groups; auto theft and credit card theft; carrying concealed weapons; taking weapons to school; assault/intimidation of victims, witnesses, and shoppers; drug trafficking; driveby shootings, and homicide, all of which frequently involve gang members and may serve as reasonably accurate "gang markers" in some communities at some points in time. (Huff, 2001, p. 403)

In many cases, the development of a gang unit occurs after a major incident. Once the incident occurs (usually with violence), the media grabs it, there's public outcry and a unit is formed. By this time, law enforcement is clearly behind the learning curve, and may very well spend the first few months of operation trying to "catch up." In addition putting together a high quality unit takes time (and in some cases additional resources) and this too puts the police in the more familiar reactive mode rather than newer proactive mode. So the question really is not whether a unit is going to exist after a major incident(s) but whether one should be developed in the normal course of business or before there is a major incident. Webb and Katz (2003) offer a different idea regarding the development of gang units:

> Our observations of police gang units convinces us that police organizations need to carefully reassess the organizational configuration of their response to gangs as well as the investment of resources in that response. The starting point is a thoughtful and careful assessment of the local gang problem and of whether or not it is of sufficient magnitude to warrant a specialized unit to address the problem. With the possible exception of Junction City, the gang units that we have observed were in communities with substantial gang problems, and specialized gang units were a reasonable response to the local problem. However, we suspect that a substantial number of gang units developed during the last decade were formed not in response to local gang problems, but were the result of mimetic processes. Mimetic processes occur when organizations model themselves after other organizations. . . . In other words, we suspect that many police departments have created gang units for reasons related to institutional legitimacy rather than to respond to actual contingencies in their environment. (p. 45)

The suspicions of Webb and Katz notwithstanding, it seems illogical to us that an agency would start a gang unit just because other agencies have one. The potential political fallout it seems would be enormous. Again, unlike a tactical unit, or a drug unit, or a bicycle unit (all of which could be explained and justified without undue citizen or politico "fear"), a gang unit clearly suggests that "our" city has a problem with gangs and that the problem is so "bad" that we must mount a specialized and focused unit to combat it. Particularly in this era of fiscal uncertainly, that just does not seem logical.

The following request appeared on a gang discussion list of which one of the authors is a member:

> Hello all,
>   I am from City [*a pseudonym*] PD in City, CA [*the state is accurate*]. We have *finally gotten our administration to admit we have a gang problem in our city (after 15 years of graffiti and shootings)*. Currently, only myself and two others are interested in being gang experts. Unfortunately, one of the others just promoted to Sergeant, leaving only two officers. I was wondering if anyone knew of POST certified gang training

coming up. My agency will send me only if it is POST certified. Any help would be
appreciated. (Emphasis ours)

Does anyone doubt that the state of California has a significant gang problem? And yet
when surrounded by cities with gang problems and gang units it took this particular
city 15 years to decide to form a gang unit. This certainly does not seem like a
mimetic process.

Katz and Webb notwithstanding, the question is not whether a unit should be
implemented after a major incident(s); rather, whether one should be developed in the
normal course of business or before there is a major incident appears obvious. "If there
is a gang problem in a city or area, one of the most important things to do is to set up a
unit that gathers and investigates intelligence about that gang problem. Only when we
understand how the gangs work can we successfully fight them" (Karlsson, 2000,
p. 60). How does one know if they have a gang problem? When the signs and signals
appear, such as graffiti appearing on buildings, an increase in drive-bys, there exists at
the very least a budding problem if not a full-scale entrenched problem. For some, if a
community has a drug problem then it is likely they have a gang problem too. While this
may not necessarily be an absolute, respondents to the 1999 National Youth Gang
Survey estimated that 46% of youth gang members are involved in street drug sales to
generate profits for the gang (Egley, 2000, p. 2). Of course the other major question is
what is a gang?

DEFINING THE GANG PROBLEM

For many, particularly academics, defining a gang is the most important element of dealing
with the gang problem. There is little consensus on the definition of the term "gang" by
either practitioners or researchers who evaluate gang activity (Wilson, 2000). Some narrow
their definition of gangs to being interchangeable with the term "street gangs" (Howell,
2000). Others define "gang" as synonymous with the term "organized crime group," and
treat them as you would the Mafia or the Triads (Orvis, 1996).

The thought is that we must define the problem before we can address the problem.
While there is certainly some truth to this, we cannot let this paralyze us. In the most sim-
plistic form, *gangs are what gangs do* (Ryan, Personal Communication, 1998; emphasis
added) and what they do is crime, in many cases violent crime. Whether the state provides a definition or not, police, probation, and district attorney agencies must have a
working definition of their own. Such a definition is important if criminal justice professionals are to effectively work together and share gang-related intelligence across jurisdictions (NAGIA, 2000). Spergel (1995) notes we don't know much about gangs in the
aggregate, because of differences in reporting across jurisdictions. Some of this, no
doubt, is due in part to the lack of a common definition. Clearly the National Youth Gang
Survey helps alleviate this problem, although they continue to focus on the "youth"
aspect of gangsterism, specifically excluding motorcycle gangs, prison gangs, hate or
ideology groups, and gangs with an exclusively adult membership (Egley, 2000). The
National Alliance of Gang Investigator's Associations has put together a definition "to
facilitate a national discussion":

Gang. A group or association of three or more persons who may have a common identifying sign, symbol, or name and who individually or collectively engage in, or have engaged in, criminal activity which creates an atmosphere of fear and intimidation. Criminal activity includes juvenile acts that if committed by an adult would be a crime. (NAGIA, 2000: http://www.nagia.org/nagia_recommends.htm#Gang Definitions).

This definition is perhaps the best that currently exists as it covers both juveniles and adults, covers the wide spectrum of ganging, from the least sophisticated to the Italian Mafia and specifies the required number of participants. It also includes the notion of criminal activity (i.e., that gangs do crime), which eliminates many "groups" who (as yet) haven't committed any crimes. Compare this definition of gang with the definition on the National Youth Gang Survey:

A group of youths or young adults in your jurisdiction that you or other responsible persons in your agency or community are willing to identify or classify as a 'gang.' (Egley, 2000, p. 1).

It is more than reasonable to assume that the police officers answering the survey are willing to identify as gangs those groups of youths that are doing criminal activity. This must be the case, as respondents are asked to estimate the proportion and types of crimes their gangs are involved in.

A benefit to the NAGIA definition is that it is applicable within and across the criminal justice spectrum, putting everyone on the same page. It can also be used for non-criminal justice agencies (e.g., education) to put them on the same page as well. Being on the same page is important as the gang problem is being assessed, as the question of do we have a gang problem is being answered; which is of course the number one question to be asked as the formation of a gang unit is being considered. Jurisdictions need a formula into which they can plug local data to determine the best approach to gang crime. It is to the development of such a formula that we now turn.

DEFINING THE GANG UNIT SOLUTION

Ideally, a comprehensive needs-analysis should be done to define the gang problem and to choose among feasible strategies to alleviate it. The community of Racine, Wisconsin, a city of 84,298 developed a gang assessment, planning, and program development team model for communities with an emerging gang problem. The long-named "Community-University Model for Gang Intervention and Delinquency Prevention in Small Cities" model involves community leaders and consists of six steps:

1) **A genuine commitment to youth** . . . The team must demonstrate a commitment to resolving local issues . . . and develop a thorough understanding of the city's social, political, and economic context, especially race and ethnic tensions.

2) **Gang problem assessment**. The team will need to investigate, observe, and document the developing gang problem while learning from neighboring jurisdictions through the exchange of information . . .

3) **Initial networking**. A task force should be formed . . . organizing community meetings and neighborhood hearings to identify solutions and develop a collaborative response to gangs.

4) **Local study of the gang situation**. The task force should identify a local college, university, or other community resource that can study the local gang problem . . . (to) provide the documentation necessary to secure external funding for the programs the task force identifies . . .

5) **Timeout** . . . the task force should publish and disseminate research findings, expand its network via conferences and other communication outlets, identify funding sources, establish political foundations for funding, and prepare grant/contract applications . . .

6) **Development of new programs**. The final stage is program development and implementation. The overall plan should include long-term-goals and a master plan . . . (Howell, 2000, pp. 28–29).

Unfortunately, there is not always enough time to engage in a comprehensive needs-analysis. The nature of gangsterism being what it is, when a gang problem arises, or when there is a perception that such a problem exists, a specialized unit should be formed to aggressively address the problem. For "if any real progress is going to be made to secure society against the threat of gangs, it will have to be through the use of sophisticated, specialized law enforcement units, specially trained in combating those elements" (Jackson & McBride, 1992, p. 93).

In determining what type of gang unit a department should develop, it would be nice if the administrator could look to research to assist in answering that question. Unfortunately there exists little or no research on the specifics of gang units or their effectiveness. Of the departments that have been studied, it would seem that departments develop gang units based on the local dynamics of "their" gang problem with an orientation that fits with the agency's orientation toward crime and disorder. What does this mean? It means that any gang problem is a localized gang problem. This does not mean, however that gangs are stationary. One gang may have "branches" throughout the country.

The FBI and local police have reported the migration of "Crips" and "Bloods" originating from Los Angeles to as many as 45 western and mid-western states (Howell, 1994). The 1999 National Youth Gang Survey reports that their respondents reported an overall estimate that 18% of their gang members were migrants from another jurisdiction. Of some interest is the variation by area that the survey reports: rural areas (34% migrants), small cities (27% migrants), suburban counties (20% migrants), and large cities (17% migrants). In addition, "83% of the respondents agreed that the appearance of gang members outside of large cities in the 1990s was caused by the migration of young people form central cities" (Egley, 2000, p. 2). However, studies of such migration suggest that the cause is family migration and local gang "genesis," and not strategic relocation (Howell, 1994). Furthermore, "most of these rural gangs appear to be primarily homegrown problems and not the result of the social migration of urban gang youth" (Esbensen, 2000, p. 1).

What is even less known is the extent to which a local gang is affiliated with gangs in other cities (e.g., Bloods, Crips, Disciples). In the area around Chicago, for example, according to one gang officer and trainer, it seems that when the Chicago-based Gangster Disciples hear of another Disciples gang, they travel to that city or town, have a "discussion(s)" with the membership and bring that gang into the larger group (oftentimes over the bodies of those that "disagreed"). Whether such a "discussion" occurs the farther one moves from Chicago is unknown. Part of the problem lies with the lack of a uniform definition, as well as the **DID** syndrome, and a general lack of communication between agencies. In many cases, as well, officers may believe that the gangster is simply trying to "be a big shot" by saying that he is from LA, or Chi-Town and not follow-up on this information.

Another reason for having a gang unit is the training and networking that occurs among gang officers. So the type of unit a department develops should be based, in part, on the extent and nature of their problem. So what kind of unit? Regardless of the type selected, a department's gang unit should be a response to their particular problem.

TYPES OF GANG UNITS

Needle and Stapleton (1983) identified three types of specialized police gang units: The Youth Services Program, the Gang Detail, and the Gang Unit (pp. 19ff). In the Youth Services form, personnel are not assigned gang control responsibilities on a full-time basis, and the "unit" is located in another section, usually the juvenile services division. In the Gang Detail approach, one or more officers are assigned exclusive responsibility for dealing with gangsters; again, usually within another unit. In the Gang Unit, a separate unit is developed to deal with the gang problem. [Author's note: This basic structure continues today, although we would submit that the youth services program approach should be scratched and that the gang detail approach should be the minimum established gang unit in any department that chooses to have a specialized gang unit. We would also submit that any gang unit should have two officers at a minimum assigned to it. Why two officers? Because regardless of the approach or mission the unit is to take, one person simply cannot do it all. When you have only one person involved, something(s) will be left undone, and the officer will be stretched and pushed to the maximum.]

In 1991, Weisel and Painter found in their study of five police departments' response to gangs the following three approaches: large specialized gang units, a more generalized approach with one or two person gang intelligence units, and a gang specialist approach located in patrol areas. Five years later they found that the three approaches were now down to two. Four of the departments had specialized gang units (although some had decreased the numbers assigned) and one had taken the specialist-generalist approach. They suggest that "the most common organizational response to gangs pairs a centralized investigative and intelligence-gathering unit with decentralized patrol personnel" (Weisel & Painter, 1997, p. 75).

While patrol cannot and should not be excluded from the "mix," the importance of a stand-alone, specialized unit cannot be overemphasized. Working and dealing with gangsters (and the information they provide) are daily activities that patrol officers simply cannot devote themselves to. As patrol officers go from call to call, reacting not responding, and involve themselves in other aspects of the total community their ability to focus on gangs and gang members is limited. The gang unit has the ability (and time) to focus their attention and efforts on gangs and gang members that patrol simply does not have. A specialized, stand-alone gang unit is the recommended approach and response to the gang problem for most departments. Once the problem has been resolved, the unit can move on to another gang-related mission, find a new mission or disband.

THE ALTERNATIVE MULTI-JURISDICTIONAL TASK FORCE APPROACH

Many communities and police agencies have gone to the "multi-jurisdictional task force" approach to handling their gang problem. Phillips and Orvis (1999) define such a task force as

> a special law enforcement organization with multijurisdictional authority created by an agreement among several government bodies to more effectively combat a delineated crime problem and using the combined resources, both human and logistical, of several law enforcement agencies to more efficiently combat the stated problem for the term of the agreement. (p. 442)

Such an agreement can include law enforcement agencies from several local governments, agencies of the state government, and even agencies of the federal government. For example, the Anti-Violent Crime Initiative begun by the Clinton administration targeted violent and drug-trafficking gangs through the use of federal, state, and local inter-jurisdictional task forces (Howell, 2000). Huff (2000, p. 345) suggests a two-tiered approach to a task force, "As preventive measures, states should consider establishing statewide intergovernmental task forces on gangs, organized crime, and narcotics . . . In addition, each large city should establish a local task force that brings together the following components (where they exist): juvenile bureau, youth gang unit, narcotics unit, organized crime unit, school security division, youth outreach project or other social service coordinating program, and juvenile court."

Of the 1238 jurisdictions reporting youth gang problems in the 1998 National Youth Gang Survey, 612 of them reported participating in a task force focusing on a youth gang problem as its primary concern. Two-thirds of the jurisdictions that participated in these gang task forces are located in large cities. These gang task forces were also found in the survey to be most prevalent in the West and least prevalent in the South, except in southern large cities (Wilson, 2000). The size of such task forces varies in the number of communities, agencies, and officers involved. Although typically a task force may include only a handful of police officers from a few jurisdictions, Minnesota has a statewide gang task force that has 40 members from local, county, and state police agencies. Its members are deputized to have statewide powers and they conduct long-term investigations using a common gang database (Howell, 2000).

As noted earlier, gangs are increasingly mobile, sometimes crossing jurisdictional lines in the accomplishment of their crimes. Members of gang task force units often have the advantage of having the legal authority to investigate and make arrests outside their normal jurisdiction. Although most task forces are in large cities, the formation of a task force allows smaller communities to conduct expensive police operations, such as undercover operations, that they could not afford individually. Several communities may share the costs for a specialized gang unit by forming a task force, and may even be eligible for federal aid in forming and maintaining the unit (Phillips & Orvis, 1999).

However, there are often "strings" tied to the federal monies, which may include oppressive reporting "red tape" and federal mandates that may realign local policy priorities. Another problem is that members of gang task forces are usually part-time in that role, thus lacking the "bonding" experience gained by working full-time in a specialized unit within a single police department, and often lack specialized training needed to deal with the gang problem. Furthermore, multi-jurisdictional task forces may suffer from a lack of adequate management due to a lack of a functioning centralized authority and/or ill-defined spans of supervisory control in the task force's initial organization (Phillips & Orvis, 1999).

MANAGING THE GANG UNIT

Given the responsibilities of the gang unit, it well almost by necessity operate independently, hence the importance of selecting both officers and supervisors very carefully. That said, there are steps that can be taken to insure a well-run unit. First, operational guidelines must be developed as the unit is being put together. These include operational definitions of gangs, gang membership, gang activity, gang-related activity, etc. In some cases, state law establishes many or all of these definitions. For example:

> "Gang" is not a historic legal term; that is, in the absence of statutory definition, gang is not a term of fixed legal meaning. For that reason, every State that has enacted a gang statute has undertaken to define gang, and these statutory definitions are similar. They state how many persons (usually a minimum of three) must be involved, what type of general activity they engage in, and he kinds of crimes involved. The type of activity is sometimes described in a separate definition of "pattern of criminal gang activity." (Johnson, Webster, & Connors, 1995, p. 2)

Where the law does not specify such definitions, the agency and unit must establish for themselves what these definitions will be. The agency should also encourage legislators to do so at the state level as well, so that all agencies and communities are sharing common definitions.

Second, in putting the unit together, it must be remembered that combating gangs requires a multi-agency approach involving all members of the community. The approach must also take into consideration suppression, intervention, and prevention. Hopefully the recognition of a potential gang problem took place *before* the development of the gang unit, so there is some intelligence already in place. If not, this must be the first approach of any unit. Those involved in the unit (as well as other members of the department, the community, etc.) must know the extent and nature of the gang problem. This can only be done through the establishment of gang intelligence files identifying as much information as possible.

Third, as with the establishment of any specialized unit, the gang unit must have the support of the "suits" (administration). The unit must be given organizational support. Those at the top must make it absolutely clear to everyone that the unit and its mission is a department priority. This must be unwavering and communicated throughout the department in as many ways as is necessary. This support must also be communicated to the politicos, the community, the media, and other important stakeholders. Hopefully the decision to form a unit was done before a media-induced hysteria occurred. If so, then unit support must be communicated in a calm, warm (i.e., not detached) manner. If not, then support for the unit must still be communicated but in a more factual way, a counter to the media-led frenzy regarding the "community's gang problem." Regardless, the unit and its establishment must be thoughtful and well coordinated and not a "knee jerk" reactive response to the media or politicians looking to be re-elected by being tough on crime. This top-level support will also go a long way in the unit's acceptance throughout the department. This support must include adequate staffing, equipment, training, and other resources necessary for the unit to accomplish its objectives. Any specialized unit needs the support and acceptance of the department, and this is better accomplished when support is acknowledged from the top.

As the department begins to explore the idea of a specialized gang unit, several questions must be addressed. The initial question is the WHY for the unit. Why is this unit being formed? The authors would suggest that the unit be formed to address the gang problem that has occurred in a community. We would also suggest that the unit may very well develop after an officer(s) have been specifically assigned to do a rudimentary needs assessment of whether and to what extent the gang problem exists. It should be remembered that when the unit is formed, the department is saying there is a need for the unit (or at least the public will perceive there to be such a need). The public and government officials may not appreciate the acknowledgment that a gang problem exists. (Indeed this may delay the formation of a unit, making the problem that much more difficult to get a handle on.)

Another question to be answered is what the unit's role will be. Will the unit function merely as an intelligence unit? Will it be involved in investigation of gangs and gang-related crime? Will it be a suppression unit, using "the full force of the law, generally through a combination of police, prosecution, and incarceration to deter the criminal activities of entire gangs, dissolve them, and remove individual gang member from them by means of prosecution and incarceration" (Howell, 2000, p. 21)? Will it be a prevention unit, designed "to prevent youth from joining gangs, but might also seek to interrupt gang formation" (Howell, 2000, p. 5)? Will it be involved with intervention programs, which "seek to reduce the criminal activities of gangs by coaxing youth away from gangs and reducing criminality among gang members" (Howell, 2000, p. 14)? Or will the unit work in all of these areas? While certainly a unit can change as the problem changes, these goals must be established at the outset as the selection of personnel will be guided by the goals established for the unit.

The establishment of the mission and goals of the unit cannot be overemphasized. All too often, specialized units are established as a public relations tool without any idea of what is to be done or how it is to be done. This idea is addressed in another chapter of this book. The mission and goals of the unit must be clearly defined and established. While all of the above (intelligence, investigation, suppression, prevention, and possibly intervention) can certainly be accomplished by and within the unit, it must be remembered that public safety is the ultimate goal. Gangs are a plague on society, gang members are criminals, in many cases violent criminals, and this fact cannot be forgotten. Despite what some may believe, and indeed despite what some gangsters themselves may believe, the authors would submit that the vast majority of gang members are not misguided youths, they are not a fraternity (or sorority), they are not disenfranchised nor are they the community's representatives; they are criminals, pure and simple. And criminality must be prevented when possible, investigated and suppressed when it occurs. Any other goal is secondary.

It must also be remembered that the gang unit is a proactive unit. Similar to Goldstein's (1990) problem-oriented policing, it responds to the gang problem, not simply reacts to it. By responding, the unit is in control, rather than the gangsters. This response employs a number of tactics and approaches, some of which may appear unorthodox when compared to other units of the police agency. The gang problem is an ever-changing phenomenon, and so the gang unit must also be evolving. As part of its tactics in addressing the gang problem, the unit must reach out to other criminal justice, education, and social service agencies to enlist their cooperation and engagement in dealing with the community's gang problem. As with many other problems, a comprehensive community-wide effort must be the approach taken.

While the police may have (and should have) the primary responsibility for addressing the gang problem they cannot do it alone. The socio-psychological and sociological ramifications of the gang problem are too immense. Coalitions must be built—agencies with complementary missions must be active members of the team. This must be something that is recognized by everyone involved in the community. However, nearly all gang task forces involve only criminal justice agencies (Wilson, 2000). An extreme example of the opposite approach is the Aurora Gang Task Force (AGTF), which won the 1992 City Livability Award from the National Conference of Mayors for its efforts to mobilize an effective response to their gang problem. Formed in 1989 in Aurora, Colorado, the AGTF's members come from government agencies, social services, churches, volunteer organizations, businesses, the military, and the media. The most amazing fact about the AGTF is that all its members volunteered for the task force (Howell, 2000).

DUTIES OF GANG UNIT OFFICERS

What does a gang unit do? Once the unit has been formed, it engages in a variety of activities, all interrelated and all related to the gang problem. The temptation to use the unit for other tasks or to address other, non-gang problems should be avoided at all costs. If there is no longer a need for the unit then disband it, or reconfigure its mission, but as long as the need exists, do not use it as a stopgap for other problems or concerns.

The unit must primarily involve itself in the following activities:

1. Profiling gangs and gangsters,
2. Tracking gangs and gangsters,
3. Obtaining, storing, and disseminating intelligence on gangs and gangsters,
4. Assist other units' investigation into gang-related cases,
5. Interact with gangs and gangsters,
6. Sweep and arrest operations,
7. Investigation of gang and gang-related crime.

The above list can be partitioned out to unit members based on knowledge, training, and expertise and clearly involve working with other members of the department, especially the patrol division. Each of these related duties is important for the overall mission of the unit and to address the gang problem. While all are important to the success of the unit and in combating the problem, perhaps the most important area is disseminating intelligence and information.

Gangs and gangsters cut across a variety of criminal and community lines. The importance of first knowing and understanding what is happening and secondly communicating that information to others is paramount. The bad guys do not just stay in one place anymore, and so law enforcement must account for and be able to explain the movements. This requires pro-activity on the part of the unit, and a good working (or at least a good communicative) relationship with a variety of agencies. Information cannot be kept to the gang units themselves; it must be shared with any and all parties that might be affected by the gangs and their members. Where information sharing is difficult, law enforcement must take the lead in smoothing out the rough spots, calling on the "bosses" and "suits" when necessary to intervene.

The importance of sharing information cannot be overemphasized. Everyone affected by the gang problem must know what's going on. This would include the media when necessary. However, there are two schools of thought on this matter. On one hand, publicity enhances the gang's image among those in their "turf" and among rival gangsters, and it might induce the gang to greater feats of criminality in order to get more publicity. On the other hand, media attention on gangs increases public awareness, and often creates public pressure on government officials to increase resources to deal with the gang problem, thus benefiting the gang unit (Jackson & McBride, 1992).

All too often, specialized units keep to themselves, not willing to let others know what they know. While this is never good for any unit, it is especially bad for the gang unit. Gangs themselves change to meet the demands "of the street" or the pressures from law enforcement, they are on the Internet, they are using schools and prisons as places of recruitment, they are traveling from city to city, they are collaborating in ways previously unheard of. If information is not shared between and among those who fight gangs, policing can be assured of always being one step behind the gangsters.

EFFECTIVE COMMUNICATION

It must be remembered that effective communication is a two-way street, requiring listening as well as talking. Often, gang criminality occurs on school property, therefore school officials, particularly school security personnel that are "trained in combating gang violence," can provide invaluable service in the investigation of such crimes and the prevention of future incidences (Jackson & McBride, 1992, p. 28). Community leaders in neighborhoods that are part of gang "turf" are invaluable sources of intelligence and investigation information as well as useful as a communication conduit to the gangs themselves, when necessary to establish gang summits or negotiate gang truces (Howell, 2000). In a broader sense, the Internet can be an invaluable source of gang information, with many websites established for that purpose. There is even a website whose specific goal is to provide web "links" useful to police gang units (Officer.com, 2001).

Gang units should communicate regularly with other law enforcement units within their own jurisdictions; maybe even through an official liaison. For example, prosecutors can educate gang units on how new criminal laws can be used to suppress gangs as well as inform police as to why some cases against gang members are not successful (Orvis, 1996). Some prosecutors operate their own computerized database and/or employ their own gang investigators (Johnson et al., 1995). In a recent survey of American prosecutors from 38 states, about a fifth received in-service training on gang awareness and a few (5%) had specialized gang prosecution units (National Gang Crime Research Center, 1995). It is also important to open communications with law enforcement agencies in other communities.

> Since gangs do not exist solely within city or county boundaries, and since gang clashes all too frequently involve more than one jurisdiction, it helps to know what gangs are active or reside in adjacent areas or jurisdictions . . . To keep informed it is necessary to maintain a good working relationship and have an effective information exchange arrangement with gang units of other law enforcement agencies. (Jackson & McBride, 1992, p. 97)

To be effective in responding to the problem, law enforcement generally and the gang unit specifically must change their old ways of doing business, and this includes hoarding of information. Egos must be put aside and collaborations must take place. The department's "brass" must make information sharing and dissemination a priority both within and across agencies. This idea of shared information, communication, and unorthodox approaches is readily seen with Operation Night Light in Boston, MA:

> The success of Night Light depends on a number of key operating principles. First, *intensive communication* and a *unified sense of mission* among all the partners [including ministers, social workers, parents, etc.] are crucial . . . Night Light is committed to a broad strategy characterized by prevention, intervention and enforcement. (Jordan, 1998, p. 2–3, emphasis added)

Regardless of which approach or combination thereof is taken by the unit, one thing must be kept in mind. The number one, most important function of the gang unit is information/intelligence. Officers in the unit must be on the streets, working the gangsters, being approachable and visible not only to gangsters, but to other citizens who may have information the gang unit needs. The gathering of intelligence fuels all of the other goals of the unit (and the department). Without up-to-date information, the ability to solve cases diminishes. Street gangs are street criminals and information sharing must be a priority within and among all agencies involved with these criminals.

It is best for the unit to use a combination of all three approaches: intelligence, investigation, and suppression. To do this, the unit should take the lead in liaison with the prosecutor's office and the probation/parole office (both juvenile and adult) to insure a coordinated approach to gangs and gangsters. If the gang unit is going to take the lead in combating gangs, the unit's personnel become very important.

PERSONNEL

The amount and type of personnel necessary for a gang unit to function properly depends on the duties assigned to it. If the unit is primarily investigative, then a cadre of detectives is necessary. If gang suppression is the main goal of the unit, then uniformed officers are important to conduct sweeps and the like; "using measurable gang activity as yardstick for the number of officers involved" (Jackson & McBride, 1992, p. 95). Logistical support staff with computer skills is important to maintain the up-to-date gang files needed to organize the intelligence gathered by a gang unit. The gang unit may need members trained in social work if the unit is also committed to prevention and/or intervention policies. If responsibilities outside the gang unit are assigned to members of the gang unit, then this too must be considered in the final analysis. Certain questions must be answered before personnel can be selected for the gang unit:

1. What will be the unit's primary function?
2. Will the unit be responsible for the criminal investigation of all gang cases?
3. Will the unit operate as a monitoring section, compiling and disseminating intelligence information?
4. Will the unit handle enforcement operations and patrol gang areas to curtail violence?
5. Will the personnel work in plainclothes or in uniform?

6. Will the unit operate covertly, with undercover operations?
7. Will the unit function as a line or support section?
8. What will be the best hours of operation, and how many days a week should the unit be on duty?
9. Where will the unit be assigned (patrol, detective, or administrative) (Jackson & McBride, 1992, p. 94).

Given the importance of the gang unit and its duties, the selection of personnel to staff it is absolutely critical. First and foremost, the officers must be above reproach. There must be nothing in their "jackets" to suggest any kind of illegal or corrupt activities. They must be considered "good" officers and not "hot dogs." Officers who are unorthodox (i.e., creative, thinking outside the box) in their methods should be considered as well, as working with gangs and gangsters oftentimes require unorthodox methods.

Officers should have at least five years of street experience and be tactically correct in their approach to the job. They must be able to get along with other officers, be tenacious, flexible, thorough, and approachable (Safe Streets, 2001). Those selected should be able to work independently and be self-motivated. Gang unit officers need excellent oral and written communication skills. They must be able to analyze and synthesize information and make decisions. Officers must be able to adapt to situations quickly, they must know how to manage red tape and how (and when) to "screw the system to make it work" (Jacobs, l993). Gang unit officers must know when to suppress their ego and must be able to take a subordinate role for the good of the ultimate job or case. Gang officers must be able to work a variety of cases, know when to "hand off," and be able (along with their families) to accept the risks inherent in working with gangs.

Officers must be able to establish rapport with gangsters without appearing weak or compromising their positions. Gangsters come from a subculture where kindness is weakness; the officer must be able to have a street persona of strength. This does not mean that the gang officer should be less than professional, but the gang officer must be perceived as being strong and fair, from the street, not of the street (Wambaugh). The gangsters must see the officer as a professional; one who knows what he or she is doing and is good at it. While not the gangster's friend (sharing a Coca-Cola and singing Kumbya), the officer must be approachable and perceived to be fair in their dealings with the gangsters. Gangsters are all about respect, they demand it, and they give it (Jackson & McBride, 1992). The officer must know this and respond accordingly.

In addition, gang officers (as should all police officers) must be well grounded and be involved in an activity(ies) outside of the job. Working a gang unit can become an all-consuming activity; the gang officer must avoid this at all cost. One way, perhaps the best way of keeping the gang unit officer "grounded" and the unit "on track" is with the proper supervision.

UNIT SUPERVISION

As important as the officer is to the unit, perhaps the most important people in the unit are the supervisors. They too must be self-directed and highly motivated. Gang unit supervisors (as should all police supervisors) need to be "hands on." They need to not be afraid to work the streets with their officers nor should they be afraid to make unpopular decisions. Consistency should be their mainstay. "There can be no question as to the identity of the team leader or the clarity of his/her orders and directions" (Safe Streets, 2001, p. 85). The unit (or sub-unit) must believe and have faith in the supervisor (and he or she in the unit/team) or it will fail. The importance of good supervisors and administrators was seen in the Los Angeles Police Department's Rampart scandal.

While the number of officers involved represent a relatively small percentage of both CRASH and the LAPD, the scandal and the NYPD's shooting of Amadou Diallo has called into question the use of specialized units. A study commissioned by the Los Angeles Police Commission indicated that, "[T]he misconduct of CRASH officers went undetected because the department's managers ignored warning signs and failed to provide the *leadership, oversight, management,* and *supervision* necessary to control this specialized unit" (Barry, 2000, p. 1, emphasis added). The supervisors and bosses (managers) of the gang unit must be especially sensitive to what's going on with their officers. They must have the ability to identify those officers who are "in too deep," and those officers who have developed "the wrong attitude."

The bosses and supervisors must communicate in specific and no-nonsense detail exactly what the mission and goal(s) of the unit are, and what the acceptable and not acceptable tactics are in accomplishing the mission. They must have the fortitude to take significant and appropriate disciplinary action when necessary. In addition, supervisors and bosses must not be captives of the politicos. They must not be afraid to stand up to the politicians and "suits" who all too often are merely "frail reeds in the winds of fickle public opinion" (Jenkins, 2000, p. 2). They must lead from the front and offer complete support for their members when they are right, and provide quick, appropriate, and correct discipline when their members are in the wrong. The gang unit is oftentimes highly regarded by those they are sworn to protect, a "welcome alternative to the tyranny of insecurity and fear caused by rampant crime" (Jenkins, 2000, p. 2). What cannot happen (and what the supervisors and bosses must be aware of) is for the unit to become as bad or worse than those they are going after. It is the supervisors and the bosses' responsibility to protect against it. As Jenkins notes, "elite units require skillful management and excellent leadership" (2000, p. 2).

SPECIAL PROBLEMS

Gang units however are unique among elite units. Gang units therefore have problems somewhat unique from other specialized units. Unlike most other specialized units, a gang unit's problems are often different from even other gang units. Thus, gang units may need special training to deal with these problems, although not all gang units need the same training. Howell (1996, p. 313) recommends a centralized federal approach to training, "Training and technical assistance for police and other local community agencies should be expanded, primarily through the addition of technical training teams, which would provide a broader range of gang diagnosis and program development expertise." Some examples of such training and technical assistance would be that provided by the National Youth Gang Center, the California Gang Investigator's Association, the National Gang Crime Research Center, etc.

As noted earlier, the mission and composition of gang units are varied. The goal or goals of the unit (i.e., gang suppression, prevention, intelligence, investigation, intervention, or some combination of these) can create problems unique to that unit. Too many goals or not enough resources allocated to a single goal can create financial problems for the unit's managers. Part-time personnel can create a division of loyalty and attention, distracting gang unit members from fulfilling their mission. A lack of a centralized unit authority or too broad a span of control for unit supervisors can create management problems as well. These logistical problems vary from gang unit to gang unit, but are similar in that they are solved if planning and support are adequate from the beginning of the unit.

Not all gangs have the same dynamics, and often this is due to the culture the gang originates from. At the turn of the twentieth century, the majority of American gangs were white with various European backgrounds. By the 1970s, four-fifths of gang members were either Hispanic or African American. Asian gangs are spreading quickly today (Howell, 1994). It was found in a recent survey that only 12% of all youth gang members are Caucasian, with 46% being Hispanic, 34% being African American, and 6% Asian. Furthermore, two-thirds of all gangs are dominated by members from a single racial or ethnic groups (Wilson, 2000). Law enforcement agencies estimate that more than 90% of gang members are male (Esbensen, 2000). However, female gang members and even all-female gangs are increasing in number, but not to a significant extent (i.e., less than 3%). Needless to say, the low social class of gang members remains constant over the years (Howell, 1994). Gang unit members must be prepared to deal with the cultural diversity or uniqueness of the gangs that they must deal with in their jurisdiction.

The gang unit must also be aware of the culture of the neighborhoods in which the gangs operates (i.e., each gang's "turf"). Some neighborhoods are continually terrorized and intimidated by gangs, and therefore afraid to cooperate with police because of gang retaliation (Jackson & McBride, 1992). However, recent emigrants from countries whose police are oppressive may fear police because of their past experiences. Further, English may not be the predominant language in some neighborhoods, so a language barrier between the gang unit and the citizens they seek to protect may have to be overcome. Gang units must be able to adapt to these contingencies, and therefore some unit members must be skilled in the languages and aware of the cultures of the neighborhoods whose cooperation they need to be successful. Gang units engaging in a strategy of "community policing" in those neighborhoods may find it easier to overcome cultural barriers:

> Among agencies that had implemented community policing for at least 1 year, 99 percent reported improved cooperation between citizens and police, 80 percent reported reduced citizens' fear of crime, and 62 percent reported fewer crimes against persons . . . Citizens in community policing jurisdictions were more likely to participate in a Neighborhood Watch Program, serve as volunteers within the agency and on agency-coordinated citizen patrols, and attend a citizen police academy. (Travis, 1995, pp. 1–2)

Study of the culture of the gang's turf is important for prevention as well as suppression programs, especially for youth gangs. The city of Westminster, California, studied youths of Vietnamese descent from neighborhood where Vietnamese youth gangs dominated in order to develop successful programs to divert the youths from joining the gangs in the future (Wyrick, 2000).

ONE APPROACH

As a department looks for a model for their developing gang unit, all too often a Chief will look toward the Los Angeles or the Chicago police departments. While certainly these cities have a plethora of gangs and their departments have responded with varying degrees of success, they have also had a number of problems associated with their units, the most public of which has been the problem with LA's Rampart Station CRASH unit. Although not LA or Chicago, the city of Westminster, California located in Orange County has established a gang unit model that with local modifications is a good one to follow.

Westminster, with a population of 83,000 has a culturally diverse population including Asians, Hispanics, Whites, and Blacks. It has also experienced a significant gang problem, which culminated in the shooting death of a 49-year-old teacher's aide (Valdez, 2000). While this shooting (in the process of a car jacking) brought media attention and crystallized the gang problem for Orange County, the Westminster Chief was ahead of the train. Recognizing the potential for gang activity and problems, he had been exploring the possibility of a unique gang unit. As is usually the case, the shooting helped those who were in the **DID** syndrome to pay attention to the Chief and ultimately to fund his grant application. The Chief's plan was to bring together the coppers, the probation department, and the district attorney's office as the Westminister gang unit. Part of the uniqueness of his approach is that the members of the unit would all be housed in the police department. This was important, because while these agencies oftentimes worked together, it was from separate offices, and as Valdez notes, ". . . information exchange is slow and seldom immediate" (2000, p. 456). Putting the three agencies together (physically and organizationally) had "a synergistic (enhancing) effect on the unit" (Valdez, 2000, p. 457). In addition to coppers, the unit was also staffed with two police service officers and an intern. The unit was named TARGET, standing for Tri Agency Resource Gang Enforcement Team. While Klein (1995) has some problem with the names attached to some gang units (representing a suppression or "war" mentality) names are important for the solidarity and cohesiveness of the unit and to identify the unit to other criminal justice and related agencies.

Now that the unit had been established and named with the appropriate Memoranda of Understanding written, how will the unit operate? [Author's note: A memoranda of understanding is a document agreed to and signed by all parties involved specifying what each agency is responsible for.] Here again, the Chief took a somewhat unique approach in law enforcement; he looked to criminology, to academia. Looking at the work of Wolfgang et al. (1972) (and similarly Hamparian et al., 1978; Shannon, 1982) the chief decided that if a small percentage of bad guys commit a large proportion of the crime, what if the same existed among the gang bangers (i.e., a small percentage of gangsters committed the majority of the gang crime). Certainly (at least in California) Wolfgang's thesis holds (cf. Humes, 1996, the 16 percenters). The thrust of the TARGET team was to focus on this group, the worst of the worst. One of the advantages of TARGET is that it puts cops, district attorneys (i.e., the court), and probation (i.e., corrections) all at the "front end" of the system. It should be noted as well that TARGET took the "soft uniform" (i.e., not wearing a "traditional" police officer uniform, wearing instead a "golf" shirt and pants) approach. The soft approach (generally favored by Klein) became the identifier for TARGET members and tended to be viewed by the gangsters in a less threatening manner than the traditional patrol uniform. Reports Valdez (2000, p. 464), "many times gangsters would tell me they knew they could not try to 'scam' us, because the type of uniforms the team wore indicated they were specialists" (i.e., professionals: people who knew what they were doing, and were good at it [Haig, personal communication, 2000]).

In addition to focusing on the "worst of the worst," and using current law and strategies, the unit was also responsible for developing new strategies to deal with the gangsters. Some of the strategies developed included multiple dwelling search warrants, honeymoon search warrants, and understanding the cooling off period. Honeymoon warrants are an interesting and effective tactic. The "honeymoon" of an incident is the first 2–12 hours after an incident (Valdez, 2000). The honeymoon warrant is a warrant obtained and served during this time period. This allows the unit to strike quickly, before the gangsters have an opportunity to shift, hide, or destroy evidence. Of course, the effectiveness of such warrants depends in large part on the quality of the intelligence gathering and information dissemination of the unit.

The cooling off phenomenon is that period of time in which evidence is hidden before it resurfaces. For example, a gang member uses a weapon in an assault or homicide. Oftentimes, the weapon will go "underground" and when the gang thinks the pressure is off, the weapon will resurface and be used again. Officers must understand this phenomenon and maintain a constant vigilance on the implements used in the gang crimes. Valdez (2000, p. 466) reports a "major decrease in violent gang crime" and an increase in the quality of life of the residents of Westminster specifically and Orange County, California generally.

THE FULL SERVICE GANG UNIT

Spergel (1995, p. 199) notes that a community-problem-solving approach to gangs seems to be slowly emerging. This is based, he avers on the idea that "an arrest and lock-em-up strategy is not sufficient." Sgt. Wes McBride, of the Los Angeles County Sheriff's Department, an author, trainer, and one of the foremost gang experts in the country, notes that a person working with gang members must understand the gang ethic, customs, and practices of the gang(s) in their area. The reason to do this is that knowing these things helps the officer develop rapport with the gangsters making dealing with them easier and allows for more and better communication with the gang and the gangster (in Valdez, 2000, p. 543). McBride (1993, p. 413) notes that when "firm but fair law enforcement, [coupled] with personal knowledge of the gang members backed by a demonstrated humanitarian concern for the status of the individual, violence within targeted gangs began to decline."

Father Greg Boyle, who runs the Delores Mission in Los Angeles has noted that "gang members need hope to help make it in life" (in Valdez, 2000, p. 543). Father Boyle suggests that many gang members have no hope and, "don't want to look in a mirror and accept what they see" (in Valdez, 2000, p. 543). Dealing with gangs and gangsters requires both a Wes McBride and a Father Boyle (and the many other types of people who work with gangs and gangsters). Dealing with gangsters needs what McBride refers to as *Full Service Gang Units*.

A full service gang unit has responsibility for suppression, intervention, and prevention. It "kicks butt" and makes arrests when necessary and it refers gangsters to a wide array of programs based on the need of the gangster. A full service gang unit is a collaborative program, using resources already in existence, while being open to new resources that become available. The key element, the cement is that all members are working toward the same goal: increasing the quality of life for the people they serve. Each of the members are working collaboratively but independently (based on their specific expertise and perspective) toward the same goal, under the auspices of a program manager.

The program manager(s) would coordinate the efforts of the various agencies involved in the full service gang unit. They would also know what each agency was doing and could insure that all agencies were "on the same page." What the full service gang unit allows for is immediate action depending on the need. When working with gangs and gangsters, all too often one cannot wait until tomorrow. When a gang crime is committed, immediate action must take place, "when a gangster comes to you and says, 'I want to change,' that is a totally different scenario and should be acted on *immediately*" (Valdez, 2000, p. 544, emphasis added). Young children need the tools and information to handle the pressures of living in gang neighborhoods and avoiding gang membership. Plus, they need the tools sooner rather than later! This, of course, would require involving schools in the full service gang unit.

Full service gang units do not reinvent the wheel. They simply take the multitude of agencies working with gangs and gangsters and put them together in a collaborative program, sharing costs, expertise, knowledge, and intelligence. The full service gang unit is a model that can be adopted by any agency and one that can be regionalized depending on the city or county. Just as gangs share some characteristics they also regionalize many practices. The full service gang unit has the flexibility to address this. The issues are time, cooperation, and commitment. Egos, rivalries, biases, and other negative influences or thoughts must be set aside (Valdez, 2000). The goal is simple: improving the quality of life of the communities in which we serve, the "achievement of 'functional communities,' that is, communities in which family life, work, religion, education, law enforcement and other institutional areas reflect and reinforce common values" (Short, 1990, p. 226).

CONCLUSION

Gangs and gangsters are a plague on our society. Gangs and gangsters are what they do, and what they do is crime and violence. When the plague descends on a community, the first response must be one of intelligence gathering, accomplished by the department dedicating an officer(s) to gather and analyze the information. Once it has been established that a gang problem (or the possibility of one) exists, the formation of a gang unit is the next step. While it should be kept in mind that the primary responsibility of law enforcement is "hooking and booking," any approach to dealing with a community's gang problem needs to be multi-faceted and involve the entire community. The approach crafted should involve intelligence, suppression, prevention, and intervention—a *Full Service Gang Unit*. Members of the gang unit must be able to shift from "rapping" with the gangster to taking a zero-tolerance approach; they must be able to communicate effectively with a wide range of community members and criminal justice personnel; and they must be able to see beyond their own agencies, egos, and communities to adequately address the gang problem.

The gang unit is a specialized unit designed to deal with the special and unique problems associated with the plague of gangs. Given the rapidity with which gangs encompass a community, their rapid mobility, their use of the Internet, and their ever-changing face, a specialized unit may be necessary to adequately address these and other concerns regarding gangs. That gang unit must adequately understand the gang ethic, customs, and practices of the gang(s) in their area. Like any specialized unit, the gang unit can lose its way, can go astray, but that is not a reason to disband it (like LAPD and NYPD). We must keep in mind that specialized units must be adequately staffed, supervised, and kept on mission. We must also keep in mind, as Larry Elder says, "cops are hired out of our flawed, human population. The process screens out numerous applicants for every one who makes it through." (http://www.townhall.com/columnists/larryelder/le000525.shtml).

[Still, things go wrong, break down, fail to work, or people fail to communicate. Planes crash. People die in car accidents. From time to time, cops act illegally, corruptly, or brutally. By all means, we must stay vigilant, and weed them out. But let's have perspective. Cops live real lives. They make mistakes. They get grumpy, surly, and sometimes lax. They sometimes get ticked—just like the rest of us. Yeah, I know the guys on "Adam 12" never made a mistake, and Joe Friday always kept his cool. "Too bad the crooks don't spend all their time watching television, instead of committing crimes. You know, like in real life" (Elder, 2000)].

Let us not throw the baby out with the bath water. Let us deal with the problems of the officers involved in the specialized units. Let us deal with the problems that the specialized units may bring about. Let us think carefully before we decide to not establish or disband a specialized gang unit. While other specialized units may not be necessary or outlive their usefulness, the plague and criminality of the gangs themselves establish the necessity of a gang unit.

RECOMMENDED READINGS

DECKER, S.H. (2003). *Policing gangs and youth violence*. Belmont, CA: Wadsworth.

GOLDSTEIN, H. (1990). *Problem oriented policing*. New York: McGraw-Hill.

GRENNAN, S., Britz, M., Rush, J., & Barker, T. (2000). *Gangs: An international approach*. Upper Saddle River: Prentice-Hall.

KLEIN, M. (1999). *The American street gang*. New York: Oxford.

KOREM, D. (1994). *Suburban gangs: The affluent rebels*. Richardson, TX: International Focus Press.

VALDEZ, A. (2000). *A guide to understanding street gangs*. San Clemente: LawTech.

REFERENCES

BARRY, D. (1996) The Columbia World of Quotations, number 9180.

——. (20 November 2000). *L.A. officials blamed for police scandal*. APBnews.com.

CURRY, G.D. (1995). *National Youth Gang Surveys: A review of methods and findings*. Washington DC: U.S. Department of Justice.

EGLEY, A. Jr. (2000). *Highlights of the 1999 National Youth Gang Survey: OJJDP fact sheet*. Washington, DC: U.S. Department of Justice.

ELDER, L. (25 May 2000). *Police scandal: Overreaction equals anarchy*. Townhall.com columnists.

ESBENSEN, F-A. (2000). *OJJDP Juvenile Justice Bulletin: Preventing adolescent gang involvement.* Washington, DC: U.S. Department of Justice.

GOLDSTEIN, H. (1990). *Problem oriented policing.* New York: McGraw-Hill.

HAMPARIAN, D., SCHUSTER, R., DINITZ, S., & CONRAD, J. 1978. *The violent few.* Lexington, MA: Lexington Books.

HOWELL, J.C. (1994). *Gangs: OJJDP fact sheet.* Washington, DC: U.S. Department of Justice.

——. (1996). Recent Gang Research: Program and Policy Implications. In C.W. Eskridge (Ed.), *Criminal justice: Concepts and issues.* Los Angeles: Roxbury Publishing Company.

——. (2000). *Youth gang programs and strategies: Summary.* Washington, DC: U.S. Department of Justice.

HUFF, C.R. (2000). Youth gangs and public policy. In B.W. HANCOCK & P.M. SHARP (Eds.), *Public policy, crime, and criminal justice.* Upper Saddle River, NJ: Prentice-Hall.

——. (2001). Youth gangs, crime, and public policy, In H.N. PONTELL & D. SHICHOR (Eds.), *Contemporary issues in crime and criminal justice.* Upper Saddle River, NJ: Prentice-Hall.

HUMES, E. (1996). *No matter how loud I shout.* New York: Touchstone.

JACKSON, R.K., & McBride, W.D. (1992). *Understanding street gangs.* Placerville, CA: Copperhouse Publishing Company.

JACOBS, M. (1993). *Screwing the system and making it work: Juvenile justice in the no fault society.* Chicago: University of Chicago Press.

JENKINS, B. (27 March 2000). *Elite units troublesome, but useful.* LA Times.com.

JOHNSON, C., WEBSTER, B., & CONNORS, E. (1995). *Prosecuting gangs: A national assessment.* Washington, DC: U.S. Department of Justice.

JORDAN, J.T. (1988). Boston's Operation Night Light. *FBI Law Enforcement Bulletin, 67*(8), 1–8.

KARLSSON, M. (2000). Keeping gangs under control. *911 Magazine, 13,* 60–63.

KLEIN, M.W. (1995). *The American street gang.* New York: Oxford University Press.

McBRIDE, W. (Summer 1990). Speech given in Gulf Shores, AL.

——. (1993). Part II-Police departments and gang intervention: The Operation Safe Streets concept. In A.P. GOLDSTEIN & C.R. HUFF (Eds.), *The gang intervention handbook*, pp. 411–415. Champaign, IL: Research Press.

NAGIA. (2000). www.nagia.org.

NATIONAL GANG CRIME RESEARCH CENTER (1995). A research note: Preliminary results of the 1995 National Prosecutor's Survey. 2 *Journal of Gang Research, 4,* 59–71.

NEEDLE, J.A., & STAPLETON, W.V. (1983). *Police handling of youth gangs.* Washington DC: OJJDP.

OFFICER.COM (2001). *Resources for gang units.* www.officer.com/special_ops/gang.htm.

ORVIS, G.P. (1996). Innovative prosecution of youth gangs by treating them as organized crime groups. In J.M. MILLER, & J.P. RUSH (Eds.), *Gangs: A criminal justice approach.* Cincinnati: Anderson Publishing Company.

PHILLIPS, P.W., & ORVIS, G.P. (1999). Intergovernmental relations and the crime task force: A case study of the East Texas Violent Crime Task Force. 2 *Police Quarterly, 4,* 438–461.

SAFE STREETS BUREAU (May 2001). *L.A. Style: A Street Gang Manual of the Los Angeles County Sheriff's Department.*

SHORT, J.F. (1990). *Delinquency and society.* Englewood Cliffs, NJ: Prentice-Hall.

SLIWA, C. (October 1987). Speech given in Birmingham, AL.

SPERGEL, I. (1995). *The youth gang problem.* New York: Oxford University Press.

SPERGEL, I. et al. (2000). *Youth gangs: Problem and response.* University of Chicago: The School of Social Service Administration.

TRAVIS, J. (1995). *Community policing strategies: NIJ research preview.* Washington, DC: U.S. Department of Justice.

VALDEZ, A. (2000). *Gangs: A guide to understanding street gangs* (3rd ed.). San Clemente, CA: LawTech Publishing.

WEBB, V.J., & KATZ, C.M. (2003). Policing gangs in an era of community policing. In S.H. Decker (Ed.), *Policing gangs and youth violence*, pp. 17–49. Belmont, CA: Wadsworth/Thomson.

WEISEL, D.L., & PAINTER, E. (1997). *The police response to gangs*. Washington DC: Police Executive Research Forum.

WILSON, J.J. (2000). *1998 National Youth Gang Survey: Summary*. Washington DC: U.S. Department of Justice.

WOLFGANG, M., FIGLIO, R., & SELLIN, T. (1972). *Delinquency in a birth cohort*. Chicago: University of Chicago Press.

WYRICK, P.A. (2000). *Vietnamese youth gang involvement: OJJDP fact sheet*. Washington DC: U.S. Department of Justice.

6

Bias Crime Units

Joseph B. Vaughn

INTRODUCTION

This chapter examines units that investigate hate crimes motivated by bias. While history is replete with crimes motivated by prejudice and bias, creation of police units specializing in these offenses is a relatively new endeavor. The first state level bias crime unit was not created until 1992 (Bureau of Justice Statistics, 2000a, p. 12). Perhaps more than any other factor, passage of the Hate Crime Statistics Act of 1990 focused attention of the police on bias crimes. Under that law, the Federal Bureau of Investigation was directed to begin compiling statistics on bias-motivated crimes through its Uniform Crime Reporting program.

Much of the existing research focuses on broad issues surrounding the definition of bias crimes and recommendations for implementation of community-based programs to address them. There is no readily available or reliable data documenting the number of departments having specialized units for this purpose.

DEFINING BIAS CRIMES

The terms hate and bias crimes are often used interchangeably. Early semantic arguments about the term "hate crime" arose as legislatures attempted to decide the parameters of hate crime laws. In general, hate crimes are crimes motivated by bias. For the purpose of data collection, the FBI defines a bias crime as a "criminal offense committed against a person or property which is motivated in whole or in part, by the offender's bias against a race, religion, disability, ethnicity/national origin, or sexual orientation" (Federal Bureau of Investigation, 1997, p. 14). In 1998 the International Association of Chiefs of Police (1999,

p. 3) suggested expanding the definition to include "a criminal offense committed against persons, property or society that is motivated, in whole or in part, by an offender's bias against an individual's or a group's race, religion, ethnic/national origin, gender, age, disability or sexual orientation."

It is important to distinguish between hate crimes and hate incidents. A hate incident is any act, whether criminal or not, where prejudice is evident as a contributing factor and the intention is to harm. While all hate crimes result from hate incidents, not all incidents are criminal in nature (Paynter, 2000a, p. 53). It is often difficult for the public to distinguish between incidents that constitute criminal conduct and those that do not. This confusion may result in demands being placed on bias crime units that they are not able to address under existing laws.

HATE CRIME STATISTICS

National data collection on hate crimes began in 1991 as a result of the passage of the Hate Crime Statistics Act of 1990. The Act required the Justice Department to collect data on crimes that "manifest prejudice based on race, religion, sexual orientation or ethnicity." In 1997 the FBI included disability-motivated crimes as a result of the passage of the Violent Crime Control and Law Enforcement Act of 1994. The Church Arsons Prevention Act of 1996 amended the Hate Crime Statistics Act (1990) and made the FBI's data collection mandate permanent by removing a sunset provision from the original act which would have required them to collect data for only four years. Several private organizations compile statistics and other information on bias crimes and hate groups.

UNIFORM CRIME REPORTING SYSTEM

The resulting data collection was integrated into the existing Uniform Crime Reporting System (UCR). The UCR is divided into the Summary Reporting Systems and the National Incident-Based Reporting Systems. The summary system is used by a majority of law enforcement agencies to report traditional Part I and Part II crimes. Agencies utilizing this system are asked to voluntarily submit data quarterly on crimes motivated by bias for the Part I offenses of murder and non-negligent manslaughter, forcible rape, robbery, aggravated assault, burglary, larceny-theft, motor vehicle theft, and arson. Data are also collected for simple assaults, intimidation, and destruction, damage, or vandalism of property (Federal Bureau of Investigation, 1999).

Those agencies using the incident-based system (NIBRS) submit bias crime data on all Group A offenses. The number of crime categories in the NIBRS system is expanded to twenty-two from the eight categories making up the Part I offenses in the summary system. This results in data being collected for more offenses under the incident-based system. The additional offenses in Group A include such crimes as bribery, counterfeiting, drug offenses, embezzlement, extortion and blackmail, fraud, and weapons law offenses (Justice Research and Statistics Association, 1995).

To ensure uniformity in the collection of data, the program provides specific definitions for the use in hate crime reporting. They limit the types of bias motivation included to those groups specifically mandated by federal legislation. Table 6-1 shows the current bias

TABLE 6-1   Bias Motivation Categories

Racial bias
    Anti-white
    Anti-black
    Anti-American Indian/Alaskan native
    Anti-Asian/Pacific islander
    Anti-multi-racial group

Religious bias
    Anti-Jewish
    Anti-Catholic
    Anti-Protestant
    Anti-Islamic (Moslem)
    Anti-other religion (Buddhism, Hinduism, Shintoism, etc.)
    Anti-multi-religious group
    Anti-Atheism/Agnosticism

Disability bias
    Anti-physical disability
    Anti-mental disability

Sexual orientation bias
    Anti-male homosexual (Gay)
    Anti-female homosexual (Lesbian)
    Anti-homosexual (Gay and Lesbian)
    Anti-heterosexual
    Anti-bisexual

Ethnicity/National origin bias
    Anti-hispanic
    Anti-other Ethnicity/National Origin

Source: Federal Bureau of Investigation (1999). *Hate Crime Collection Guidelines*. Washington, DC: Author.

categories. In addition to bias motivation, information is also collected on the type of offense (shown in Table 6-2), type of location, victim type (individual, business, or other), number of offenders, and the suspected offender's race (Federal Bureau of Investigation, 1999).

UCR HATE CRIME STATISTICS

The latest figures compiled by the Federal Bureau of Investigation (2001) reveal that in 2000 there were 8063 bias-motivated criminal incidents resulting in 9430 separate offenses. Individuals were targeted in 82.1% of the 9430 offenses with the remaining victims being businesses, religious organizations, or other unspecified targets.

TABLE 6-2   Bias Crime Annual Report Offense Categories

Crimes against persons

  Murder and non-negligent manslaughter
  Forcible rape
  Aggravated assault
  Simple assault
  Intimidation
  Other

Crimes against property

  Robbery
  Burglary
  Larceny-theft
  Motor vehicle theft
  Arson
  Destruction/damage/vandalism
  Other

Crimes against society

  Anti-physical disability
  Anti-mental disability

Source: Federal Bureau of Investigation (1999). *Hate Crime Collection Guidelines*. Washington, DC: Author.

Agencies reported 7530 offenders were known to be associated with 8063 of the incidents. Whites accounted for 64.4% of the offenders while blacks accounted for 18.7%. Police were most likely to identify offenders with crimes against persons rather than crimes against property. Offenders involved in religious-bias crimes were among the most difficult to identify because most of those crimes were crimes against property.

The most common motivation for single bias incidents is race, accounting for over half (53.8%) of the 8.063 incidents with known offenders. Religious and sexual orientation bias accounted for 18.3% and 16.1% of the incidents respectively. Ethnicity/national origin bias accounted for 11.3% of the incidents. Less than 1% of the incidents were caused by disability bias (Federal Bureau of Investigation, 2001).

Most frequently hate crime incidents occurred at residences (32.1%), while incidents occurring on highways, roads, alleys, or streets accounted for 17.9%. Less than 12% (11.4) occurred at schools or colleges.

The data collected by the UCR are believed to be inaccurate and incomplete. A government study of the first ten years of data collection under the Hate Crimes Statistics Act (Bureau of Justice Statistics, 2000b) illustrates the problems. While nearly 12,000 agencies now participate in the UCR hate crimes program, roughly 83% of those agencies report no bias crimes occurring in their jurisdiction. In 1998, fifteen states had ten or fewer agencies submitting incidents. Almost one-third of the states reported 50 or fewer incidents for the entire state (Bureau of Justice Statistics, 2000b, p. 4). Of the approximately 17,000 agencies submitting data to the UCR for 2000, only 11,690 submitted hate crime data. Of those submitting data, 83.8% reported no hate crimes had occurred in their jurisdiction. Many of the agencies did not submit data for the entire year. The report cautions that ". . . the reports from these agencies are insufficient to allow a valid national or regional measure of the volume and types of crimes motivated by hate . . ." (Federal Bureau of Investigation, 2000, p. 2).

The accuracy of the submitted reports has also been called into question. As part of the evaluation of hate crime data collection by the UCR, a survey of 2657 law enforcement agencies was conducted. In agencies identified as not participating in the UCR's hate crimes program, 37.1% responded that their agency had investigated one or more hate crimes. Perhaps even more troubling, in agencies identified as participating but reporting no hate crimes in their jurisdiction, 31% of the respondents indicated their department had investigated one or more such incidents.

INTERPRETING THE DATA

The only accurate conclusion that can be drawn from the data at this point in history is that we do not know how many hate crimes occur in the United States. Most observers agree that the current UCR program is inaccurate due to the lack of participation by local law enforcement agencies. Given this lack of reporting, any meaningful trend analysis is all but impossible.

Comparison among sources is difficult at best, given the manner in which data is collected. For example, in 1998 the UCR reported there were 1081 anti-Jewish incidents and 1260 incidents that were motivated by bias against sexual orientation. Statistics compiled by the Anti-Defamation League (2001, January) indicate there were 1611 such incidents. The report issued by the National Coalition of Anti-Violence Programs (2001) documents 2512 anti-gay incidents, almost twice as many as reported in the UCR. The UCR relies on crimes reported to it by local police agencies. Both the Anti-Defamation League and the National Coalition of Anti-Violence Programs rely on self-reports from victims and/or police reports.

STATUTES REGULATING HATE CRIMES

Legislative efforts to address hate crimes raise many issues. Some individuals question whether such laws are appropriate. Others would argue that some groups should be protected while others should not. Determining the type of motivation required for a crime and the applicable categories of behavior to be controlled has serious implications for the constitutionality of the law.

Justifications for Bias Crime Statutes

At various levels society is still struggling with the issue of whether it is appropriate to legislatively control bias crimes. The arguments range from the pragmatic to the philosophical. Opponents of the legislation would argue that punishing bias crimes amounts to censorship of ideas. They would maintain that there are already laws that prohibit conduct without punishing a person for their thoughts or beliefs. In their view, society should not afford special treatment to victims of certain types of crimes. Bias crime statutes are viewed as placing more value on some people's lives than others.

Proponents would argue that bias crime statutes punish conduct and not speech. The enhancement of penalties is warranted because bias crimes cause greater harm than similar acts not motivated by bias. Crimes motivated by bias toward a particular group reinforce social divisions and hatred while legislation reinforces the desired messages of tolerance and inclusion. Bias crimes are believed to have a direct impact on other members of the targeted group. Fear and intimidation often spread through the target group because they see the crime as an attack on themselves (Lawrence, 1999).

Federal Statutes

Prior to the passage of the Civil Rights Act of 1964, protection of individuals was primarily the province of local governments. That Act, coupled with the murders of civil rights workers Michael Schwerner, Andrew Goodman, and James Chaney in Mississippi in June of 1964 provided the impetus for a sustained federal effort to protect civil rights (Operation Lone Wolf, 2001).

One of the primary federal laws used by the Justice Department to investigate and prosecute hate crimes is the Conspiracy Against Rights section of the United States Code. The law makes it a crime for two or more people to conspire to "injure, oppress, threaten, or intimidate any person . . . in the free exercise or enjoyment of any right or privilege secured to him by the Constitution or laws of the United States" (18 U.S.C. 241), or to travel in "disguise" on the highway or property of another with the intent to deprive him of a federally protected right. The law applies to every citizen and is not limited to any specific categories as are most hate crimes.

Crimes motivated by bias based on race, color, religion, or national origin are prosecutable if the offender was attempting to prevent the victim from exercising a federally protected right (18 U.S.C. 245, 1969). The Prevention of Intimidation statute (42 U.S.C. 3631, 1996) makes it a crime to injure, intimidate, or interfere with anyone on the basis of race, color, religion, sex, handicap, familial status, or national origin in an effort to prevent them from buying, selling, renting, or occupying a dwelling unit. It also makes it a crime to disrupt protests opposing the denial of fair housing.

In 1994, Congress passed the Gender Motivated Violence Act (42 U.S.C. 13981) to address the problem of interstate domestic violence and stalking. Although the legislative history makes clear the statute was designed to address violence against women, its wording would make it applicable to both males and females. At least one federal appeals court (*Schwenk v. Hartford*, 2000) held that it applied to males, females, and transsexuals as well. The Supreme Court has overturned a provision that allowed civil remedies for victims (*United States v. Morrison et al.*, 2000).

In 1995, sexual orientation was added for the first time to a federal hate crimes law. The Violent Crime Control and Law Enforcement Act of 1994 (28 U.S.C. 994) contained a section that directed the U.S. Sentencing Commission to develop guidelines to enhance the penalty for anyone who intentionally selected their victim because of the victim's actual or perceived race, color, religion, national origin, ethnicity, gender, disability, or sexual orientation. The statute applies only to crimes committed in national parks or on other federal property.

Motivated by a dramatic rise in arsons that were occurring at African American churches, primarily in the south, Congress passed the Church Arsons Prevention Act of 1996 (18 U.S.C. 247). It provides authority to investigate arson, property damage, and interference in the exercise of religion. If the crimes committed were motivated by race, color, or ethnic characteristics, prosecutors were not required to show they affected interstate or foreign commerce.

State and Local Laws

Laws regulating hate crimes can be grouped into three general categories: (a) those that prohibit specific activities, (b) those that punish behavior motivated by bias, and (c) those that enhance penalties for otherwise criminal acts (Bureau of Justice Statistics, 1999). In 1981 the Anti-Defamation League (Hate Crimes Laws, 2001, March) proposed model legislation to address all forms of hate crimes. The model primarily uses the sentencing enhancement approach that was upheld by the Supreme Court (*Wisconsin v. Mitchell*) in 1993. It also contains provisions addressing institutional vandalism, civil penalties, mandatory reporting, and mandatory training for law enforcement.

Recent studies suggest 40 states and the District of Columbia have enacted laws that address bias-motivated violence and intimidation (Anti-Defamation League, 2001, March). Thirty-eight states and the District of Columbia have statutes prohibiting institutional vandalism while 21 entities criminalize interference with religious worship. The most commonly protected categories are race, religion, and ethnicity (40 states and the District of Columbia), while the least protected category is sexual orientation (12 states and the District of Columbia). The District of Columbia and 21 states prohibit crimes based on disability. Gender is included as a protected group by 19 states and the District of Columbia. Seven states have no statutes regulating hate or bias crimes. Some states, such as Texas, have a general statute that refers to victims selected because of a bias or prejudice against a person or group without specifying the targeted groups.

Court Decisions

The ultimate arbiter of the government's authority to regulate behavior is the United States Supreme Court. In recent years the Court has addressed relatively few cases arising out of the enactment of bias crime laws.

The authority of the federal government to enact hate crime statutes has been limited by the Court's narrowing of what may be justified under Congress' power to regulate interstate commerce (*United States v. Lopez*, 1995). Absent a direct connection to interstate commerce, their authority would appear to be restricted to crimes occurring on federal property or those that interfere with a federally protected activity (*United States v. Morrision,* et al., 2000). This however, does not apply to crimes motivated by racial bias that may be regulated under the 13th Amendment (*Jones v. Mayer*, 1968; *McDonald v. Santa Fe*, 1976).

Court decisions indicate that laws aimed at controlling certain kinds of conduct which may be offensive to target groups are possibly unconstitutional. The danger is that the court may construe the conduct to be a form of speech. With very limited exceptions, the First Amendment prohibits regulation of speech (*R.A.V. v. City of St. Paul*, 1992). Far easier to defend are laws which provide penalty enhancements for otherwise criminal actions which are motivated by bias (*Wisconsin v. Mitchell*, 1993). The bias motive must be established beyond a reasonable doubt (*Apprendi v. New Jersey*, 2000).

BIAS CRIME UNITS IN THE UNITED STATES

Federal Agencies

The Federal Bureau of Investigation is the primary federal agency for enforcement of civil rights laws. The civil rights program in the FBI consists of four sub-programs: (1) racial/religious discrimination, (2) color of law/police misconduct, (3) involuntary servitude/slavery, and (4) freedom of access to clinic entrances. While a majority of their investigations are in the area of color of law/police misconduct, the program places its highest priority on hate crimes (Operation Lone Wolf, 2001, January).

The Bureau of Alcohol, Tobacco, and Firearms (January, 2001) has the primary responsibility for investigating crimes resulting from bombings and arsons. As with the FBI, their jurisdiction is restricted by federal statutes and court decisions.

The Community Relations Service (1998) of the Department of Justice provides assistance in settling destructive conflicts and disturbances in communities. Working with state and local officials, they assist in identifying the sources of violence and conflict and provide crisis management assistance. The Community Relations Service is not a law enforcement agency and has no power to investigate or prosecute cases. By law their activities are confidential and are limited to hate crimes motivated by the victim's race, color, or national origin.

In response to a sharp increase in the number of church arsons, particularly among African-American churches in the south, The National Church Arson Task Force was created in 1996. The Task Force is comprised of the Bureau of Alcohol, Tobacco, and Firearms, Federal Bureau of Investigation, Community Relations Service, Federal Emergency Management Agency, Housing and Urban Development, and the Department of Justice. They are authorized to investigate racially motivated arsons without having to demonstrate the incident was connected to interstate commerce.

State and Local Agencies

Relatively little research exists which documents the existence of bias crime units. In a 1997 study, Reaves and Goldberg (1999) surveyed law enforcement agencies employing 100 or more officers for the Bureau of Justice Statistics. Nearly half of the state agencies (49%) and almost one-fifth of the local agencies (19.94%) had no policy, procedure, or personnel designated to address the issue of hate crimes.

The most common method of addressing hate crimes was through the use of special policies or procedures. Other agencies designated specific personnel to investigate hate crimes as needed. Only two states and 52 local agencies had special units with personnel assigned on a full-time basis. The Los Angeles Police Department has the largest bias

TABLE 6-3   Local Law Enforcement Agencies With Bias Crime Units of Five or
More Sworn Personnel

Agency	Sworn	Civilian
Los Angeles Police	90	
New York City Police	23	3
Chicago Police	16	
Philadelphia Police	11	
Cincinnati Police	10	1
Riverside Police	10	2
Kansas City Police	9	
Phoenix Police	7	
Boston Police	7	1
Union Police	5	
Nassau County Police	5	
Charlotte-Mecklenburg Police	5	
Tucson Police	5	2

Source: Adapted by author from Reaves, B. and Goldberg, A. (1999). *Law enforcement
management and administrative statistics, 1997: Data for individual state and local agencies with
100 or more officers*. Washington, DC: Bureau of Justice Statistics.

crime unit in the United States with 90 sworn personnel assigned. There were only
13 local agencies with five or more sworn personnel. (See Tables 6-3 and 6-4.) When the
survey was re-done two years later (Reaves & Hart, 2000) the information for bias crime
units was not collected.

New Jersey is generally recognized as the first state to establish a bias crime unit. In
February of 1992, a unit was created in the State Attorney General's Office. Six people were
assigned to the unit responsible for examining patterns in bias crimes (Bureau of Justice
Statistics, 2000b, p. 12). The New Jersey State Police began collecting bias crime data in
1988. The State's Office of Bias Crime and Community Relations assists agencies in investigation and prosecution of crimes (Bureau of Justice Statistics, 1999).

In 1992 the State of Maine began the Civil Rights Officers Project to coordinate
enforcement with local, county, campus, and state police agencies (Bureau of Justice
Statistics, 2000a). Each department designates a civil rights officer to report hate crimes to
the Attorney General's Office. The Attorney General then assists the law enforcement
agency in their investigation, coordinates prosecution with the county district attorney, and
seeks involvement of the U.S. Attorney if federal prosecution is warranted.

Many of the bias crime units in local agencies were originally created to address
racial problems. In 1978 the Boston Police Commissioner established the Community
Disorders Unit to oversee all racial violence cases (McLaughlin, Brilliant, & Lang, 1995,
p. 281). In addition to performing investigations, the unit takes a proactive approach by

TABLE 6-4   Bias/Hate Crime Enforcement Strategy for State and Local
Agencies with 100 or More Officers

Strategy Employed	State Agencies		Local Agencies	
	Number	Percent	Number	Percent
Special unit with personnel assigned on a full-time basis	2	4.08	52	7.97
Specially designated personnel assigned to address the problem as needed	7	14.29	204	31.29
Agency has special policies or procedures that address this problem, but no specially designated personnel	16	32.65	264	40.49
Agency has no policy, procedure, or personnel designated to address this problem	24	49.00	130	19.94
Agency did not provide the information			2	.31
Total	49	100.00	652	100.00

Source: Adapted by author from Reaves, B. and Goldberg, A. (1999). *Law enforcement management and administrative statistics, 1997: Data for individual state and local agencies with 100 or more officers*. Washington, DC: Bureau of Justice Statistics.

using decoys and covert surveillance. Where possible they pursue not only criminal charges, but injunctive relief through the courts as well. Recently the unit was subjected to public criticism after a series of articles in the *Boston Globe* charged it was pursuing civil rights violations with less vigor than it had in years past (Boston PD fires back at critics over its handling of hate crimes, 2000).

The Bias Investigating Unit, now the Hate Crimes Task Force, of the New York City Police Department was created to coordinate the department's response to hate crimes (McLaughlin et al., 1995, p. 281). The unit performs an initial evaluation of the incident to determine appropriate police action to stabilize volatile situations and determine bias motivation on the basis of race, ethnicity, religion, or sexual orientation. If bias is suspected, the department requires the response of a patrol supervisor and a commanding officer. The unit assumes responsibility for completion of a bias incident report and for making the final determination if an incident is bias-motivated or not.

In addition to investigating bias-motivated criminal acts, the Baltimore County Police Department's bias unit is responsible for providing victim services. They are trained to respond to the needs of victims, families, and other community members who are impacted by hate crimes (McLaughlin et al., 1995, p. 281). Members are expected to take a pro-active role in promoting peace and harmony in the county.

ADMINISTRATIVE CONSIDERATIONS

Unit Structure

Administrators must determine what approach is to be used for investigation of bias crimes. The options include: (1) protocol approach, (2) case management approach, (3) general assignment approach, and (4) task force approach. Primarily the number of hate crimes experienced in a jurisdiction should dictate selection of the department's approach. It would be impractical to establish full-time units with dedicated personnel in jurisdictions that experience a small number of hate incidents or crimes each year.

Protocol Approach   At a minimum, every agency should adopt a protocol approach to address bias crimes. This requires the creation of department policies that clearly set forth the criteria for identification, reporting, and investigation of hate crimes. The policy should detail the responsibilities of the responding officer, investigators, and supervisors. The method of complying with any requirements created by state law or local ordinance should be specified. These requirements may include notifying victims of available resources, reporting to another agency or commission, notification of the local prosecutor or state attorney general, or provision of specific services to victims.

Case Management Approach   This approach is an intermediate step between operation by policy and establishment of a full-time unit. Officers are designated to function as case managers. They assume responsibility for case reviews, provide assistance to investigators, coordinate victim services, maintain liaison with prosecutors, ensure required notifications to outside entities are made in a timely manner, and perform other activities required to address hate crimes. The advantage of using a designated person is that he or she will be familiar with the required protocol and more aware of specific issues that arise in these types of cases.

General Assignment Approach   In jurisdictions with a significant number of hate crimes it may be preferable to create a unit with sworn personnel assigned on a full-time basis. This level of specialization will increase the ability of the investigators to address the issues arising in hate crimes with victims, prosecutors, and the community at large. It demonstrates the department's firm commitment to prevent this type of occurrence and aggressively seek prosecution. The disadvantage of this approach, absent a fairly large unit, is that while the investigators will be specialists in bias crimes, they will of necessity have to be generalists with the ability to investigate all the underlying criminal offenses. Those offenses may range from harassing phone calls to homicide.

Task Force Approach   The task force approach can be used as the primary protocol or can be used in combination with the other three approaches. When a bias crime or series of bias crimes occur, officers are selected based on their skills to conduct the investigation. The Sacramento Police Department (Bureau of Justice Assistance, 1997) effectively employed this strategy to address a series of arsons in 1993. Between July and October of that year there were four arsons and three attempted arsons committed by a white separatist who called himself the "Aryan Liberation Front." Targets of the arsonist included the local office of the NAACP, a Jewish temple, offices of the Japanese American Citizens League, the home of an Asian-American city councilman, and the State Office of Fair Employment and Housing.

The investigation originally involved investigators from the local police, arson investigators from the fire department, one FBI agent, and lab technicians from the ATF. It was being hampered because there were conflicting demands and directions placed on the investigators from their own agencies. When the task force was created it included local police and fire officials, FBI agents, ATF agents, the California Attorney General's Office, and DOJ investigators all under the supervision of a lieutenant from the police department. A federal grant was obtained to purchase a sophisticated vehicle for covert surveillance and a geographical information system. By combining this expertise to focus on these serial hate crimes authorities were able to identify and arrest a suspect approximately one month after the task force was created.

Unit Authority

Administrators must determine the areas of responsibility and authority for the bias crime unit. Of primary importance is identification of target groups that will be included in the unit's activities. There are several options to select from: (1) criteria established by the International Chiefs of Police (IACP), (2) criteria used by the FBI in the UCR, and (3) criteria established by state statutes or local ordinance. The unit should be guided by clearly stated objectives. In addition, the administrator must determine if the unit will investigate only hate crimes or will include hate incidents that may not constitute a criminal offense as well.

The IACP's criteria for inclusion are the broadest. It includes "a criminal offense committed against persons, property or society that is motivated in whole or in part, by an offender's bias against an individual's or a group's race, religion, ethnic/national origin, gender, age, disability or sexual orientation" (International Association of Chiefs of Police, 1999, p. 3). The FBI's criteria would not include gender as a target group (Federal Bureau of Investigation, 1997, p. 14). The most restrictive category may be state statutes or local ordinances.

Selection of the target groups has operational, legal, and political implications. As the categories of targeted groups are increased you increase the workload of the unit. Restriction of categories to those specified in state law or local ordinances is a sound administrative policy, if you do not factor in the political ramifications. The administrator can justify the unit's authority based on social policy set by the legislature. Those who disagree with inclusion or exclusion of any group can be referred to the appropriate legislative body. The political implications will become important if excluded groups feel they are not being treated fairly by the department or if others feel the department is inappropriately giving special treatment to targeted groups. If the unit provides victim services, the excluded groups will feel further alienated.

Objectives for the department should be clearly spelled out by the administrator. It is suggested that the following be included (McLaughlin et al., 1995, p. 209):

- To prevent the occurrence of bias crimes,
- To identify and investigate crimes that are possibly motivated by bias,

- To work cooperatively with neighborhood and community groups to reduce the impact and possible repercussions of these crimes,
- To help bring the offenders to justice.

The policy and objectives for the department and unit may also include the justifications for addressing hate crimes in a different manner. The justifications might include the deep emotional distress experienced by victims, increased recovery time and feelings of vulnerability for victims, fear of future victimization based on their group identity, fear generated among the targeted group in the community as a whole, or establishing a climate of tolerance in the community for diverse groups (Sonnenschein, undated; Herek, Gillis, & Cogan, 1999; Lawrence, 1999). The 1987 policy of the Boston Police Department provides an illustration of this:

> It is the policy of the Metropolitan Police department to support and defend the civil rights of all persons and to see that they are protected against interference by acts of threats of violence . . . The Department recognizes the serious impact of such crimes and their intimidating effect on their victims, and the community as a whole . . . Each officer must be sensitive to the feelings, needs and fears that may be present in the community as a result of incidents of this nature. (Boston Police Department, 1987)

The decision must be made whether the department will investigate hate incidents that do not constitute crimes or will be limited only to criminal offenses. As was previously discussed, a hate incident is one where prejudice is evident as a factor and the intent is to harm, regardless of whether there is a criminal offense (Paynter, 2000a, p. 53). While criminal charges may not result from these investigations the department may realize other benefits. The public is often confused as to what constitutes a crime. Investigation of these incidents with a conclusion by the police or prosecutor that charges cannot be filed may relieve some of the anxiety victims of hate incidents feel. Resolution of incidents by the police may prevent them from escalating into a criminal offense. Inclusion of hate incidents may allow the department to identify "hot spots" where future problems are likely to occur. Identification of those committing hate incidents may provide useful intelligence information for future criminal investigations. Addressing issues of hate through means other than criminal prosecution may help send a message to the community that those types of activities are taken seriously by the department and will be aggressively pursued.

Training

If a department is to respond effectively to bias crimes, the entire department must be involved in the training process. All employees from the dispatcher to the chief must be trained to recognize hate crimes and to initiate the proper response. Training should include cultural diversity, indicators of bias crimes, nature of the perpetrators they may encounter, identification and collection of bias crime evidence, victim sensitivity, legal issues, department policies, case preparation, victim's services, and community collaboration (Paynter, 2000b). Departments should not take a one-time approach to training. It is important for employees to periodically undergo refresher training. In agencies with low numbers of hate crimes this refresher training may become more important because the employees will be less familiar with what is required due to a lack of experience.

Zero Tolerance

Police departments exist to protect the rights of all citizens. Officers who engage in discriminatory behavior based on bias have no place in law enforcement. Their actions are inconsistent with the purpose of the profession. Administrators must create an agency culture that has zero tolerance for discrimination. Law enforcement agencies lead the community by example. If the agency acts inappropriately it sends a message to the community that hatred and bias are acceptable behavior. It discourages victims from coming forth to seek assistance for fear of being mistreated by the police.

The message of zero tolerance for offensive speech and behavior should begin on the employee's first day. Training programs should stress the department's policy of non-discrimination in provision of services. Agencies should establish policies that define the boundaries of acceptable conduct and hold each member of the department accountable for their enforcement. Appropriate penalties for violation of the policy should be applied on a continuum consistent with the infraction and employee's history (Paynter, 2000b).

Failure to do so can have negative consequences for an agency and the profession as a whole. Almost every police agency in America has been impacted by the recent focus on racial profiling. State legislatures have rushed to enact laws imposing data collection and reporting requirements on agencies under threat of criminal penalties or loss of funding (Soule, 2000). Incidents of abuse by individual officers or agencies receive national attention and are attributed to all officers and agencies by the public.

One example may help illustrate the seriousness of the problem. The National Coalition of Anti-Violence Programs (1999), as part of its data collection on bias crimes motivated by sexual orientation, documented the victim's perception of the police response in 821 of the 1010 incidents reported. The officers were characterized as being verbally abusive to the victim in 13.03% of the cases. Victims report the police as being physically abusive in 7.06% of the cases. Officers appeared to be indifferent in 27.04% of the cases while victims rated them as courteous in only 52.86% of the cases. By 2001 (National Coalition of Anti-Violence Programs), the percentage of victims reporting the police had verbally or physically abused them dropped to 12.33% and 4.82% respectively. The number of victims reporting the police as treating them courteously dropped to 41.15% while indifference increased to 29.19%.

The Glendale, California Police Department's (1988) policy addresses these issues and reinforces the agency's zero-tolerance approach:

> It shall be the policy of this Department to bring the investigative and enforcement forces of the police department into quick action following any and all reported or observed incidents of violence or threats . . . motivated, all or in part, because of race, ethnicity, religion, or sexual orientation . . . It must be remembered that the actions taken by this agency in dealing with incidents of racial, religious, ethnic, or sexual oriented bias are visible signs of concern and commitment to the community on the part of Glendale government and its police department.
>
> The proper investigation of reported incidents . . . is the responsibility of all Glendale police officers. Each officer must be sensitive to the feelings, need, and fears that may be present in the community as a result of incidents of this nature.

Developing Accurate Data

It is generally accepted that hate crime statistics are inaccurate and under-reported. This is caused by law enforcement's inability or unwillingness to collect and report the data and the reluctance of victims to report crimes to the police. Increasing the accuracy of agency statistics can be accomplished through policy and procedure. It is somewhat more difficult to address the reluctance of victims to report crimes to the police. Table 6-5 shows the recommended key decision points for hate crime reporting.

In addition to the established general reasons for not reporting crimes to the police other factors may apply for bias crimes (Bureau of Justice Statistics, 2000b, pp. 29–30). Victims may not realize that the act was motivated by hate, particularly in property crimes where there is no accompanying graffiti or indication of bias. They may suspect bias but may be uncomfortable discussing it with the police. The targeted groups may perceive the police as being biased toward that group and fear they will be mistreated or that the police would not believe them.

Differences in culture and language may inhibit victims from contacting the police. First generation immigrants or foreign citizens may anticipate that American police officers are as corrupt or abusive as those in their native country. Ethnic minorities who are in this country illegally may fear deportation if they report a crime to the police. Gays and lesbians may fear that reporting the crime will result in their being "outed" to their family, friends, or coworkers. If they live in a state with sodomy laws they may fear criminal charges would be filed against them for engaging in homosexual activity.

Agencies should establish a system of secondary review for reports to make a final determination on bias classification (Bureau of Justice Statistics, 2000c, p. 11). This function serves a twofold purpose: to review the field officer's determination of bias for compliance with established agency standards, and to identify cases of bias which may have been improperly classified by the original officer. The agency size will usually determine whether this second level of review is outside the normal supervision process. In larger

---

TABLE 6-5　Key Decision Points in Reporting Bias Crimes

1. Victim understands that a crime has been committed
2. Victim recognizes that hate of their real of perceived status or attribute may be a motivating factor
3. Victim or another party solicits law enforcement intervention
4. Victim or another party communicates with law enforcement about the motivation of the crime
5. Law enforcement recognizes the element of hate
6. Law enforcement documents the element of hate and, as appropriate, charges suspect with civil rights or hate/bias offense
7. Law enforcement records the incident and submits the information to the Uniform Crime Reports, Hate Crime Reporting Unit

---

Source: Adapted by author from Bureau of Justice Statistics (2000c). *Improving the quality and accuracy of bias crime statistics nationally: An assessment of the first ten years of bias crime data collection*, p. 6.

agencies the secondary review of bias incidents may be outside the normal chain of command to include review by the bias crime unit.

Establishing Community Partnerships

The community-policing strategy is particularly suited to addressing bias crimes. The department should seek to form partnerships with those in the community who are affected by or are committed to reducing the problem. Any of the strategies used for community policing can be applied to the issue of hate crimes. The agency will want to identify advocacy organizations for the various targeted groups. These may include both national and local organizations. The advocacy groups can facilitate communication between the department and the community, provide diversity training for law enforcement officers, encourage victims to report crimes, and facilitate victim assistance.

Partnerships with local schools should be developed using existing relationships forged through school resource officers, DARE programs, police athletic leagues, etc. Educators should be trained to recognize hate crimes and report them. Curriculum programs for students should include diversity training and conflict resolution. Educators and police officers should routinely share information designed to prevent hate crimes and identify potentially dangerous trends before they become a problem (Paynter, 2000a, p. 60).

The list of potential partnerships is limited only by the department's imagination and willingness to seek their assistance. Other examples may include use of citizen academies directed at community advocacy organizations. The department gains additional knowledge about the advocates, who in turn are exposed to the true nature of law enforcement with all its inherent legal restrictions and limitations. The medical community can be solicited to identify victims of suspected hate crimes and encourage reporting and/or referral to community support groups.

The Provincetown, Massachusetts Police Department's plan and the State of Maryland's model response program serve as two examples of successful partnerships (McLaughlin et al., 1995; Paynter, 2000a). Provincetown is a resort area that attracts a significant number of gay and lesbian visitors. In response to community advocates' concern over an increasing number of bias incidents, the city adopted the community-oriented policing model to address the issue. The police, community leaders, victim assistance agencies, and the City's Human Services Department cooperated to develop new policies to guide their response to bias crimes.

The State of Maryland created an advisory committee made up of community organizations, human relations specialists, and law enforcement officials to develop a model for local jurisdictions to use in dealing with racial, religious, and ethnic violence. The model establishes best practices for prevention, detection, reporting, and response strategies. It includes involvement with the schools, community at large, police, victim assistance programs and a community support program.

OPERATIONAL CONSIDERATIONS

Investigation of a bias crime follows the normal procedures for investigation of any criminal act. Recall, that most bias crimes are penalty enhancement statutes based on acts that are already criminal, e.g., homicide, robbery, assault, property damage, etc. There are differences, however, in a bias crime and other criminal offenses. Clearly establishing the motive becomes crucial to a successful prosecution. Officers need to be trained to identify indicators of bias crimes and how to properly document the evidence for prosecution. Familiarity with offender profiles and the ideology of different organized hate groups will facilitate the investigation. Availability of victim assistance programs will vary between jurisdictions. In any given location there may or may not be community resources to which the victim can be referred, such as advocacy or support groups. The ability to deal compassionately with the victim increases the likelihood they will cooperate with the police in identifying and prosecuting the offender.

Indicators of Bias Crimes

Establishing a bias motive is admittedly subjective. Merely because a suspect is biased against a particular group does not in and of itself establish the crime was motivated by bias. A combination of facts and circumstances are generally used to support a finding of bias by responding officers, investigators, and reviewing officials (Federal Bureau of Investigation, 1999, p. 4). The most frequently cited indicators of bias motivation (Federal Bureau of Investigation, 1999, pp. 5–6; International Association of Chiefs of Police, 1999, p. 7; U.S. Department of Justice, 2001, March) can be grouped into victim characteristics, offender characteristics, and modus operandi.

Victim Characteristics

- The victim has a different religious, racial, ethnic/national origin, disability, or sexual orientation than that of the offender.
- Victim is a member of a group that is overwhelmingly outnumbered by members of another group in the area where the incident occurred.
- Victim has previously received harassing mail, phone calls, or verbal abuse based on affiliation with a targeted group.
- Victim was engaged in activities promoting or closely identified with his/her group.
- Victim was in the company of, or married to a member of a targeted group.
- Victim is a member of an advocacy group that supports a targeted group.
- Animosity has existed historically between the victim and offender's groups.
- The victim or witnesses perceive the incident was bias-motivated.

Offender Characteristics

- Suspect was previously involved in a similar incident or has a history of previous crimes with a similar modus operandi, or there have been multiple victims of the same target group.
- Suspect is a member of or associates with an organized hate group.
- Bias-related comments, gestures, or written statements were used by the suspect.

Modus Operandi

- The crime occurred at or near an area commonly associated with or frequented by targeted groups, e.g., social clubs, religious facilities, gay bars.
- Several crimes occur in the same area with members of a targeted group.
- Bias-related drawings, markings, symbols, or graffiti were left at the scene.
- Object or items representing organized hate groups were observed by witnesses or left at the scene, e.g., white hoods, burning crosses, swastikas.
- An organized hate group claims responsibility for the crime.

All of these factors should be taken into account when making a determination as to whether a crime is bias-motivated. In some instances, the motive is not identifiable, which itself may be an indicator of bias. One of the factors often considered is the lack of any motive, economic or otherwise for a particular act which is committed against a member of a targeted group.

Preservation of Evidence

Once the immediate concerns of obtaining medical assistance for victims, locating and identifying witnesses and suspects, providing for officer safety, and securing the crime scene have been addressed, attention should be focused on collection of evidence. The initial officers and investigators should take steps to preserve and document the evidence that is necessary to establish the elements of the crime and prove bias. The standards and procedures for evidence collection are the same as any other crime, with one exception. In some jurisdictions physical evidence would not be collected or photographed for relatively minor offenses or where the dollar loss was low. Because motive is an essential element in successful prosecution of hate crimes, all physical evidence in bias crimes should be documented, photographed, and where possible collected.

Offender Profiles

Compared to other types of offenders, relatively little has been published on the profiles of individuals who commit bias crimes. Profiles can provide useful information that can be used by the police to focus the investigation, identify suspects, and conduct interviews and interrogations. Demographic data can be gleaned from the Uniform Crime Reports and other sources. Researchers at Northeastern University created a typology based on the characteristics of the offender, victim, and modus operandi of the crime.

Data from the Uniform Crime Reports is used by the Federal Bureau of Investigation to identify several characteristics of hate crimes (Operation Lone Wolf, 2001). Hate crimes are typically more violent and involve a higher level of assaults than other crimes. While only 10% of all crimes involve assaults, 45–55% of hate crimes are personal assaults. Physical injury occurs in 74% of the bias crime assaults as opposed to 29% of the non-biased assaults. A series of confrontations and incidents that escalate in severity often precede attacks. Hate crimes are more likely than other types of criminal activity to be committed by multiple offenders. Unlike most crimes against persons where the victim knows the offender, bias crimes are most commonly committed by strangers.

Levin and McDevitt (1993) of Northeastern University identified three categories of offenders; thrill seeking, reactive, and mission. The thrill-seeking offender is generally young, not associated with an organized hate group, and engages in the activity to gain a thrill or acceptance by his peers. Crimes are generally committed away from the area where the offenders live, are random in nature, and often involve property destruction. Victims may be members of a group the offenders view as inferior or vulnerable. Bias toward the victim is often superficial.

Reactive offenders focus on individuals or group of individuals that are believed to be interfering with their way of life. Most often the victims are chosen on the basis of race. The crimes occur in the offender's own neighborhood because they are attempting to send a message to the unwanted individual or group. Fear and intimidation will be used to force the outsiders to leave. The offender will generally have no prior criminal history. Once the "threat" is perceived to subside, the criminal behavior will be reduced or stop. The offender will feel little guilt or remorse as they view themselves as being victimized by the outsiders.

In the weeks following the September 11, 2001 terrorist attack on the World Trade Center Towers and the Pentagon, reactive type hate crimes rose dramatically against Muslims and Arabs. The Council on American-Islamic Relations reported 625 anti-Muslim incidents. The American-Arab Anti-Discrimination Committee said it had confirmed 315 anti-Arab threats or incidents of violence. In the 18 days following the attack the FBI began investigating 90 hate crimes related to the incident (CNN, 2001, September 29). Less than two months after the incident the number of hate crimes aimed at Arabs, Muslims, and Sikhs in the United States being investigated by the FBI had risen to 160 (CNN, 2001, October 16).

Mission offenders are believed to be the rarest type. They are often psychotic or suffering from other mental illnesses. They will be withdrawn from other people. The victim is perceived as evil or subhuman. The mission offender has been instructed by God, the Fuhrer, or some perceived authority figure to eliminate the victim or target group. The crimes will be concentrated in the areas where members of the targeted group are most likely to be found. The crimes are of a particularly violent nature. Often the offender will ultimately commit suicide.

Victim Assistance

Agencies may find it beneficial to establish a network of victim assistance resources to encourage victims to report crimes and address their physical and psychological needs. This may include liaisons to the police, prosecutors, courts, human rights commissions, other government agencies, victim advocacy organizations and support groups (Bureau of Justice Statistics, 1999, p. 40). Provisions need to be made for use of bilingual officers when the victim does not speak English as their primary language. Other cases may require the use of someone capable of speaking sign language. Where applicable assistance should be provided in filing victim compensation claims.

CHAPTER SUMMARY

The terms hate and bias crimes are often used interchangeably. Various groups adopt different definitions resulting in the inclusion or exclusion of victims. A distinction must be made between hate crimes and hate incidents. While all hate crimes result from hate incidents, not all incidents are crimes.

There is no reliable data on the number of hate crimes and hate incidents that occur each year. Efforts by the UCR have thus far not produced reliable data that can be used to compare trends over time. Data collected by private organizations often uses different definitions and is not collected consistently. They often include incidents that are not criminal in nature.

Legislative efforts to control hate crimes have raised both legal and political issues. Laws that provide penalty enhancements for otherwise criminal acts motivated by bias are more likely to be viewed by the appellate courts as constitutional than those prohibiting specific conduct. The courts have limited the role of the federal government under the Interstate Commerce Clause of the Constitution. Absent a direct connection to interstate commerce, their authority appears to be limited to crimes occurring on federal property, those that interfere with a federally protected activity, or those motivated by race that can be regulated under the 13th Amendment.

The political issues raised by legislation focus on whether hate crimes should be treated differently than other crimes and the groups that should be protected. There is a disagreement as to the federal government's role in prosecuting these offenses. Inclusion of gender and sexual orientation as protected groups has generated the most opposition to proposed legislation.

Relatively little research exists on bias crime units in law enforcement agencies. It was suggested that agencies adopt one of four approaches to bias crime investigations; protocol, case management, general assignment, or task force. Selection of the model would be dictated by the type and number of hate incidents occurring in a jurisdiction. Once a model is selected the administrator must determine the unit's authority, scope of activities, training, and community involvement. These decisions should take into account the operational, legal, and political implications for the agency.

The administrator must address the standard of conduct for their employees, including a zero-tolerance policy. Accurate data collection must be undertaken with reporting requirements clearly established. A system of secondary review should be implemented to ensure bias incidents are properly recorded and responded to.

SUGGESTED READINGS

BUREAU OF JUSTICE STATISTICS (1999). *A policymaker's guide to hate crimes*. Washington, DC: U.S. Department of Justice.

——. (2000). *Improving the quality and accuracy of bias crime statistics nationally: An assessment of the first ten years of bias crime data collection*. Available at: http://www.ojp.usdoj.gov/bjs/abstract/iqabscn.htm.

FEDERAL BUREAU OF INVESTIGATION (1999). *Hate crime collection guidelines*. Washington, DC: U.S. Department of Justice.

McLAUGHLIN, K.A., BRILLIANT, K., & LANG, C. (1995). *National bias crimes training for law enforcement and victim assistance professionals*. Washington, DC: U.S. Department of Justice.

REFERENCES

ANTI-DEFAMATION LEAGUE (2001, January). *Audit of anti-semitic incidents.* Available at: http://www.adl.org/frames/front_99audit.html.

———. (2001, March). *Hate crimes laws.* Available at: http://www.adl.org/99hatecrime/intro.html.

APPRENDI V. NEW JERSEY, No. 99-478 (U.S. June 26, 2000).

BOSTON PD FIRES BACK AT CRITICS OVER ITS HANDLING OF HATE CRIMES (2000). *Law Enforcement News, 26*(536), 1, 8.

BOSTON POLICE DEPARTMENT (1987, November 19). *New departmental civil rights policy and procedures.* General Order Number 87–27.

BUREAU OF JUSTICE ASSISTANCE (1997). *Stopping hate crime: A case history from the Sacramento police department.* Washington, DC: U.S. Department of Justice.

BUREAU OF JUSTICE STATISTICS (1999). *A policymaker's guide to hate crimes.* Washington, DC: U.S. Department of Justice.

———. (2000a). *Addressing hate crimes: Six initiatives that are enhancing the efforts of criminal justice practitioners.* Washington, DC: U.S. Department of Justice.

———. (2000b). *Improving the quality and accuracy of bias crime statistics nationally: An assessment of the first ten years of bias crime data collection.* Available at: http://www.ojp.usdoj.gov/bjs/abstract/iqabscn.htm.

———. (2000c). *Improving the quality and accuracy of bias crime statistics nationally: An assessment of the first ten years of bias crime data collection, executive summary.* Available at: http://www.ojp.usdoj.gov/bjs/abstract/iqabscn.htm.

CHURCH ARSONS PREVENTION ACT OF 1996, 18 U.S.C. 247.

COMMUNITY RELATIONS SERVICE (1998). *Hate crime: The violence of intolerance.* Washington, DC: U.S. Department of Justice.

CONSPIRACY AGAINST RIGHTS, 18 U.S.C. 241. (1996).

COUNCIL ON AMERICAN-ISLAMIC RELATIONS (2001). The status of Muslims civil rights in the United States 2001. Retrieved November 24, 2001, from the World Wide Web: http://www.cair-net.org/civilrights/body_index.html.

CNN (2001, September 29). *Advocates say reports of hate crimes slowing.* Retrieved November 24, 2001, from the World Wide Web: http://cnn.com/2001/US/09/29/rec.attacks.backlash/index.html.

CNN (2001, October 16). *Ashcroft, Muslim leaders to discuss hate crimes.* Retrieved November 24, 2001, from the World Wide Web: http://cnn.com/2001/US/10/16/rec.just.antimuslim/index.html.

FEDERAL BUREAU OF INVESTIGATION (1997). *Training guide for hate crime data collection.* Washington, DC: U.S. Department of Justice.

———. (1999). *Hate crime collection guidelines.* Washington, DC: U.S. Department of Justice.

———. (2000). *Hate crime statistics 1998.* Washington, DC: U.S. Department of Justice.

———. (2001). *Hate crime statistics 2000.* Washington, DC: U.S. Department of Justice.

FEDERALLY PROTECTED ACTIVITIES, 18 U.S.C. 245 (1969).

GENDER MOTIVATED VIOLENCE ACT, 42 U.S.C. 13981 (1994).

GLENDALE, CALIFORNIA POLICE DEPARTMENT (1988, August 20). *Crimes motivated by race, religion, ethnicity, or sexual orientation.* General Order Number 715.

HATE CRIME STATISTICS ACT of 1990, 28 U.S.C. 534.

HEREK, G.M., GILLIS, J.R., & COGAN, J.C. (1999). *Journal of Consulting and Clinical Psychology, 67*(6), 945–951.

INTERNATIONAL ASSOCIATION OF CHIEFS OF POLICE (1999). *Responding to hate crimes: A police officer's guide to investigation and prevention.* Washington, DC: National Criminal Justice Reference Service.

JONES V. ALFRED H. MAYER, 392 U.S. 409 (US, 1968).

JUSTICE RESEARCH AND STATISTICS ASSOCIATION (1995). *Report on federal record keeping relating to domestic violence*. Washington, DC: National Institute of Justice. Retrieved November 24, 2001, from the World Wide Web: http://www.ojp.usdoj.gov/ocpa/94Guides/FedRec/welcome.html.

LAWRENCE, F.M. (1999). *Punishing hate: Bias crimes under American law*. Cambridge: Harvard University Press.

LEVIN, J., & MCDEVITT, J. (1993). *The rising tide of bigotry and bloodshed*. New York: Plenum. In K.A. MCLAUGHLIN, K. BRILLIANT, & C. LANG (1995), *National bias crimes training for law enforcement and victim assistance professionals*. Washington, DC: U.S. Department of Justice, pp. 136–148.

MCDONALD V. SANTA FE TRAIL TRANSPORTATION CO., 472 U.S. 273 (US, 1976).

MCLAUGHLIN, K.A., BRILLIANT, K., & LANG, C. (1995). *National bias crimes training for law enforcement and victim assistance professionals*. Washington, DC: U.S. Department of Justice.

NATIONAL COALITION OF ANTI-VIOLENCE PROGRAMS (1999). *Anti-lesbian, gay, bisexual and transgender violence in 1998*. Available at: http://www.avp.org/publications/reports/1998ncavpbiasrpt.pdf.

——. (2001). *Anti-lesbian, gay, bisexual and transgender violence in 2000*. Available at: http://www.avp.org/publications/reports/2000ncavpbiasrpt.pdf.

OPERATION LONE WOLF (2001, January). Washington DC: Federal Bureau of Investigation. Available at: http://www.fbi.gov/majcases/lonewolf/lonewolf2.htm.

PAYNTER, R.L. (2000a). Healing the hate. *Law Enforcement Technology*, 27(4), 52–61.

PAYNTER, R.L. (2000b). Protecting all the people. *Law Enforcement Technology*, 27(4), 62–66.

PREVENTION OF INTIMIDATION, 42 U.S.C. 3631 (1996).

R.A.V. V. CITY OF ST. PAUL, 505 U.S. 377 (US June 22, 1992).

REAVES, B.A., & GOLDBERG, A.L. (1999). *Law enforcement management and administrative statistics, 1997: Data for individual state and local agencies with 100 or more officers*. Washington, DC: Bureau of Justice Statistics.

REAVES, B.A., & HART, T.C. (2000). *Law enforcement management and administrative statistics, 1999: Data for individual state and local agencies with 100 or more officers*. Washington, DC: Bureau of Justice Statistics.

SCHWENK V. HARTFORD, No. 97-35870 (9th Cir. February 29, 2000).

SONNENSCHEIN, F.M. (undated). *Hate crime. A training video for police officers. Discussion manual*. New York: Anti-Defamation League.

SOULE, H.M. (2000). *Contemporary issues and trends in racial profiling: Is it a police tool or a civil rights violation?* Unpublished manuscript, Central Missouri State University, Warrensburg.

UNITED STATES V. LOPEZ, 514 U.S. 549 (U.S. April 26, 1995).

UNITED STATES V. MORRISON et al., No. 99-5 (U.S. May 15, 2000).

VIOLENT CRIME CONTROL AND LAW ENFORCEMENT ACT of 1994, 28 U.S.C. 994.

WISCONSIN V. MITCHELL, 508 U.S. 47 (US June 11, 1993).

U.S. DEPARTMENT OF JUSTICE (2001, March). 1998 national victim assistance academy, chapter 21, section 1, hate and bias crimes. Available at: http://www.ojp.usdoj.gov/ovc/assist/nvaa/ch21-1hb.htm.

7

Volunteer Units: A Case for the Use of Citizen Volunteers in Sundry Police Tasks

Peter W. Phillips

INTRODUCTION

Volunteer units are integral to policing, having both a rich history and a significant contemporary presence. Before organized policing, whatever law enforcement there was consisted of able-bodied men gathered as the Shire Reeve's (sheriff's) "army." Tithings (groups of ten) and other volunteer configurations evolved to advance the "Hue and Cry" when trouble arose in the community. On the American frontier, posses were hastily assembled by sheriffs or town marshals to pursue bank and stage coach robbers; in the growing cities, the Night Watch consisted of citizen volunteers (Klockers, 1985; Lane, 1992; Monkkonen, 1992).

In a case study prepared in support of the National Evaluation of the COPS Program, Coles (2000) reports that for Civil War duty, three-quarters of the St. Paul (MN) police enlisted in the Army, city revenues dropped precipitously, the night police were disbanded, and a force of 200 volunteers organized themselves to take the place of the police—even going so far as to divide into four companies for the purpose of patrolling different sections of the city. In a review of the literature on volunteer units, Garry (1980) notes that in several places volunteers were both the first police officers and the organizers of our first police agencies, such establishment in America predating what is considered by some to be the first modern police department, the NYPD (see, e.g., Klockers, 1985). Today, volunteer units are well-established parts of police agencies everywhere—but far expanded in role and scope.

The purpose of this chapter is twofold. The first purpose is to identify and explain the varied roles of volunteers in contemporary American policing, with particular emphasis on their deployment in departments managed by community-policing objectives. The second purpose is to examine some of the special considerations important to police agencies in the management of volunteer units. Because of the coverage of reserve and auxiliary police by Shernock in Chapter 3 and because of the growth of volunteer units supporting many police functions besides law enforcement, the clear emphasis in the following discussion will be on volunteers performing duties other than those of reserve or auxiliary police officers. (For those readers particularly interested in reserve and auxiliary units, however, the following references are suggested: Berg & Doerner, 1988; Cooper & Greenberg, 1997; Dow, 1978; Wallace & Peter, 1994.)

DEFINING THE VOLUNTEER UNIT

The definition of a volunteer unit is not as obvious as may appear at first blush. Three questions must be answered: (1) What is volunteerism? (2) Who is a volunteer? (3) When do volunteers constitute a police unit?

What is Volunteerism?

Volunteerism by its very nature implies a symbiotic relationship between the organization and the individual, each deriving benefit from the other. The organization benefits from both the physical and intellectual contributions of the volunteer. Volunteers contribute time and talent, and often spend a significant amount of their own money in the process (such as vehicle expenses, or, in the case of reserve officers, normally all uniform and personal equipment expenses).

Benefits to volunteers are primarily intrinsic (psychological or emotional). Intrinsic benefits derive from making civic contributions, including various kinds of recognition associated with the contribution (Gidron, as cited in Sundeen & Siegel, 1986). Some volunteers fulfill a need for identification with a source of authority (Berg & Doerner, 1988).

Who is a Volunteer?

For the purposes of this chapter, any person who provides services to an organization—in this case a police agency—without direct compensation in the form of salary or wages is a volunteer. Volunteerism is not limited by age (within reason, excluding young children), ranging from teens to elder citizens. Youth may volunteer in programs sponsored by police agencies such as Explorer Scouts and Teen Court. Elder citizens volunteer for a far wider range of duties.

When do Volunteers Constitute a Police Unit?

This part of the definition is a little trickier. A review of the literature, as this chapter demonstrates, reveals a large number of reports depicting the activities of volunteer units in policing. A precise definition of what constitutes a *unit*, however, was not discovered.

Should the definition be quantitative or qualitative in nature? Does one volunteer constitute a unit? Is it three or more? Is it appropriate to consider a volunteer group a unit when the department provides unity of command by assigning a permanent employee as unit supervisor? What role, if any, does the existence of a formal, unit mission statement play in determining the unit's status?

Definition of a Volunteer Unit

In response to the above questions, the author offers the following definition of a volunteer unit in policing:

> A volunteer unit in policing exists when persons contribute services to a police agency without direct compensation in the form of salary or wages, are assigned specific duties in support of the organization's mission, and are supervised by permanent employees of the agency. It is advantageous if the volunteers are provided with written position descriptions and if the volunteer unit has a well-defined mission statement that shows a clear correspondence with a subset of the police agency's goals and objectives as articulated in the agency's mission statement.

Examples of a volunteer's position description and a volunteer unit's mission statement are shown in Appendix A.

RATIONALE FOR USING VOLUNTEERS

Since the 1970s, increasing pressure has been placed on publicly funded agencies to practice cutback management (Arreola & Kondracki, 1992), also known, as noted below, as "doing more with less." The accountability movement that started in public education then hit other tax supported institutions, including the police. As social change also affected the institution of policing, particularly in urban areas (Moore & Stephens, 1991), the public clamored for increases both in efficiency and in effectiveness in dealing with problems of disorder. By the early 1980s, policing had entered the "community era" (Kelling & Moore, 1988) and the concept of "problem-oriented policing" (Goldstein, 1979, 1990) was beginning to take hold. Progressive police administrators, looking for ways to meet these economic and social forces head on, began to open their agencies to the communities they served. One way was to accept, then increasingly solicit volunteer assistance (see, e.g., the ten organizational change case studies prepared in support of the National COPS Evaluation [Roth, Ryan, Gaffigan, Koper, & Moore, 2000]).

In a 1994 study of the use of volunteers in policing, the American Association of Retired Persons (AARP) conducted interviews of 1030 individuals from municipal police agencies and county sheriffs' offices who were responsible for supervising volunteers. Sixty-one percent of the respondents said they started volunteer programs because they needed extra help. In addition, 24% said they started a volunteer program because of budget cuts and constraints, 18% said they were asked by a private organization or approached by individuals to start a volunteer program, and 14% said they started a volunteer program to encourage community involvement in law enforcement.

According to Sundeen and Siegel (1986), the rationale for using volunteers in police agencies may be categorized as follows:

1. Cost effective[ness]. In terms of "doing more with less," volunteers can fill [personnel] shortages, provide additional help, or provide services at reduced budgetary costs.
2. Better services. The use of volunteers enables police departments to free up sworn officers to devote greater time to activities considered more essential to crime prevention or reduction. Also, greater citizen involvement in crime prevention, particularly neighborhood watch and citizen patrols, appears to contribute to greater community security.
3. Citizen participation. The use of volunteers increases the extent to which local citizens can participate in and build bridges with local police. The consequence of this interaction is said to be improved police–community relations and greater input of citizens' views into the police organization.
4. Source of recruits. Some departments have come to view their reserve units and explorer scout posts as a source of potential sworn officers. In these cases, the reserve or explorer role takes on the casting of a screening process through which the most outstanding candidates for full-time employment may be identified (pp. 49–50).

Today, under the rubric of community policing, "co-production" must be added to Sundeen's and Siegel's list. Co-production is a political science concept meaning citizens and government working together to accomplish goals that neither can accomplish alone (Frank, Brandl, Klorden, & Bynum, 1996; Parks, Baker, Kiser, Oakerson, & Ostrom, 1981). The concept is even broader than the notion of good police–community relations. As Palumbo (1994) notes:

> For some problems, such as crime control, there is a need for people themselves to become involved in finding solutions. . . . Without the active participation and help of citizens, there is little that government agencies can do to control crime or solve the many other problems facing the country. (p. 14)

This is an important concept for police to grasp. Inasmuch as citizens need their police to intervene in situations when "something-ought-not-to-be-happening-about-which-something-ought-to-be-done-NOW!" (Klockers, 1985, p. 16), the police need citizens' information. To elicit information, to encourage citizens to volunteer information or any other form of support, requires a certain partnership between the police and the public. That partnership is considered the very essence of community policing: "Community policing's future is dependent on the government and the governed coalescing to identify needs that can be addressed through a combination of government resources and citizen activism and *volunteerism*" (Trojanowicz, 1994, p. 258; emphasis added).

THE COST-BENEFIT OF VOLUNTEER USE

One might suppose that volunteers provide "free labor," but "free" is a relative term. The use of volunteers may (a) result in cost avoidance (otherwise compensating additional employees to produce the same products or services), thereby paying for added services with the "saved" funds; or (b) simply provide products and services which, without the "free" volunteer, could not be provided within budget allocations. As Sundeen and Siegel (1986) note, however, there are always organizational costs associated with volunteers. There are direct costs involved in recruiting, training, supervising, and rewarding volunteers. There are indirect costs, as well, particularly if the use of volunteers causes dissention among the ranks—discussed further below.

In their definitive study on the use of volunteers in policing, Sundeen and Siegel (1986) demonstrate average departmental benefits and costs of five volunteer types: paid reserves officers, non-paid reserve officers, neighborhood watch volunteers (tantamount to today's "citizens on patrol" groups), Explorer Scouts (vis-à-vis paid interns), and clerical volunteers. Minus the direct costs of recruiting, training, supervising, and rewarding volunteers, the average avoided cost among eight departments, valued in 1984 dollars, for non-paid reserves was $13,983; the average among seven departments for neighborhood watch patrols, $9504; the average among 14 departments for Explorer Scouts, $2815; and the average among seven departments for clerical volunteers, $3666. Surprisingly, the authors found a low but negative cost-benefit to the seven departments employing paid reservists, minus $1769 as a departmental average. Departments very well may have justified this expense, however, simply by pointing to the fact that this asset existed and could be mobilized in any emergency or other demand situation.

For fiscal year 2001, one police chief whose department serves a population of approximately 78,000, reports that if he were to compensate civilian employees at $10 per hour, its Pawn Shop Detail data entry volunteers, working a total of 1337 hours, would have cost $13,370 for the year; and if its Citizens on Patrol volunteers, working 5069 hours, were converted to CSOs (Community Service Officers, non-sworn), the cost would have been $50,690. If its reserve officers, working 942 hours, were to have been compensated at $15 per hour, the department cost would have been $14,130. For FY 2001 (September 1, 2000 through August 31, 2001), therefore, this department avoided costs of $78,190 by deploying volunteers in just these three areas alone (Key, 2001). For calendar year 2000, based on an average hourly wage for civilian employees in all categories of $14.30, another municipal police department with 178 sworn officers calculated that its volunteers saved the city $210,639 (Lynch, 2001).

In the study of 30 Texas police agencies, one researcher reports that:

> [d]epending on the extent of use and the number of volunteers recruited, a department can provide human resources for certain police services free of charge thus saving city or county funds. The estimated monthly value of services performed by policing volunteers ranged from $400 to $160,000, with an average of $15,961.
>
> The average annual cost for volunteer maintenance (e.g., providing uniforms and other equipment), ranged from $10 to $500 a volunteer. . . . While there is concern about the cost of maintaining volunteers, the figures indicate that it is not substantial compared to the value of services provided. (Brock, 1996, p. 3)

OPPOSITION TO THE USE OF VOLUNTEERS

In the AARP (1994, p. 12) study mentioned earlier, 95% of the respondents rated the quality of volunteers' work as "excellent" or "very good." Even though the positive aspects of deploying volunteers overwhelmingly outweigh the negative aspects of their use, from time to time there may be expressions of dissatisfaction with the use of volunteers. Criticism, if it comes, is usually internal to the department—although some may be external. We shall consider the internal first.

Intra-agency objection to the use of volunteers tends to center within the lower police ranks and is often an issue with the bargaining unit (union) or, if none is recognized, a more informal, but nonetheless formidable police benevolent association (PBA). Burden (1988) notes that many police officers consider volunteers to be a job threat, particularly in periods of fiscal exigency; at the very least, they suggest, volunteers take jobs away from disabled or "limited duty" officers.

It is suggested, however, that any adverse rank-in-file reaction to the infusion of volunteers in a police agency is negatively correlated to the efforts of management to announce its intentions to implement a new program, to consult with the employee unit, and to practice participative decision-making in the process. The following anecdote is offered as evidence in support of this conclusion.

In the Fremont, California study, Thatcher (2000) reports that there were police union concerns about the use of volunteers. Communication was a concern in that the union wanted to be part of the decision-making process when volunteer positions were created. When new job descriptions for volunteer positions are drafted, therefore, the union is provided with a copy and asked for comments. For example, a question once arose regarding a volunteer assisting with filing in the Chief's office. That this person might have access to personnel records was objectionable to the union. The objection being voiced, the job description was modified to allow the volunteer to help with all filing except that relating to police personnel documents.

The California Police Officers' Association president in the Fremont department is reported as having stated that his association has no fundamental objection to volunteers as "job thieves." "If a volunteer is actually replacing a position, yeah, I think the complaint is justified and it needs to be addressed. [However] if the volunteer is merely helping an overburdened position, then God bless them" (quoted in Thatcher, 2000, p. 37). To this, the department's volunteer coordinator noted that success lay in carefully delimiting the volunteers' responsibilities. It is most important for volunteers to know what their boundaries are; the idea of the volunteer program is not to replace staff but to augment the services needed. "[B]ecause of the volunteers, we are able to accomplish a lot more and are able to offer a lot more services to the community" (quoted in Thatcher, 2000, p. 37).

Other internal criticisms of volunteer programs focus on the availability of volunteers to deal with peak workload periods (AARP, 1994; Sundeen & Seigel, 1986) and volunteer turnover (AARP, 1994; Brock, 1996). Although not all contingencies can be foreseen, workload analyses in different police divisions should provide a modest amount of information regarding peaks and valleys in the demand for certain tasks. To the extent possible, supervisors should schedule volunteers to meet these demands. During recruitment and in discussions with potential recruits regarding job placement within the department, these needs should be discussed. The turnover problem, however, may well relate to motivation. As Sundeen and Seigel (1986, p. 59) suggest, "some volunteers lose interest in their endeavors and drop out after an initial surge of enthusiasm." Motivation is discussed in a later section of this chapter.

External criticism of volunteer programs and activities tends to come from citizens who believe that they are being policed by non-police, that is, by persons without authority to pass judgment on their behavior. Anecdotal evidence for this conclusion comes from critiques regarding Citizens on Patrol (COP) kinds of programs. In COP programs, generally, citizen volunteers are charged with the responsibility of reporting violations of law committed by their fellow citizens. COP activities concentrate on neighborhoods, with citizen activists patrolling the area and reporting alleged violations to the police.

In another version of the COP program, citizen volunteers monitor traffic either at assigned intersections or at random. Sometimes called "traffic monitors," at intersections the volunteers record traffic violations, turn their reports over to the police department, and the department sends a letter to the registered owner of the vehicle noting the observation, the section of traffic law allegedly violated, and a plea for future safe driving. When operating "at random," the volunteer makes observations whenever he or she is "out driving." Notes may be taken as described above, or the volunteer may call in violations to a special traffic desk number as they occur. Whether stationary posts or mobile observations, running red lights is the most common violation reported (Pappas, 2001).

Much of the public objection to COP relates to traffic safety programs. Three recent comments of citizens in one city where a "Citizens for Traffic Safety" program was initiated are as follow:

> Citizen committees or citizen spies? Communist countries and Nazi Germany effectively used citizens to spy on fellow citizens in order to extend their control of everyday lives and perceived breaches of authority. Why not us?
>
> I see the Tyler PD got the go-ahead for the Traffic Safety Program where 20 volunteers will be snooping around out there to see if we commit some traffic violation! All you're going to have is some unqualified citizen "wanna-be traffic cop" snitching on folks.
>
> I hope I am not the first person to receive a letter from [the police chief] and [the Traffic Safety Board chairman] admonishing me for an alleged traffic infraction reported to them by an untrained volunteer. Safety awareness is not the issue: Harassment is. ("Residents divided . . . ," 2001)

Never can all criticism be avoided, nor should it be. When initiating new programs, however, police administrators should assure that both internal and external stakeholders are consulted. In the case of programs such as COP that affect the public so directly, it would seem prudent to achieve consensus among the municipality's elected officials, thence engage in an aggressive public education campaign regarding the program before it is actually implemented. Because two of the three letter writers above appear to believe they are being observed by untrained persons (implying high error rates or the potential thereof), the pre-program publicity should include information regarding the volunteers' selection, training, and testing related to the tasks undertaken. Periodic reinforcement is important, too, in matters that are potentially controversial. Where evaluation is built into program design—as it should be—periodic reports to the public regarding the outcomes of police programs serve as reminders of the reasons these programs were created and the conditions under which they continue.

TYPES OF VOLUNTEER UNITS

Developing a typology of volunteer units is somewhat problematic—there are many and they are varied in function. In 1986, Sundeen and Siegel surveyed a stratified random sample consisting of 18 police agencies in Los Angeles (LA) County. There and then, the emphasis was on reserve and auxiliary police (89%).

In the study conducted by the American Association of Retired Persons (1994) to which reference already has been made, the agencies in the survey reported that three-quarters of their volunteer programs used volunteers for clerical work. Nearly 60% used volunteers for program support, over 50% for administration and general office assistance, and slightly more than one-third as receptionists. About 15% of the agencies used volunteers for program management and supervision.

In a study of Texas police agencies, Brock (1996) found that 73% of the departments used volunteers. As shown in Table 7-1, clerical duties appear to be the most frequently reported volunteer activity (57%) and data processing appeared to be the most frequently performed function (33%). "The responses clearly indicate that volunteers are useful resources for departmental management, as well as other tasks more traditionally thought of

TABLE 7-1   Use of Volunteers in Texas Police Agencies

Function	Number of departments	Percent of departments
Clerical assistance	17	57
Data processing	10	33
Neighborhood watch	9	30
Citizen patrol	8	27
Victim assistance	7	23
Public education	7	23
Translation services	7	23
Reception	6	20
Communication	4	13
Motorist assistance	4	13
Citations	3	10
Explorer post	3	10
Vehicle maintenance	2	7
Other	9	30

Source: Adapted from Brock (1996), p. 2.

as police officer duties (e.g., citizen patrol, assisting with victims, etc.)" (Brock, 1996, p. 2). Over the period of the three studies cited (1986–1996), an interesting redistribution among the categories of volunteer activities may be noted; namely, the number of volunteers supporting non-law enforcement police functions.

A more detailed description of how volunteers may be used is shown in Appendix B. The appendix contains abridged job descriptions of volunteers in policing. It has been constructed as a sample of the varied job-tasks to which volunteers can be assigned. It is not intended to be conclusive; it is intended, however, to stimulate innovation in the use of volunteer assistance in police agencies of all sizes.

Somewhat parenthetically, it should be noted that small police agencies may be most in need of help because of limited resources and they may suffer two limitations not faced by large organizations: (1) Because the communities are small, the pool of citizens willing to volunteer is small; (2) because the agency is small, there are fewer higher-level sworn or civilian administrative personnel to supervise (train, lead) the volunteers.

NATIONAL COPS EVALUATION

The organizational change case studies prepared for the Urban Institute in support of the National COPS Evaluation (Roth et al., 2000) are insightful regarding contemporary applications of volunteer units in policing today. The cities studied were Albany, NY; Colorado Springs, CO; Fremont, CA; Knoxville, TN; Lowell, MA; Portland, OR; Riverside, CA; St. Paul, MN; Savannah, GA; and Spokane WA. Particularly instructive are the studies of Colorado Springs, St. Paul, and Fremont.

Colorado Springs

In a case study of Colorado Springs, Sheingold (2000a) documents several of the volunteer unit activities, all designed to forge a closer relationship between the department and city's citizens. For example, an all-volunteer interfaith Chaplaincy Corps was organized to provide counseling assistance following traumatic incidents and a group of senior citizen volunteers was organized to respond to elderly crime victims who might need crisis intervention services. In addition to several storefront operations staffed by volunteers, since 1996 the Colorado Springs Police Department (as have many other departments) has established a handicapped enforcement unit that trains volunteers, many of whom are themselves handicapped, to write tickets to violators of handicapped parking regulations.

> In 1996, 336 volunteers worked more than 43,000 hours on all of the various projects. Myers [a departmental spokesperson] believes that the volunteers play an important role in the department's commitment to community policing. "By working here, getting a feel for the department, knowing what is going on, knowing how the department operates, the volunteers carry [that information] back out to the community," Myers said. "I think that it also . . . says [that] this department is willing to go a step further to help people in the community, beyond what officers are trained to do or have the time to do." (Sheingold, 2000a, pp. 50–51)

St. Paul

In St. Paul today, volunteers are recruited primarily from the alumni of the Police Department's Citizen Police Academy. As Coles (2000) reports, however, volunteerism in that department actually began in 1976 with replication of the Dayton, Ohio Neighborhood Assistance Officer (NAO) program. Co-production is the concept driving this Neighborhood Watch-type program where trained citizens make vacation and other home security checks; conduct residential security training; monitor neighborhood gatherings, such as garage sales; and might also monitor troublesome youth in neighborhoods. These citizens provide feedback to the police regarding their observations, clearly adding needed eyes and ears both to patrol officers and to investigators. For the calendar year 1996, approximately 110 volunteers worked with the department; 45–50 doing 85% of the work. One volunteer worked nearly 2500 hours in one year alone (Coles, 2000).

Fremont

In its transition to community-oriented policing and problem-solving (COPPS), the most visible way in which the Fremont, California, Police Department reached out to its citizens was through a planned expansion of its volunteer program. Donna Gott, a department project manager, in an interview with the author of this case study, David Thatcher (2000, p. 35), said, "[The administration] found that in order to bridge the gap between the police department and the community, what better way to do it than get the citizens involved in what we do to make our job easier? Break down those barriers by getting [the citizens] involved." With a core of eight volunteers, the number grew rapidly to exceed the department's goal of 100 after Gott prepared a pamphlet describing volunteer opportunities available in the department.

SOURCES OF CITIZEN VOLUNTEERS

The primary source of volunteers in policing appears to be alumni of citizen police academies (Liddell, 1995; McDowell, 2001; Pappas, 2001; Sheingold, 2000b). These citizens have undergone an average of 36 hours of training related to the tasks associated with policing—not as if to be police, but to understand the nature of police work. Graduates tend to be significantly more empathetic regarding police procedures and the outcomes of many police–citizen interactions than they were before receiving the training. In the main, they have developed a clear understanding of the critical relationship existing between the public and their police and have become advocates for the police in their respective spheres of influence. Not all are converted, but even those who remain critical of the police admit that they are far better educated than when they began the program (Peverly & Phillips, 1993).

Other sources of volunteers come from departmental publicity campaigns to start programs or to increase the numbers of volunteer participants, from command level officers announcing volunteer opportunities when addressing civic groups and when making presentations to neighborhood forums, and from word-of-mouth recruitment conducted by persons who are already volunteers in policing and are finding the experience rewarding.

Police officers can be effective recruiters as they encounter good prospects both in their official capacity and as citizens themselves interacting with others through personal business, social relationships, and civic and religious associations. Two volunteer coordinators interviewed during the preparation of this chapter stressed word-of-mouth recruiting by satisfied volunteers and positive citizen encounters with police officers coincident with departmental publicity campaigns—possibly ranking second and third after citizen police academy graduates (McDowell, 2001; Pappas, 2001).

SELECTION, TRAINING, AND SUPERVISION OF VOLUNTEERS

Selection

Some volunteer positions may not require the same considerations of knowledge, skills, and attitudes (KSAs) as for sworn or particular non-sworn paid positions, but certain tests of honesty, integrity, and confidentiality should be considered as standard in the selection of volunteers. Although the overwhelming majority of volunteers are truly civic minded, it should not be assumed that every volunteer applicant will have the same motives.

On the dark side, the author recalls that one centralized state police academy was so concerned about infiltration by the Weather Underground, it randomly placed listening devices under seats in its main lecture hall to capture off-hand remarks by recruits. On a somewhat lighter, but nonetheless embarrassing note, the author also recalls recommending volunteer work to an outstanding student as a means of further enhancing his résumé. The police agency that accepted the recommendation did so without a background check on the student and while he was seated one day entering data in the detective division, a Secret Service Agent entered and arrested him on an outstanding warrant for credit card fraud.

In Brock's (1996) study of 30 Texas police departments, an alarming 27% did not perform a background check on volunteer applicants. Also disturbing was Brock's finding that only one department utilized drug testing as a form of selective screening. Two departments conducted polygraphs, but reported that volunteers were difficult to obtain as a result of this requirement.

It is recommended that criminal history, traffic history, employment references, and drug screening constitute the minimum number and kinds of background checks performed before a volunteer applicant is accepted. Following is a list of actions or attributes that should automatically disqualify any person from volunteer service in a police agency:

1. Been convicted of a felony or any offense that would be a felony if committed in this State, or of any crime involving unlawful sexual conduct, or of any crime related to domestic violence.
2. Used ("tried") marijuana in the past six months.
3. Sold marijuana.
4. Used ("tried") any dangerous drug or narcotic including cocaine, crack, heroin, LSD, etc.

5. Sold narcotics or dangerous drugs.
6. Had/has a pattern of abusing prescription medication.
7. Chronically abuses alcohol.
8. Been dishonorably discharged from the United States Armed Forces.
9. Had excessive traffic violations in the last three years.
10. Been previously employed as a law enforcement agent and since has committed or violated federal, state or city laws pertaining to criminal activity.
11. Lied during any part of the volunteer selection process.
12. Falsifed his or her questionnaire or application.

The next list is of those actions or attributes that some departments may deem discretionary disqualifiers for acceptance as a volunteer in policing:

1. A physical or mental disability that would substantially impair an individual's ability to perform his/her duties.
2. Excessive traffic violations.
3. Commission of any crime.
4. Any discharge from the U.S. Armed Forces other than an honorable discharge.
5. Bad debts or a demonstrated unwillingness to honor fiscal responsibilities.
6. Any other conduct or pattern of conduct that would tend to disrupt, diminish, or otherwise jeopardize public trust in the law enforcement profession.

As opposed to those on the previous list, the reason that these items are classed as "discretionary" is because the significance of some may depend on the task to which the volunteer is assigned. For example, a good driving record is critical to the qualifications of a person ferrying squad cars between the police station and the city garage, but not critical to the task of the volunteer front desk receptionist. An accumulation of bad debts may be particularly troublesome in positions where the volunteer has access to petty cash, but not as much so in others.

To avoid allegations by volunteer applicants of unequal treatment, it is recommended that departments decide from the list of discretionary disqualifiers which items they believe should be moved to the list of automatic disqualifiers and do so. A rough guideline may be found in what actions, omissions, or attributes a police department would, or would not, tolerate by prospective paid employees assigned the same job-tasks.

Additional recommendations include:

1. Development of a comprehensive application form that will provide enough personal information about the volunteer applicant to serve as an outline for a complete background check.
2. Inclusion in the applicant packet of a/an
   a. authorization to release information,
   b. emergency contact list (in case the volunteer becomes ill or is injured on the job), and
   c. confidential information agreement form.
3. Development of a volunteer agreement form that precisely stipulates the volunteer's responsibilities to the police agency and the agency's responsibilities to the

volunteer. An example of one police department's Volunteer Agreement Form is included as Appendix C and is particularly instructive regarding these mutual responsibilities.

All forms should be signed and/or acknowledged as having been received by the volunteer. A personnel file should be established in which to accumulate records pertaining to the volunteer's performance, both for his or her benefit and for the agency's.

Training

The tasks to which volunteers may be assigned are so varied as to preclude recommendation of a "standard curriculum." One to ten training hours was the most common range reported in Brock's (1996, p. 5) study (42%), although 37% of his departments trained longer. Again, a disturbing figure: 21% reported zero hours of training. Perhaps those agencies reporting no training were not counting informal or on-the-job training experiences of the new volunteers working alongside experienced employees or other volunteers, but truly to provide no training is both irresponsible and demotivating.

Employers have a clearly defined responsibility to provide training for all workers, however formal or informal. How else should one learn the agency-defined method for satisfactorily completing a particular task? If tasks are not completed satisfactorily according to agency standards, critique may turn to criticism and the latter is a disincentive to continued effort. More important, the official attention given to a volunteer in a training situation confers significance to the position and the job-tasks involved. By devoting time and attention to training, the agency demonstrates its investment in the person; taking the time to fully train a volunteer adds value to the person, and employees in any category need to feel valued.

Most volunteers are older adults, including many senior citizens, who bring previously acquired job skills to the volunteer position. Primarily as older adult learners, volunteers require less classroom oriented training than do younger persons without comparable work and life experience, and relatively more hands-on training experiences using the actual tools of the agency, whether those be forms and filing systems or internal communication networks and specially written databases. Training should be seen as an ongoing process of continued skill building, as this too is intrinsically rewarding. Generally, training responsibilities are vested in the person designated as the volunteer's supervisor.

Supervision

In the main, volunteers should not be supervised or evaluated differently from paid civilian employees who are performing the same function in the department or who otherwise might have been employed to perform that function. Supervisory personnel should remember, however, that even though the volunteer may have demonstrated a commitment to the task as a matter of personal work ethics, he or she does not necessarily have the same commitment to the agency as a paid employee dependent on income from the position.

Lowered morale among employees may lead to a loss of productivity; lowered morale among volunteers leads to their loss, period.

Pappas (2001) charges volunteer supervisors with taking occasional breaks with volunteers for the purposes of both getting to know them and providing them with management recognition. She notes that supervisors constantly should be on the alert for volunteers' contributions and suggestions and to recognize them in a timely manner; further, to recognize special events in volunteers' lives, such as birthdays, anniversaries, births of grandchildren, and so forth. The important point here, we believe, is to establish acceptance of the volunteer in the same manner as any other employee is accepted by the agency. Because volunteers are not paid *per se*, in no way however slight should they question their status as "first class citizens."

When creating volunteer positions, supervisors should remember that volunteers want to be involved in projects they consider worthwhile. They become frustrated when repeatedly given boring, purposeless jobs. Work that regular employees simply do not want to do is *not* the work to assign to volunteers. Volunteers will do their share of mundane work, but only if they perceive some equity in its distribution. Supervisors making assignments to subordinates in any work environment should be creative and mix the mundane with more interesting tasks, tasks for which subordinates can see an organizationally important outcome. This is especially true when dealing with volunteers—for as the saw goes, "They can vote with their feet."

INCENTIVES AND REWARDS

Speaking of old saws, it very well may be true that "nobody works for nothin'." Volunteers, by definition, do not work for the extrinsic motivation of monetary reward—the rewards sought are intrinsic. They are motivated to volunteer, to give of their time and effort, for the inner satisfaction of making a valued civic contribution; of self-development and achieving greater personal meaning (Sundeen & Siegel, 1986). Volunteers' pay, the extrinsic equivalent of salary or wage is in the form of official, and often public, recognition. We already have spoken about the importance of an agency's demonstrating that it values its workers as persons. Valuing volunteers' contributions, however, can take many forms.

In his study, Brock (1996, p. 5) found that 60% of the departments held formal ceremonies to recognize volunteers for their contributions; 40% provided volunteers with awards and certificates, one-third gave gifts such as jackets and caps, 23% organized group outings, and 12% treated volunteers to social events such as ball games and plays. To this list we should add a particular type of recognition, an annual Volunteers of the Year banquet or formal reception with an invited media presence, usually held in conjunction with National Volunteer Week (in 2001, the last week of April). In one department as part of the ceremony, the City Council receives a check representing the cost savings to the police department based on volunteer contributions for past year. Several volunteers who have made extraordinary contributions are specially recognized, with their "stories" being reported by local newspapers and television stations (Lynch, 2001).

SUMMARY AND CONCLUSION

The trend for at least the last decade clearly has been an increase in the deployment of volunteers in policing. As contingency management techniques have been forced on police administrators, the "do more with less" theme means that innovative strategies must be employed to meet the increased demand for community-policing functions in addition to traditional law enforcement. Developing a qualified volunteer corps is an important, contemporary innovation in policing.

Volunteers can both supplement and complement the activities of regular departmental personnel. Supplemental actions include engaging in the same activities as regular employees, such as clerical tasks; complementary actions include bringing specialized experience and expertise to the department, for example, professional counseling.

In well-managed programs, there is little internal resistance to volunteer input. Where it exists, it usually can be reduced to nil by good-faith negotiations with employee groups. External resistance usually can be predicted and pre-empted through a planned program of public education.

The primary key to a successful volunteer program is the quality of supervision. Volunteers wish to be treated equally with paid employees. The jobs that regular employees do not want to do are not the jobs to disproportionately assign to volunteers. In lieu of pay *per se*, volunteers seek signs and symbols that the organization values them both as persons and for their donations of time and effort. Volunteers can be a great asset to police agencies, but never should be a "hidden asset."

RECOMMENDED READINGS

AMERICAN ASSOCIATION OF RETIRED PERSONS (1994). *The 1994 study of the use of volunteers in police agencies*. Washington, DC: Author [Monograph].

ASSOCIATION FOR VOLUNTEER ADMINISTRATION. *The Journal of Volunteer Administration*. Available: P.O. Box 32092, Richmond, VA 23294. [A current journal]

BOTTOF, J., & KING, M. (1996). *Volunteer Management Made Easy Series*. Available: PROVAL, 3106 S. Extension Rd. Mesa AZ. 85210 [pow.cas.psu.edu/Final/5FResources.html].

FRANK, J., BRANDL, S.G., WORDEN, R.E., & Bynum, T.S. (1996). Citizen involvement in the coproduction of police outputs. *Journal of Crime and Justice, XIX*(2), 1–30.

KULP III, K., & NOLAN, M. (2000). Trends impacting volunteer administrators in the next 10 years. *Journal of Volunteer Administration, 19*(1), 10–19.

VINEYARD, S. (1993). *Megatrends and volunteerism*. Downers Grove, IL: Heritage Arts Publishing.

REFERENCES

AMERICAN ASSOCIATION OF RETIRED PERSONS (1994). *Report on the 1994 study of the use of volunteers in police agencies*. Washington, DC: Author.

ARREOLA, P., & KONDRACKI, E.N. (1992). Cutback management, cost containment, and increased productivity. *Police Chief, 59*(10), 110–112, 115–118.

BERG, B.L., & DOERNER, W.G. (1988, Spring). Volunteer police officers: An unexamined personnel dimension in law enforcement. *American Journal of Police, 7*(1), 81–89.

BROCK, D. (1996, August). Volunteers. *TELEMASP Bulletin, 3*(5), 1–7. [Huntsville, TX: Bill Blackwood Law Enforcement Management Institute of Texas.]

BURDEN, O.P. (1988). Volunteers: The wave of the future? *Police Chief, 55*(7), 25–29.

COLES, C. (2000). *National COPS evaluation organizational change case study: St. Paul, Minnesota* [On-line]. Available at: *http://www.ncjrs.org/nij/cops_casestudy/stpaul2.html.*

COOPER, K., & GREENBERG, M.A. (1997). Auxiliary police help prevent delinquency. *Police Chief, 64*(10), 116–118.

DOW, R.E. (1978). *Volunteer police: Community asset or professional liability.* New York: New York Conference of Mayors. Available: NCJRS paper reproduction, Box 6000, Dept. F., Rockville, MD 20849.

FRANK, J., BRANDL, S.G., WORDEN, R.E., & BYNUM, T.S. (1996). Citizen involvement in the coproduction of police outputs. *Journal of Crime and Justice, XIX*(2), 1–30.

GARRY, E.M. (1980). *Volunteers in the criminal justice system: A literature review and selected bibliography.* Washington, DC: National Institute of Justice [NCJ 65157].

GOLDSTEIN, H. (1979). Improving policing: A problem-oriented approach. *Crime and Delinquency, 25,* 236–258.

——. (1990). *Problem-oriented policing.* New York: McGraw-Hill.

KELLING, G.L., & MOORE, M.H. (1988). The evolving strategy of policing. *Perspectives on Policing,* No. 4, 1–15. Washington, DC: National Institute of Justice.

KEY, A.J. (2001). Personal correspondence, December 11.

KLOCKERS, C.B. (1985). *The idea of police.* Beverly Hills, CA: Sage [Vol. 3, Law and Criminal Justice Series].

LANE, R. (1992). *Urban police and crime in nineteenth-century America.* In M. Tonry and N. Morris (Eds.), *Modern Policing* (pp. 1–50). Chicago: University of Chicago Press.

LIDDELL, R.J. (1995, August). Volunteers help shoulder the load. *Law Enforcement Bulletin, 64*(8), 21–25.

LYNCH, G. (2001, April 27). Police salute 95 volunteers. *Tyler Morning Telegraph,* Sec. 1, p. 2.

McDOWELL, F. (2001). Personal correspondence, December 11.

MONKKONEN, E.H. (1992). History of urban police. In M. Tonry and N. Morris (Eds.), *Modern Policing* (pp. 547–580). Chicago: University of Chicago Press.

MOORE, M.H., & STEPHENS, D.W. (1991). *Beyond command and control; The strategic management of police departments.* Washington, DC: Police Executive Research Forum.

PALUMBO, D.J. (1994). *Public policy in America.* (2nd ed.). Ft. Worth: Harcourt Brace.

PAPPAS, A. (2001). Personal correspondence, May 22.

PARKS, R.B., BAKER, P., KISER, L., OAKERSON, R., & OSTROM, E. (1981). Consumers of public services: Some economic and institutional considerations. *Policy Studies Journal, 9,* 1001–1011.

PEVERLY, W.J., & PHILLIPS, P.W. (1993). Community policing through citizen police academies. *Police Chief, LX*(8), 88–89.

RESIDENTS DIVIDED OVER CITIZENS ON PATROL ISSUE (2001, April 30). *Tyler Morning Telegraph,* Sec. 1, p. 5.

ROTH, J.A., RYAN, J.F., GAFFIGAN, S.J., KOPER, C.S., & MOORE, M.H. (2000). *National evaluation of the COPS Program—Title I of the 1994 Crime Act.* Washington, DC: National Institute of Justice [NCJ 183643].

SHEINGOLD, P.M. (2000a). *National COPS evaluation organizational change case study: Colorado Springs, Colorado.* Available at: http://www.ncjrs.org/nij/cops_casestudy/colorado.html.

——. (2000b). *National COPS evaluation organizational change case study: Spokane, Washington.* Available at: http://www.ncjrs.org/nij/cops_casestudy/spokane.html.

SUNDEEN, R.A., & SIEGEL, G.B. (1986). The uses of volunteers by police. *Journal of Police Science and Administration, 14*(1), 49–61.

THATCHER, D. (2000). *National COPS evaluation organizational case study: Portland, Oregon* [On-line]. Available at: www.ncjrs.org/nij/cops_casestudy/portlan2.html.

TROJANOWICZ, R.C. 1994. The future of community policing. In D.P. Rosenbaum (Ed.), *The challenge of community policing: Testing the promises.* (pp. 258–262). Thousand Oaks, CA: Sage.

WALLACE, H., & PETER, A.P. (1994). Police reserves: Rights and liabilities. *FBI Law Enforcement Bulletin, 63*(5), 20–23.

APPENDIX A
SAMPLE VOLUNTEER UNIT MISSION STATEMENT

The goals of Volunteers in Policing (VIP) are as follow:

Enhance our police-community alliance by establishing a closer relationship within the community through the VIP Program.

Spread the message of the police department through the community by the actions of volunteers. In turn, law enforcement personnel will have the opportunity to hear the viewpoint of the citizen-volunteer. This will strengthen the partnership of law enforcement and community working together.

Implement a volunteer network to expand existing programs, and initiate new ones utilizing community resources for the enrichment of our citizens.

Augment staff with volunteers to achieve departmental goals. Recently, law enforcement has had to curtail or not implement many community services and programs with paid employees because of fiscal constraints. Volunteers have helped significantly to provide these services without interruption.

Source: Tyler, TX, Police Department (2001)

SAMPLE POSITION DESCRIPTION

Job Title:	Tyler Citizens For Traffic Safety
Purpose:	Assisting citizens in helping to make the city streets safer.
Description:	Record the license plate, make, and color of vehicles seen violating traffic laws. A letter then follows from the Police Chief and Traffic Safety Board Chairman, encouraging drivers to voluntarily comply with traffic laws.
Qualification:	Be at least 21 years of age, possess a valid Texas License, provide minimum vehicle insurance on personal vehicle, and meet any other requirements that might be established by the appropriate law enforcement agency.
Training:	One two-hour training session on traffic safety laws.

Time Commitment: Varies, depending on the Volunteer's weekly schedule, four hour shifts or at random.

Responsible To: Volunteer Coordinator and Traffic Sergeant.

Source: City of Tyler, Texas, Municipal Police Department (2001)

APPENDIX B
TYPICAL TASKS PERFORMED BY VOLUNTEER UNIT PERSONNEL

Accreditation Assistant—Maintains file folders, highlighting applicable areas in written policy and assist with light computer entry duties in the existing computer program.

Alarms Clerk—Assists with notification of homeowner of city alarm ordinance and the penalty for non-compliance.

Auto Theft—Answers phones, picks up paperwork in Records, filing, working front desk and weekly overdue supplement list.

Career Development Clerk—Assists in data entry, filing and typing of Completion Certificates.

Children's Safe House Attendant—Assists investigators with child interviews by baby-sitting and answering the phones.

Citizen Contact—Volunteer makes daily phone calls to elderly citizens at their request.

Crime Prevention Clerk—General office duties, compiling information packets for Neighborhood Crime Watch and Business Crime Watch programs.

"Crime Watch Live" Operators—Telephone screeners for live call-in TV show.

Criminal Investigations Clerk—Filing, phones, copying, and light computer work for the following areas: burglary, criminal mischief, auto theft, forgery, homicide, gang unit, personal violence, intelligence, administration.

Communications Clerk—Clerical support for communications director. Computer and organizational skills.

Crisis Counselor Assistant—Clerical support and phones for crisis counseling and victim's assistance programs.

Disabled Parking Violation Monitor—Volunteer will issue parking citations and take polaroid pictures of violators in disabled parking spaces. A copy of the ticket and photo will then be sent to municipal court.

Equipment Clerk—Sizing, labeling and sorting of police uniforms, removing sleeve patches from discarded uniforms.

File Clerk/Typist—Filing of confidential information documents into police employee folders, filing index cards numerically, use of copy machine.

Fingerprint Technicians—Fingerprinting of the public at the station and of children at the schools.

Front Desk Receptionist—Answering telephones to screen calls, typing letters and documents, use of copy machine.

Jail Clerks—Clerical support for detention staff, phones, light data entry.

Mini-station/Storefront Receptionist/Office Assistant—Volunteer will answer telephone, take messages, greet visitors, provide general information, perform other light clerical duties.

Narcotics—Enter narcotic complaints and Crime Stoppers tips into computer system, help maintain files, and typing of search warrants.

Park Patrol—Volunteer daily inspects conditions of city parks, looking for problem areas to report.

Pawn Ticket Entry Clerk—Data entry into pawn ticket computer assisting investigators with reported stolen merchandise.

Property Room Assistant—Data entry and general support for quartermaster.

Records Clerk—Data entry, filing, public contact at service window.

"Santa Cop" Volunteer—Organization and distribution of donated toys during the month of December.

Sign Language Instructor—Volunteer will instruct classes to officer's and civilians, teaching the basic's in communicating the sign language.

"Speedwatch" Volunteer—Operates radar and speedboard on residential streets at the request of citizens and the traffic division.

Squad Car Inspector—Inspection and supply for all fleet vehicles. Delivery of vehicles for service or repair.

Traffic Clerk—Clerical support for Traffic Division. Phones and light computer work.

Traffic Safety Monitor—Volunteer will record the license plate number, make, and color of vehicles observed violating traffic laws at designated intersections.

"Turkey Drive" Volunteer—Collection and distribution of food to the needy during the month of November.

"Vacation Watch" Patrol—Checks homes of citizens at their request during their absence.

Victim Assistance—Answers phones, picks up paperwork, assists with clerical duties, limited counseling, informing victims about victim compensation.

VIP Secretary—Clerical support for volunteer coordinator.

Volunteer Chaplain—Volunteer will ride with officers in patrol units, counsel and assist employees with various functions, and participate in police events as needed by representing a cross section of religious communities in the area. Auction, phone referral transfers, entering and updating the computer juvenile ID records, assist the missing persons clerk in "call back" verifications of new runaway and missing persons reports.

Warrants Clerk—Clerical support for warrant office.

Weapons Range—Sorting and cleaning shell casings, reloading ammunition, repairing used targets, answering telephones, cleaning of weapons and bullet trap areas, etc.

Word Processing Assistant—Word processing of police documents and proof-reading.

Youth Bureau—Assists the bicycle officer with the job of getting the bicycles ready for auction, phone referral transfers, entering and updating the computer juvenile ID records, assists the missing persons clerk in "call back" verifications of new runaway and missing persons reports.

———————

Source: Consolidated from Tables 2 and 3 (Brock, 1996) and Tyler (TX) Police Department Volunteer Opportunity List (Pappas, 2001).

APPENDIX C
THE CITY OF TYLER, TEXAS
TYLER POLICE DEPARTMENT VOLUNTEER AGREEMENT

I, _____ agree to give _____ hours each week as _____

(Position)

I understand that I will be regulated by all City of Tyler and Police Department Rules and Regulations, and will be expected to conduct myself in the same manner as a staff member. I will be expected to assume the responsibilities as listed in the job description of the job for which I have volunteered.

I agree that I will:

1. Be punctual, reliable and aware of the need to offer friendship and assistance to fellow volunteers and staff members.
2. Work at least _____hours per week.
3. Notify my supervisor if unable to report for regular assigned shift at least one (1) hour before the shift begins.
4. Keep all departmental data confidential and comply with the Texas Open Records Act. Events that occur within the Police Department are extremely confidential, and not to be discussed outside the Department.
5. Request assistance from staff member or volunteer coordinator if I have any questions pertaining to my duties or if I am unsure of a procedure.
6. Make suggestions! They are welcome as a means of improving our program.

The Tyler Police Department agrees to:

1. Provide volunteers with an orientation course as well as adequate job training.
2. Provide adequate space, working conditions and privileges as due paid staff members.
3. Provide a I.D. badge to the volunteer.
4. Review volunteer performance on a regular basis, keep account of volunteer hours, and provide a letter of recommendation for satisfactory performance when requested after completing six months of service.

I understand that I am a volunteer, and will receive no financial compensation, no special dispensation regarding traffic citations, legal counsel or City of Tyler Employee Benefits for my services. I understand I will receive no special consideration with regard to regular paid employment with the City of Tyler.

I understand my participation in the Tyler Police Department Volunteer Program may be terminated at any time for the good of the Department or the volunteer.

_____   _____
Volunteer   **Date**

_____   _____
Volunteer Coordinator   **Date**

8

From SWAT to Critical Incident Teams: The Evolution of Police Paramilitary Units

Robert W. Taylor
Stephanie M. Turner
Jodi Zerba

There are several controversial issues facing police forces today that employ highly specialized, paramilitary units (or what have been referred to as PPUs—Police Paramilitary Units in the literature). Most of these issues focus on complaints that police officers in these units are too aggressive, too heavily armed, and too "scary" for the general public (Macko, 1997). This discussion is an especially "hot" topic coming on the heals of high profile incidents such as Waco and Ruby Ridge and the utter debacle of using PPUs to secure the safety of individuals during these hostage and crisis situations. Further, a recent nationwide survey of 690 law enforcement agencies (serving cities with populations of 50,000 or more) revealed that 90% of American police departments have active police paramilitary units (Kraska & Kappeler, 1997). This is a dramatic increase from 60% in the 1980s. Kraska and Cubellis (1997) also found that PPUs are being used much more than ever before. Fostered by increased budgets, more violent and isolated criminal activity, and the "war on drugs," many police agencies now call on their PPUs to serve high-risk search warrants, conduct proactive and directed uniform patrol, enforce gang ordinances, *and* to act as the containment and tactical group during hostage and barricaded situations. Whereas PPUs were being mobilized just a few times a year, they are now being used extensively throughout the policing function. Kraska and Cubellis (1997) found that the lure to a more proactive paramilitary subculture fulfilled greater internal needs for the police:

> The techno-warrior garb, heavy weaponry, sophisticated technology, hyper-masculinity, and "real-work" functions are nothing less than intoxicating for

paramilitary unit participants and those who aspire to work in such units. (Kraska & Cubellis, 1997, p. 625)

By virtue of rational expansion and extension, police executives and managers then, often use these PPUs as a way to visibly show how their departments are "tough on crime and criminals." These are particularly troubling observations as most of modern law enforcement has recently adopted the tenets of community policing—a movement fostering partnerships between the police and the community. It involves a mutual commitment to crime prevention, police accountability, and the tailoring of police services to fit the needs of specific communities. At the heart of the movement, rests the concept of increasing communication and building mutual respect between the police and the community through eliminating the number of incidents of police use of force—especially in minority communities (Swanson, Territo, & Taylor, 2001).

While this debate is interesting, it only sets the stage for this paper. Our purpose is much more descriptive in nature, reflecting the historical development of police paramilitary units to their most sophisticated and logical end. We argue that today's PPUs are much more than their early SWAT counterparts and that their success in handling hostage and barricaded situations may be more dependent upon their ability to have well-trained negotiators, than expert marksmen. Today's PPUs must evolve, as they have in many jurisdictions, to highly sophisticated teams of individuals aimed at containing a situation and ending it—peacefully.

HISTORICAL DEVELOPMENT

In most cases, the general public refers to the police paramilitary unit as a SWAT Team—Special Weapons and Tactics Team. The acronym conjures up images of a camouflage-clad, heavily armed group of law enforcement officials who arrive at the scene of a situation too big to be handled by ordinary street officers. They quickly arrive in their specially marked van and quell the situation, more often than not with the precision of a highly skilled surgeon, extracting the "bad guy" with one far-away sniper shot or a simultaneous entry of five Uzi-armed officers rappelling from the building side. Such were the themes of movie and television scripts depicting a bigger than real life SWAT activity. In fact, during the 1970s, there was a television show called "SWAT" which depicted the activities of the Los Angeles SWAT team, and it is in Los Angeles that we see the first modern-day approach toward the development of a police paramilitary unit.

According to several authors, the first full-time SWAT team was indeed, established in Los Angeles, California in 1967 (Albanese & Mohandie, 1995; Gelles, Hatcher, Mohandie & Turner, 1998). The Los Angeles Police Department developed groups of officers to swiftly curtail hostage and barricaded incidents using military tactics. Officer John Nelson, with the LAPD, told an up-and-coming inspector Darryl Gates, about the idea of specialized teams of officers that were highly trained to deal with these types of incidents. The individuals would be trained to use specialized weapons and strategies to control and overcome an incident. Hence the birth of the Los Angeles Police Department's SWAT team formally developed in 1971 as a full-time military-oriented group to primarily deal with all escalating violent and protracted conflict incidents. Its stated duty was to assist and support patrol officers in handling violent incidents such as hostage and barricaded situations in a quick and efficient manner.

The composition of the first LAPD SWAT teams consisted of fifteen teams of four men. The agency chose volunteers from the patrol and other areas within the department. Each man (there were no women assigned to SWAT) had a military background and some type of specialized physical skill (e.g., excellent marksmanship, intimate knowledge of bombs and explosives, experience in rock climbing). Each team was trained on a monthly basis and was "called out" when a critical incident required their services. In most cases, patrol sergeants and mid-level managers had the capability to call-out the LAPD SWAT team when needed.

While certainly the first modern day SWAT team developed in Los Angeles under the direction of Darryl Gates, the first documented "specialized" weapons team was established in New York City over a hundred years earlier, in the mid-1880s. These teams were known as "strong-arm squads" (Ayoob, 1975). These "strong arm squads" were given nightsticks to protect themselves and effect an arrest when they dealt with deviant individuals. Their role was to arrest only the most "armed and dangerous offenders" under the most difficult of situations. This changed in the 1920s when the New York Police Department re-organized their "strong-arm squads" and renamed them Emergency Service Units (ESU). These units were armed with handguns, rifles, and submachine guns ostensibly to curtail the violence of organized crime during the Prohibition Era. In the 1960s, New York City created a "Stakeout Squad" where police sharpshooters or snipers were employed as support for high-risk surveillance situations. These individuals were given military training and as a result, became excellent marksmen. During the early 1970s, the police sharpshooters became part of the newly revised and reorganized, New York City ESUs. The NYPD Emergency Service Units have eventually evolved into the modern day Crisis Incident Team employing the combination of hostage negotiators and tactically skilled individuals. These teams have continued to expand their roles and are highly skilled in a variety of areas to include bombing and arson incident management, domestic and international terrorism, domestic crisis mediation, felony in-progress incidents and hostage situations.

As SWAT and specialized teams were evolving in Los Angeles and New York City, similar changes were occurring elsewhere. For instance, in August of 1966, Charles Whitman, a very disgruntled resident of Austin, Texas, stabbed and shot his mother to death then continued home and stabbed his wife to death. The next morning, he packed up some personal items along with an arsenal of long rifles, pistols and ammunition and climbed to the top of the University of Texas's Clock Tower on a personal suicide mission. He killed three individuals upon entering the clock tower and once he reached the top he continued to randomly shoot into the crowds of people on the ground below. In 96 minutes, Whitman shot and killed 15 people and wounded 31. Police officers on the ground were helpless with pistols that were virtually worthless in attempting to return fire at Whitman. Their weapons simply could not accurately cover the distance, and they were not able to confront Whitman. At one point, an airplane was flown near the tower in an attempt to gain a tactical advantage. The ploy did not work, and it was not until three officers rushed the tower under direct gunfire that Whitman was finally neutralized. The University of Texas Tower incident was one of the first major, protracted conflicts in a major U.S. city. It had a dramatic impact on police agencies throughout the country, as law enforcement began to realize their own incapability to handle such situations in a quick, safe and effective manner. No longer were SWAT teams viewed as necessities for *only* big cities, but now almost every jurisdiction could justify the development of such a team under the "what if scenario." The federal Law Enforcement Assistance Administration Technology Program was more than happy to assist in the development of these teams through the purchasing of equipment and other resources for local jurisdictions. It was not uncommon to see small cities purchase vast amounts of weaponry, ammunition, training, and even armored cars under this program.

The 1960s and 1970s were turbulent times for the United States. Riots marked the Civil Rights and Anti-Vietnam experience domestically, and several international incidents and trends (e.g., Middle East terrorism and hijacking, the Munich Olympic disaster in 1972, the Israeli Army raid at Entebbe) became commonplace in the news media. Most police agencies were quick to follow the lead of the Los Angeles Police Department and develop their own police paramilitary unit. At that time, the Los Angeles Police Department may arguably have been one of the most professional, well-respected and widely accepted leaders in American policing. Several law enforcement agencies developed SWAT Teams tailored after the LA model.

OPPOSING PHILOSOPHIES OF ASSAULT

The prevailing approach to handling hostage and barricaded situations was one of forceful assault. There were three primary methods used when police agencies confronted a hostage or barricaded incident:

1. They relied on the responding officer's verbal skills to influence the hostage taker or suspect in surrendering to police authorities.
2. They simply left the incident, unresolved, and hoped for the best. On more than one occasion, the police simply left the area during a domestic crisis in a veiled attempt that time would solve the issue.
3. They displayed a show of force by massing tactical SWAT teams in full view of the suspect hoping that intimidation would result in surrender. If surrender did not ensue, then a tactical assault followed.

This was clearly the predominant approach at Munich and during other international crisis wherein governments (and police agencies) were loathe to "negotiate with terrorists" (Gelles et al., 1998).

Time and again, the assault philosophy resulted in bloodshed and civil litigation. Due to the unpopularity of the "force only" approach afforded by SWAT teams, two New York Police Department officers developed an alternative. Lieutenant Frank Bolz and psychologist/police officer Dr. Harvey Schlossberg tested this alternative on January 19, 1973 at "John and Al's Sporting Goods" in a hostage situation that had developed from a failed robbery. Bolz and Schlossberg believed that they could talk the suspects into surrendering, given the one element of time. They opted for an extended, de-escalation of the situation versus a police assault of the building. It was successful in that no lives were lost and thus the practice of "hostage negotiation" was born.

In response to the failed tactical assaults and the civil litigation that followed, the FBI also developed hostage negotiation training at their academy in Quantico, Virginia in 1973. This was a huge leap toward acceptance from local agencies, in offering a model different than the

Los Angeles SWAT assault and force philosophy. McMains and Mullins (1996) asserted that with its emphasis on behavioral science knowledge and specialized teams, hostage negotiations represented an extension of the early 20th century reforms in policing. The use of negotiations was viewed as an attempt to professionalize the police, resulting in the restoration of the public's confidence in the police. Crisis negotiations results in the safe release of hostages, without injuries to anyone involved, 95% of the time. If a mental health consultant is employed, the success rate rises to 99%. These are certainly more acceptable figures than the 78–85% injury and/or death rate afforded by tactical assault (McMains & Mullis, 1996).

The use of hostage and crisis negotiations was also an attempt to squelch the onslaught of civil litigation that often followed unsuccessful tactical assaults from the use of SWAT teams. Post-incident litigation generated the landmark *Downs v. United States* (1975) decision. The Downs case evolved from an airplane hijacking incident. The FBI launched a tactical assault against the hostage taker, resulting in the deaths of the hostages. This incident was heavily scrutinized because the hostage taker had already released hostages prior to the tactical assault by police. The Supreme Court believed that there was a better suited alternative to tactical assault and that the goal of the FBI SWAT team should have been the safety and security of the hostages, not the capture and arrest of the suspect. After all, the negotiation team had successfully negotiated the release of one of the hostages and surrender of one of the hijackers previous to the police assault. Further negotiations would likely have secured the peaceful release of the remaining hostages. It is important to note the concept of negotiations in working with tactical teams under highly stressful protracted conflict and crisis situations such as hostage and barricaded incidents. Negotiations are *not* bargaining for position, but rather emotion management. It is an alternate resolution technique that attempts to de-escalate a crisis, moving the focus of the incident from emotion to logic. Taylor (1984) states, "The prevailing philosophy behind negotiations is that human life is the main concern—including the life of the hostage taker or suspect."

Thus, we contend that the beginning of SWAT teams in American policing led to the development of a tactical, assault alternative when faced with a crisis or protracted conflict situation. These SWAT teams emerged as a result of several ongoing developments during the 1960s and 1970s:

1. Highlighted violent criminal events that involved international terrorism and massive civil unrest in many parts of the world, including the United States;
2. A growing fear or perception of escalating violence during protracted conflicts that *might* occur in almost any jurisdiction, and the ancillary fear of police agencies having little capability to handle such an incident;
3. The ease of acquiring necessary equipment and training for such teams from the acquisition of federally sponsored programs;
4. The misconception of police agencies to view SWAT operations as a visible "get tough on crime" manifestation to their communities;
5. The error of many agencies to follow the Los Angeles Police Department model emphasizing a tactical assault-force option rather than a negotiation alternative expressed in the New York model.

At their best, SWAT teams in their early years, were very important and highly specialized units within their police community. They were called upon to undertake highly

control, barter and other concepts, tolls and devices in an attempt to ease the crisis, *not escalate it*. Assault tactics employed by the unit are used as a very last resort, and referred to as past police strategies of handling and dealing with situations that went awry. In turn, the concept of "apprehension of the suspect" is accomplished through the primary discipline of approaching each situation as a "life saving mission." The goal of the operation should be to accomplish each mission without injury or death to *any* person resulting from a tactical police intervention. Foremost in the mind of decision-makers and critical incident team members are the preservation of life and the safety of the general public.

CRITICAL INCIDENT TEAM COMPOSITION AND SELECTION

Essentially, two separate units comprise the Critical Incident Team: the negotiation unit and the tactical unit. Each unit requires a different set of personnel skill and development. A negotiator should be an individual who is emotionally mature and stable. He or she should be an excellent listener. Many people mistakenly believe that negotiators are good talkers; that they are persuasive speakers, but in actuality, it is the concept of listening that is most important. The negotiator must have patience, flexibility, and must be able to communicate to all kinds of people in a myriad of situations. Negotiators must be able to handle responsibility without having any authority. They must be comfortable with handling the awesome responsibility of establishing and maintaining communication with the hostage taker while lacking the authority to make decisions about demands and tactical actions. Good negotiators must be team players and should not be competitive with other team members. They must be physically and mentally fit, and a healthy lifestyle is a must for negotiators since they are often exposed to high levels of stress and long work hours.

Similarly, individuals on the tactical side must be team players. They are often called upon to wait long hours in the rain, sleet, snow, or hot sun, without exact knowledge of the situation. They must understand that their role is secondary, and that the probability of their use of deadly force will most likely be a rarity, if ever. They must profoundly understand that their role is to contain and control a specific geographical location, *not* to develop a tactical solution to a problem. Skill in marksmanship gives way to attitude and proper understanding of the desired role for the tactical unit. Arduous physical training and rope work (rappelling) focusing on tactical assault should be kept to a minimum, as such a requirement will rarely be utilized. In like manner, competitions such as the "Best in the West" which pit police tactical teams against one another should be discouraged. These types of competition focus on tactical solutions and strongly reinforce the aggressive, hypermasculine, and authoritarian personality traits often exhibited in the police subculture (Bittner, 1970; Manning, 1977; Van Maanen, 1978; Taylor, 1983). These competitions focus on officer endurance courses, "sniper" shooting, force entry assaults, combined weapons courses, shotgun courses, and live-fire team assaults. Teams from various police departments are then judged on the basis of their precision and timing (Taaffe, 2000). It is precisely these types of tactical assaults which are extremely high risk, placing officers, hostages and the general public in direct danger; and which must be avoided at all costs. By emphasizing the control and containment objectives and missions associated with the tactical unit, a department often eliminates the aggressive, "hypermasculine" attitude that characteristically described the SWAT teams of the past. Like negotiators, tactical unit personnel must also be mentally and physically fit.

Both negotiation and tactical personnel should apply for positions on the Critical Incident Team, as opposed to being drafted because management thinks he or she would be good in that position. An individual who is made to do a job, instead of choosing to do it, will most likely not do the job as well as someone whose desire and heart is in their work. The entire team suffers when someone is drafted or "forced" to work within the team environment.

Individual candidates should be interviewed by other team members and the critical incident commander for their ability to work in a team environment. Physical tests can also be used to ascertain physical fitness for duty while under the pressure and stress of actual working conditions. Once the interviews and tests are completed the candidates are ranked based on their total performance. Depending upon their ranking, the officer is sent to the police department psychologist for a complete psychological work-up. Final selection should be discussed openly between the team members and potential incident commanders, with highest regard given to the past working characteristics of the individual under review.

In selecting a negotiator to the team, the demographics of the area must be considered. The negotiation team should be somewhat representative of the area served. In an area where a good part of the community is Hispanic, the negotiation unit must have Spanish-speaking negotiators. In contrast to the "all-boy" SWAT teams of the past, female negotiators and officers should be important additions to the Critical Incident Team. However, selection should not be overly dependent on meeting demographic needs to the detriment of an individual candidate's personality and other qualifications.

Policy

One of the first steps in training for a member of the Critical Incident Team is to understand the mission of the team during an incident and the role of each unit. Critical to this understanding is a clear and existing policy that sets forth a supportive agency philosophy. McMains and Mullins (1996) suggest that the policy should be developed by those who know most about the process: the Critical Incident Team itself. After each incident, the policy should be reviewed for improvement. The comprehensiveness of the negotiation aspects as compared to the tactical aspects of the policy should be a clear indication of the department's attitude and philosophy about tactical force options and assault. They must be strongly discouraged by policy, in favor of a negotiation and de-escalation position. The policy should also be detailed, especially relating to command structure, to avoid potential problems in a crisis situation. Yet, flexible enough to allow for unusual circumstances.

A "critical incident" must be defined by policy as any event where protracted conflict is a potential. These will usually include suicidal and barricades subjects, hostage situations, and the like. The primary decision-maker during a critical incident must be clear to everyone at all times. Control is essential to guarantee a coordinated response that maximizes the saving of life as well as makes the most efficient and effective use of resources during an incident (McMains & Mullins, 1996).

Training

Much of the critical incident training is in crisis simulation formatting where persons, sometimes hired actors, act as the hostages and suspects. The Critical Incident Team then responds to such a crisis. Schultz and Sloan (1980) conducted a study on crisis simulations training where the participants, which included various civilian police agencies, military police agencies, and personnel from a hostage negotiation course, were told that the critical incident would be a routine drill. Unfortunately, they found that in almost all of the simulations, the overriding pattern that characterized the teams who where responding to the crises was an "action orientation." That is, the trainees were all heavily employed in finding a tactical solution involving direct assault or force while attempting to engage in effective negotiations. There was little trust in the effectiveness provided through time and negotiation strategies.

Traditionally, officers are trained to control situations, to be assertive, and to exhibit authority. This is the exact opposite of the role played as a negotiator, or as a tactical officer in a Critical Incident Team. Schultz and Sloan's study (1980) found a reliance on traditional police methodologies that employed aggressive and authoritative behavior. The negotiators were poor listeners, and tried to order the hostage takers to do as they wished. Courses on crisis negotiation, often times conducted by mental health professionals and other mediation experts, attempt to remedy these shortcomings immediately. Exercises and instructions are given in an attempt to effect an attitude change. Training exercises on active listening and less adversarial communications hope to achieve a shift in the communication focus and process. While the simulation training process has come a long way in the last 20 years, the subcultural traits of police officers, so colorfully depicted by Kraska and Cubellis (1997) still exist, and are powerful nemesis to effective negotiation strategies during a critical incident.

We suggest that training should focus on four areas:

Advanced formal education. Every negotiator should have post-secondary education emphasizing the recognition and handling of abnormal persons. Advanced classes in abnormal behavior centering on the psychopathic personality, alcoholism, drug addiction, domestic crisis and conflict, and domestic/international terrorism are especially important, as well as a variety of effective communication courses. Ideally, negotiators will have formal education that includes the study of mediation techniques aimed at easing the psychological and social motivations of human being under stressful conditions.

Hostage negotiation schools and seminars. All negotiators should attend at least one special training school on hostage negotiations and the critical incident management. These classes are offered a variety of times by the FBI Academy, various state police academies, local police academies, and private training organizations and institutes. Additionally, refresher courses should be attended on at least a bi-yearly basis.

Departmental training. The entire Critical Incident Team should refresh their skills constantly through intermittent role-playing. Departmental training should take the form of videotaped dramatizations and role-playing situations supplemented with training films that emphasize tactical procedures and negotiating strategies together. Additionally, the team should constantly critique their past performances and attempt to develop refined method for resolving critical incidents through teamwork.

Coordination. The entire team consisting of both tactical and negotiation units must train together. This is a critical factor in building organization, discipline, and teamwork with the negotiating strategy. A strong emphasis on trust and reliability within the team must be encouraged.

Culley (1974) reaffirms the need for self-assessment and debriefings of every significant crisis event that takes place. During such critiques, "Monday morning quarterbacking" and speculation are encouraged. From the situation under study, officers gain new insights and learn new techniques. Mirabello and Trudeau (1981) report that major hostage incidents are infrequent in most jurisdictions; therefore, they recommend that agencies that have a hostage (or critical) incident meet with others for the sharing of experiences on a regional basis. Such activity builds trust and awareness with other teams in a specific location, which may be very important in the unlikely event that a major protracted event (such as the siege on the Branch Davidian compound in Waco, Texas) occurs. A strong relationship should also be forged with local members of the FBI Hostage Rescue Team and Negotiating Team for similar purposes.

Unfortunately, Mirabello and Trudeau (1981) also report that it is not uncommon for a police department to train members of their respective Critical Incident Team unit and then assume that they will remain proficient as the result of everyday police work. Police departments must understand that critical incident training is an ongoing process; and that continued training is a necessity not a luxury for the success in a critical incident. We strongly advocate continued education and training for those involved in the Critical Incident Team format as discussed.

Team training should focus on the many roles used in the negotiation unit. Ideally, there are usually at least five separate roles, and sometimes even six, during a major crisis incident. Some individuals in the negotiation unit may be better qualified to perform in certain roles than others. However, special circumstances of a crisis, such as the race or gender of the hostage taker/barricaded subject may make a role change necessary. Training should teach each team member to function in each role. Each team member brings with him or her a background of experiences, training, education, and his or her personal strengths and weaknesses. Team members should learn to compensate for the known weaknesses of team members while learning to play on their particular strengths. Simulations training within the team will bring out the strengths and weaknesses of the team members.

Although some crisis incidents do not necessarily require that the tactical team be present, major critical incidents obviously do. Inter-team training is crucial for the successful resolution of a crisis incident. Cultivation of relations between the negotiations unit and the tactical unit is part of the focus of the inter-team training. Although each unit operates with a completely different focus, the team has the same goal: the peaceful resolution of the crisis incident. Without inter-team training, essential solidarity cannot be achieved. Schultz and Sloan (1980) state the there is a pervasive tendency for tactical officers to feel that their job, by its nature, does not require them to develop the skills of communications and psychological sensitivity. Negotiators may share similarly parochial feelings that they need not understand the basics of tactical control and containment. Each unit should be specialized in its particular field, but integration is essential for successful operation. Crisis and hostage negotiators have tactical duties and tactical members have negotiation duties. In the absence of inter-team training, misconceptions about the other unit are unavoidable. Once again, Schultz and Sloan (1980) note that negotiators are often perceived by the tactical unit as having the tendency to be soft on the suspect(s), while the negotiators may view their counterparts as being too willing to seek a tactical solution through force. In some cases, this lack of integration and trust has led to disaster. Lanceley (1999, p. 4.) agrees and asserts

> It is not enough to know and respect the role of the other crisis management components. It is essential that all components view themselves as part of a single, unified effort and convey only one message to the subject if the desired outcome is to be achieved.

Any discrepancies in the message communicated to the hostage taker or barricaded subject by the Critical Incident Team will often result in his or her distrust of the negotiator and, consequently, will hinder the crisis resolution.

Communications during the actual incident between the negotiation unit and the tactical unit must also be a training consideration. If the negotiation unit has a tactical liaison, then this person(s) needs to be familiar with tactical procedures and should have a good working relationship with the members of the tactical unit. The liaison should be able to communicate critical information obtained during negotiations to the tactical unit, to be used in the event of last resort—a tactical force solution. The liaison should also be able to communicate to the negotiation unit the needs of the tactical unit, specifically, the procedures for hostage release, and hopefully, the hostage taker surrender procedure. Any miscommunication in these procedures could be disastrous, possibly leading to the death of the hostages, hostage taker, or police personnel. The only conceivable way that the liaison would be versed in tactical procedures, and able to effectively communicate for the negotiation unit, is to regularly train with the tactical unit in a simulation environment.

Further, we suggest that training should include the entire Critical Incident Team comprised of the negotiation unit and the tactical unit, with incident command staff, executive police and city staff, the media, utility personnel, and any other outside agencies that will be involved in any potential critical incident. Swanson et al., (2001) assert that training for protracted conflicts must include top-level decision-makers as well as operational commanders and chiefs. Attorney Generals, governors, city managers, mayors, councilperson, and top police executives must be trained in coping with such conditions.

Unfortunately, the command staff in most departments is not adequately trained in the management of crisis incidents. There are few comprehensive training courses in this area. However, the FBI offers a course on hostage negotiations and tactics for commanders. This lack of understanding of the negotiation effort could lead to poor decision-making on the scene of a crisis incident. Although, a commander will not likely ever be negotiating for the team, he or she should understand the process. This is a critical element of effective decision-making during a protracted conflict—that the decision-maker can draw upon their knowledge of the resolution process coupled with the accurate and timely information provided to them from both tactical and negotiation units.

Although it is unlikely that the media, outlying agencies, and other support agencies are included in training sessions with the Critical Incident Team, such events are of immense value. Several cities have developed mock training events (usually focused on a natural disaster) which include all of the emergency responders such as fire, EMS, police, and hospital staff within a specific geographical region. While these events are very difficult to coordinate and plan, they offer a real-life simulation of the problems often associated with inter-departmental coordination and communication during a major crisis. An invaluable experience if the unthinkable type of event does occur in a specific jurisdiction.

COMMAND AND STRUCTURE

Several issues surround the involvement of the command staff overseeing the crisis incident. The location of management with relation to the incident speaks volumes about the independence of the Critical Incident Team from micro-management. McMains and

Mullins (1996) indicate that the location of the command post should be in close proximity to the entire team (negotiators, tactical unit, and representatives of other agencies), but not commingled. This will allow the on-scene, incident commander to monitor the negotiations without distracting the negotiators, to keep updated on intelligence, to coordinate tactical issues, to resolve potential disputes between team members, and to coordinate with other agencies.

Another command issue that warrants consideration is the use of non-management personnel as negotiators. There is an old negotiation maxim: "Negotiators don't command and commanders don't negotiate" (Lanceley, 1999). In any kind of critical incident, negotiators should never be the final authorities or portray themselves as the final authority. Before any deals or decision are made, the negotiator must confer with command to ensure uniformity of thought among all parties, and to make sure that the situation has not changed. On the other hand, commanders should never negotiate because if they become too intimately involved, no one has the authority to remove them from the negotiation process. McCrystle and Poland (1999) suggest the negotiator be a neutral third party in the process. If the negotiator is lower ranking, it allows for bargaining between the hostage taker (or barricaded suspect) and the commander. It is very difficult, if not impossible for someone to negotiate and manage an incident simultaneously and properly. The incident commander will have considerable difficulty building or maintaining rapport if he or she is not able to stay with the hostage taker and focus solely on the negotiation process.

The structure of command between units is an important consideration when determining levels of support for each. There is little reported on this matter in the literature, however, we suggest that each unit should have supervisors that are equally ranked and more importantly, who have an equal voice when conferring with the incident commander.

Of paramount importance is the negotiator's ability to make decisions based on the negotiation unit's assessment without pressure from command. Particularly if the negotiator believes he or she is making progress in the negotiation process, negotiations should continue until the negotiation unit (and supervisor) believes that peaceful success cannot be achieved with this method. Unfortunately, incident commanders sometimes press for a quick resolution due to unfavorable media coverage, public scrutiny, and staffing constraints, hence the need for ongoing, simulation training involving the incident commanders.

In an effort to assist the media, the entire Critical Incident Team must provide the departmental public relations officer with accurate and timely information. This should be the role of one member of the negotiation unit. If this does not occur, media coverage may become negative, simply because the media has little information to report to the public. In such instances and with the absence of information, rumors and stories have a tendency to be exaggerated and believed. Often time, the competency of the police department and, particularly, the negotiations unit is called in to question. Neither the public nor the media understands the concept of de-escalation. Slowing down, acting deliberately and prolonging the incident increases the likelihood of a peaceful resolution. The incident may have caused street closures or other inconveniences to the public that may have prompted negative public opinion on the handling of the incident by the police. Patrol staffing shortages created by the crisis incident may also tempt management to pressure the team for a quick resolution. Crisis incidents can last for many hours or even days, forcing an agency to pay extensive overtime. At these times, it is important to remember that the amount of overtime paid during a crisis event is most likely much less that the litigation expenses that may follow a rushed and faulty decision to use tactical assault and force where lives are lost. Again, significant training and awareness with the media can be invaluable previous to such an incident. The utilization of experts in discussing the negotiation process with the media during a crisis is also an effective strategy in remedying miscommunication and improving the police reputation with the public.

A department's adherence to non-management negotiations and a non-interference approach is an indication of healthy support and confidence in the negotiations process generally, and their Critical Incident Team specifically. An agency's support of a program can also be gauged by the number of personnel it assigns to the program. An agency's willingness to pay more to bring in outside professional help is another indication of program support. Over half of the police agencies with Critical Incident Teams utilizing the negotiation process have mental health consultants in some fashion, of which 88% are psychologists as opposed to psychiatrists, social workers, or others (Gelles et al., 1998). This practice is becoming more popular. Butler, Leitenberg and Fuselier (1993) found that the use of a mental health consultant to assess the suspect resulted in fewer hostage incidents leading to serious injury or death of a hostage, more negotiated surrenders, and fewer incidents in which the tactical force had to be used. Using a mental health consultant tends to keep the negotiation process professionally focused and informed.

However, the use of a mental health consultant has the potential for problems if the consultant is not versed in police matters. Greenstone (2000) asserts that a mental health consultant should be a police officer or at least experienced in the criminal justice field with an educational background focusing in the field of criminal justice. Many police officers will not initially trust a mental health consultant. This trust may be gained one case at a time. Again, most police officers are accustomed to traditional policing focusing on a more assertive approach. The mental health consultant is perceived as a "bleeding heart" who is too kind to and too trusting of the hostage taker/barricaded subject. Often time, the traditional police officer does not believe that the hostage taker/barricaded subject is deserving of such kind treatment and will not be open to the use of a mental health consultant, nor will he or she recognize the benefits of the consultant's expertise. The use of a mental health consultant can also create problems with hostage takers. Sometimes hostage takers will resent the mental health consultant as a negotiator because it implies that they are mentally ill. This resentment will hinder the negotiation process, therefore, flexibility in the use of such professionals is important.

An agency employing a mental health consultant portrays an image of openness to outside professional help congruent with the tenets of community policing. It demonstrates an effort to professionalize the police and reduces an agency's liability by taking measures to ensure that the best help is available to ensure the safety of all involved. Mental health consultants may also be used to debrief the Critical Incident Team officers and hostages involved in the crisis incident. Addressing the stress and psychological impact of the situation on those involved also improves the image of the agency.

TWO MODELS OF STRUCTURE: EAST COAST AND WEST COAST

While we advocate the concept of a merged team composed of both a negotiations unit and a tactical unit, this should not apply to the roles of individuals within the team itself. Each team member can only belong to one unit, and hence each function is clearly separated.

McMains and Mullins (1996) indicate that it is unrealistic to expect a person to spend hours negotiating and then prepare to physically take the suspect into custody. Likewise it is unrealistic to expect an officer to perform the role of a negotiator in one situation and then act as "point" on the assault unit in the next. Each role is significantly different and requires different training and skills. The separated concept was the original mode implemented in 1973 in the New York Police Department, and therefore is coined the "East Coast Model". In the separated concept, tactical members cannot become negotiators because it is not feasible that the person can switch these roles. In this arrangement, each team is specialized in its own practice. This is not to say that the two units (and functions) cannot act as one team or that they cannot train together. This is precisely the goal—to make two groups with two different functions (negotiation and tactical) to work in concert with each other as one team.

In the other model, commonly referred to as the "West Coast Model" exhibited by the Los Angeles Police Department, all negotiators are former tactical unit members. Proponents for this kind of team suggest that a negotiation member's familiarity with tactical operations fosters a more cooperative effort if negotiations are unsuccessful. They fear the negotiator without a tactical background will tip off the hostage taker of an impending assault, although there has never been a reported incident of such behavior. We suggest that this merging of functionality clouds the process toward a peaceful outcome during an incident, and is more the product of an historically aggressive tactical team traditionally associated with the LAPD. However, on a more positive note, Albanese and Mohandie (1995) report that the integrated negotiation/tactical process was slow to be accepted initially in Los Angeles, however that cooperation, rather than competition between the two functions was a natural outcome of this structural arrangement. It seems LAPD tactical officers felt more comfortable with negotiators who had once been part of their unit.

No matter which structure exists, most negotiation units consist of a primary negotiator, secondary negotiator, an intelligence officer, tactical liaison, and a mental health consultant. The only person actually negotiating is the primary negotiator. The other members serve as support for the primary negotiator. Members of the unit change roles with each incident depending on the circumstances. As an example, LAPD uses 22 of their 60 officers as negotiators. They also have seven on-call psychologists who rotate their consulting responsibility. In similar manner, most teams have at least eight individuals assigned to the tactical unit, with four to six officers on the entry or assault group and two officers in marksmanship roles. Most agencies have abandoned the classification of "sniper" on their teams. The word connotes a highly aggressive, offensive military action that is unpalatable to most civilian communities. As an option, Critical Incident Teams assign a minimum of two individuals to the role of "containment group." One individual may be trained as a marksman, while the other acts as an observer. Given the events of Ruby Ridge, it is safe to say that most police agencies should use police marksman as a very last resort.

There is little research concerning differences in effectiveness associated with the two different models of structure. However, it is important to re-emphasize that the current shift in police paramilitary units nationwide is to a more critical incident management orientation; with tactical and force options giving way to a strong negotiation strategy.

Most large police agencies designate a Critical Incident Team as full-time or on-call 24-hours a day. Obviously, this is not the case for all agencies, especially small ones, where regional, integrated teams may be more financially feasible.

MULTI-JURISDICTIONAL TEAMS

Most large police organizations have the resources and continuous need for a Critical Incident Team that can readily respond to incidents of escalating violence and protracted conflict. They have the financial and human resources. However, a small agency must weigh many factors before committing to such an undertaking.

In a critical incident, numerous officers will be needed to establish a perimeter, provide a sizable containment team, negotiators, as well as other support functions. In addition, a number of patrol officers will still be needed to respond to other calls for service while the incident unfolds. Some agencies depend on their counterparts in large metropolitan areas, or on state police organizations to assist during these crisis incidents. However, they are not always available. A recent concept gaining popularity is the multi-jurisdictional team comprised of personnel from a number of agencies that together can provide the resources needed to adequately respond to a critical incident (Green, 2001).

While the concept has merit, numerous hurdles need to be overcome when undertaking such a project. The fact that immediate response and containment is possible within a specific jurisdiction is one of many factors that will need to be used to convince agency administrators that such a cadre of officers is needed. Other obstacles that will have to be addressed are differences in command structure, training, equipment, and tactical capabilities of each participating officer and agency. Then too, considerable planning and legal agreement must be undertaken considering the ultimate liability, jurisdictional arrest powers, and worker's compensation issues that may arise during an incident. A unified agreement would need to be in place as to who would serve as ultimate commander of the team, and who would be in charge of the ultimate decisions regarding negotiation and tactical strategies (Green, 2001).

ALTERNATIVE METHODOLOGIES FOR SMALLER AGENCIES

However, there are alternatives to multi-jurisdictional Critical Incident Teams for smaller departments, those policing populations of less than 50,000 people. One such example may be to address the potential issues from a department-wide, macro perspective. The paradox still remains for smaller agencies as for larger ones: How do we strike a balance between reducing the paramilitary images of the police while still being prepared to isolate sophisticated and well-armed criminals during protracted conflict incidents with limited personnel and equipment resources? Some smaller agencies have attempted to meet the challenges of each side of the paradox.

Realizing that the size of a city has limited relevance to the potential for a violent critical incident, and that almost all jurisdictions have vulnerable locations such as daycare facilities, schools, sporting events, and commercial business, is an important first step in understanding the need for critical incident preparedness. Then too, no jurisdiction is immune to domestic violence that may also escalate to a protracted conflict. For those smaller departments attempting to develop a complete critical incident team, the experience has been fraught with much difficulty both philosophically and financially. Small departments embracing community-policing tenets have a very difficult time convincing their city council that they require significant resources to maintain a traditional SWAT team or a newly adopted Critical Incident Team. Such teams are typically of a paramilitary nature, armed with expensive automatic weaponry, requiring extensive and ongoing training. Even for the small, resource-rich department, problems in developing a team are numerous. First, team members often have an extended response time due to the need for call-in procedures. Second, training is difficult for the entire team due to schedule conflicts, officer attrition, and the virtual size of the agency. Finally, many departments opt only for a traditional paramilitary unit without the development of several hostage negotiators, essentially, the negotiation strategy often gives way to the force concept. The facts simply do not go away for small departments: It is a very time-consuming and expensive proposition to develop an exclusive critical incident team, even to participate in a multi-jurisdictional team. Gathering enough sworn personnel from on-duty and call-back resources to provide a secure perimeter should an incident occur would be difficult enough for most smaller departments, much less execute a full paramilitary and/or negotiation operation.

Maybe the worst of all plans for any agency rather small or large, is to develop a team and *not* provide the necessary training and hostage negotiation strategies fast becoming legally incumbent upon such units. Even worse, may the deployment of a poorly trained team that results in harm to officers and to the general citizenry. Were this to happen, the issues of vicarious liability surely would loom large.

The alternative is to provide new training and equipment to *all* officers within the small department. All officers, regardless of assignment, receive new training and equipment for proper responses to violent incidents that emphasize, once again, a negotiation versus a tactical solution. Considering that most incidents have resolution prior to the arrival of a Critical Incident Team, and that such an incident would necessarily involve the entire department anyway, it may be better to have everyone properly trained rather than just a few.

Premier training and weapons can be provided to each officer. Officers should be issued standardized, semi-automatic side arms (.40 caliber or 9 mm), as well as each marked vehicle should be equipped with a semi-automatic, tactical shotgun. Every officer receives significant training and becomes intimately knowledgeable of the shotgun's function and use. A mix of shotgun slugs and tactical 00 buckshot must be available to each unit and officer. These ordinance function well in short to moderate range engagements, as well as provide a very diverse capability to almost any critical incident. They are also preferred because they reduce some of the inherent dangers incurred in a suburban setting of discharging a .223 caliber or rifled 9 mm round that have ranges well over one mile. This training and equipment gives officers the resources to establish a perimeter until assistance arrives, or in exigent circumstances, make an immediate approach and intervention.

Standardized tactics of confinement and methodologies for critical incident approach and resolution, and negotiation strategies are then taught to all members of the department, not just a select few. All departmental members embrace the no-force negotiation strategy, and several individuals are called upon to take specialized training in negotiations. In-service training emphasizes this philosophy.

This overall shift allows each member of a small department to function in their primary role and to maintain the philosophical ideals often found in community policing. Should an incident occur, all members of the department respond and are knowledgeable of their important role knowing that resolution will most probably not include escalation to a force option but rather end peaceably through a negotiated strategy.

CONCLUSION

Many agencies are exploring new concepts concerning the use of Critical Incident Teams. Indeed, much research and attention is being given to the use of less lethal weapons. The concept involves the use of incapacitating gases and plastic or rubber bullets, and even bean-bag projectiles. The rapid demand for such equipment by law enforcement agencies has spurred development in the less lethal technology arena, creating a demand for numerous types of non-traditional weapons that provide alternatives to deadly force, particularly in protracted conflict situations. Such advancement in the development of these weapons systems combined with advanced training of Critical Incident Teams, offer credible alternatives to the use of lethal tactical force.

But for every step forward, there are often several back. Some departments are now using "Rapid Response Deployment" or "QUAD," which stands for Quick Action Deployment. The design of this concept is to train patrol officers in the principles and tactics of rapid deployment in critical incidents, especially reacting to incidents of school violence. In theory the concept has merit—train immediately responding officers to enter buildings where in-progress, life-threatening situations exist, or where loss of life to innocent victims is imminent. The concept argues that such action cannot wait for the Critical Incident Team to respond and police action must be taken immediately. This type of rhetoric plays directly to the aggressive subculture of traditional policing. No doubt that such a condition could present itself; however, the vast majority of incidents (including school shootings) are de-escalated by time. We suggest that the best methodology for responding patrol officers to a critical and violent incident, especially one involving potential hostages, is to take action only when appropriate and that their primary mission is to control and contain the scene. Let the negotiation strategy work!

Non-violent police responses to violent incidents demonstrate a desire to be first and foremost "peace" officers. It is an important shift from the traditional role of arrest and capture. The more modern role, indeed the community-policing role, suggests a focus on prevention and protection. We have attempted to document the shift from paramilitary, tactical and force-oriented SWAT teams to a more progressive, negotiation and containment strategy. The concept of de-escalation, crisis intervention, and negotiation *over* tactical assault and force during critical incidents is not only reflective of national trends, but also supports modern community-policing goals by reducing the number of police involved shootings.

RECOMMENDED READINGS

ALBANESE, M., & MOHANDIE, K. (1995). The development of the West Coast crisis negotiation model. *The Journal of Crisis Negotiations, 1*, 39–41.

GELLES, M.G., HATCHER, C., MOHANDIE, K., & TURNER, J. (1998). The role of the psychologist in crisis/hostage negotiations. *Behavioral Sciences and the Law, 16*, 455–472.

KRASKA, P.B., & KAPPELER, V.E. (1997). Militarizing American police: The rise and normalization of police paramilitary units. *Social Problems, 44*, 1–18.

LANCELEY, F.J. (1999). *On-scene guide for crisis negotiators*. Boca Raton, FL: CRC Press.

MCMAINS, M.J., & MULLINS, W.C. (1996). *Crisis negotiations. Managing critical incidents and hostage situations in law enforcement and corrections*. Cincinnati, OH: Anderson Publishing.

REFERENCES

ALBANESE, M., & MOHANDIE, K. (1995). The development of the West Coast crisis negotiation model. *The Journal of Crisis Negotiations, 1*, 39–41.

AYOOB, M.F. (1975). Special weapons and tactics: The New York City approach. *Law and Order, 3*, 56–60.

BITTNER, E. (1970). *The functions of police in modern society.* Chevy Chase, CA: National Clearinghouse for Mental Health.

BUTLER, W.M., LEITENBERG, H., & FUSELIER, G.D. (1993). Use of mental health professional consultants to police hostage negotiation teams. *Behavioral Sciences and the Law, 11*(2) pp. 213–221

CULLEY, J.A. (1974). Defusing human bombs – hostage negotiations. *FBI Law Enforcement Bulletin, 43*, 10–14.

GELLES, M.G., HATCHER, C., MOHANDIE, K., & TURNER, J. (1998). The role of the psychologist in crisis/hostage negotiations. *Behavioral Sciences and the Law, 16*, 455–472.

GREEN, D. (2001). Implementing a multi-jurisdictional SWAT team. *Law and Order, 1*, 68–72.

GREENSTONE, J.L. (2000). Personal interview at the Fort Worth, Texas Police Department, October 13, 2000.

KRASKA, P.B., & CUBELLIS, L.J. (1997). Militarizing mayberry and beyond: Making sense of American paramilitary policing. *Justice Quarterly, 14*, 607–629.

KRASKA, P.B., & KAPPELER, V.E. (1997). Militarizing American police: The rise and normalization of police paramilitary units. *Social Problems, 44*, 1–18.

LANCELEY, F.J. (1999). *On-scene guide for crisis negotiators.* Boca Raton, FL: CRC Press.

MACKO, S. (1997). SWAT: Is it being used too much? *Emergency Net New Service, 1*(196), 1–4.

MANNING, P. (1977). *The police: The social organization of policing.* Cambridge, MA: MIT Press.

McCRYSTLE, M.J., & POLAND, J.M. (1999). *Practical, tactical, and legal perspectives of terrorism and hostage-taking.* Lewiston, NY: The Edwin Mellen Press.

McMAINS, M.J., & MULLINS, W.C. (1996). *Crisis negotiations. Managing critical incidents and hostage situations in law enforcement and corrections.* Cincinnati, OH: Anderson Publishing.

MIRABELLO, R., & TRUDEAU, J. (1981). Managing hostage negotiations. An analysis of twenty-nine incidents. *The Police Chief, 48*, 45–47.

SAFFLE, J.L. (2000). Emergency plans. *Oklahoma Department of Corrections* OP-050401, pp. 1–53.

SCHULTZ Jr., R.H., & SLOAN, S. (1980). *Responding to the terrorist threat. Security and crisis management.* Elmsford, NY: Pergamon Press.

SWANSON, C.R., TERRITO, L., & TAYLOR, R.W. (2001). *Police administration: Structures, processes, and behavior.* Upper Saddle River, NJ: Prentice-Hall.

TAAFFE, L. (2000). Competition sharpens S.W.A.T. teams. *Los Altos Town Crier,* at www.losaltosonline.com/latc/arch/2000/38/News/2swat.html.

TAYLOR, R.W. (1983). Critical observations on the police subculture. *Law Enforcement News, IX*, 3–5. 4.

——. (1984). Hostage and crisis negotiation procedures: Assessing police liability. In L. TERRITO (Ed.), *Police Civil Liability.* Columbia, MD: Hanrow Press.

VAN MAANEN, J. (1978). The Asshole. In P.K. MANNING, & J. VAN MAANEN (Eds.), *Policing: A View from the Street.* Chicago: Goodyear.

9

Specialized Drug Enforcement Units: Strategies for Local Police Departments

David Olson

INTRODUCTION

The purpose of this chapter is to examine the role that specialized drug enforcement units play in the array of strategies used by police departments in their drug control efforts. In doing so, the prevalence, characteristics, role, evolution, and impact which specialized drug enforcement units have across jurisdictions in which they operate will be examined from both national and state perspectives; the latter through assessment of selected, specialized drug enforcement units operating in Illinois.

For purposes of this chapter, specialized drug enforcement units can take two forms. First are units established within individual local police departments and consisting of full-time officers who focus their efforts specifically on drug enforcement. The second type of specialized drug enforcement unit to be examined in this chapter is referred to generally as a multi-jurisdictional, or multi-agency, drug task force. These drug task forces are generally defined as units that include (a) full-time officers, (b) from a variety of different law enforcement agencies, (c) within a specific geographic region, (d) that conduct drug investigations and drug enforcement activities, (e) across a geographic region that spans individual departmental jurisdiction. Although the overall prevalence of these two types of specialized drug enforcement units is relatively low nationally, as will be seen later, they play a critical role by ensuring that the full spectrum of the drug market is targeted by law enforcement. However, to fully understand the important role that these specialized drug enforcement units play in the United State's drug control strategy, it is important first to understand certain aspects of the illegal drug market and to appreciate the full continuum of law enforcement effort related to drug control measures.

BRIEF HISTORY OF THE PROBLEM

Although the enforcement of drug laws in the United States dates back to the passage of the Harrison Narcotics Act in 1914, during the past two decades, unprecedented resources and attention have been focused on drug control efforts in the United States. As a result of sweeping legislation and increased public concern, dramatic changes have occurred in response to drug-law violations by the criminal justice system. Record numbers of people have been arrested, prosecuted and incarcerated for drug-law violations, as well as admitted to substance abuse treatment programs. Between the mid-1980s and late 1990s, the number of arrests made by law enforcement agencies in the United States for drug offenses more than doubled, from approximately 811,000 in 1985 to more than 1.5 million by 2000 (Federal Bureau of Investigation, 2001). Dramatic increases in the numbers of persons incarcerated and the proportion of persons imprisoned for drug offenses *vis-à-vis* other offenses also occurred during that period. By the end of 2000, more than a quarter million adults were incarcerated in state or federal prisons for drug offenses (Office of National Drug Control Policy, 2000), and more than 900,000 adults convicted of a drug offense were on probation in the United States, accounting for one out of every four probationers nationally (Bureau of Justice Statistics, 2001b). Finally, during 1998, more than 1.1 million people were in some type of drug treatment program (National Drug Control Policy, 2000). Most important, much of the increase in arrests occurred during a period when drug use among the general population, as measured through the National Household Survey on Drug Abuse, was decreasing. For example, between 1985 and 1992, drug arrests in the United States increased 31% (Federal Bureau of Investigation, 2001), but drug use among the general population decreased almost 50%, from an estimated 23.3 million current users in 1985 to 12 million in 1992 (Office of National Drug Control Policy, 2000). From these data, two points arise. First, even during a period when overall drug use among the general population in the United States was decreasing, the ability of law enforcement agencies to increase the number of people they arrested for drug law violations was not inhibited. The second point, which has become quite obvious, is that while law enforcement appears to have an ability to make more and more arrests for drug-law violations, the rest of the justice system has struggled under the weight of the ever increasing caseloads and populations of these offenders who need to be processed, punished, and/or rehabilitated. Specifically, some have suggested that "prosecutors, judges, and corrections officials are also likely to adapt to a sudden burst of not-very-exciting retain drug cases with cheap plea bargains, time-served sentences, and early release policies" (Kleiman & Smith, 1990, p. 87). Despite the increased focus on drug-law violators by police departments throughout the United States, however, what constitutes drug enforcement in the minds of many people is skewed toward stereotypes about drug offenders and drug arrests made by law enforcement agencies. The next section will provide information to clarify these dimensions of drug enforcement, such as who is responsible for different functions, and how the characteristics and nature of illegal drug markets shape their responses.

POLICE, U.S. DRUG CONTROL STRATEGY, AND DRUG MARKETS

Drug control efforts in the United States are typically categorized as being demand reduction or supply reduction in focus. Demand reduction, usually takes the form of prevention efforts (e.g., Drug Abuse Resistance Education (DARE) and various public service announcements) or substance abuse treatment programs. Clearly, law enforcement plays an important role in the provision of some demand-reduction/prevention programs, such as DARE.

Traditional supply reduction strategies range from international efforts, such as working directly with source countries to reduce production and cultivation, to interdiction efforts at U.S. ports and across U.S. borders, and to street-level enforcement targeting those selling illegal drugs to consumers (see e.g., www.whitehousedrugpolicy.gov). Federal agencies, such as the U.S. State Department, U.S. Drug Enforcement Administration (DEA), U.S. Customs Service, the U.S. Border Patrol, and the Federal Bureau of Investigation are involved in many national supply reduction efforts; particularly those related to international and interstate trafficking. When drugs are smuggled into the country and make it to intended points of sale—the street—much of the work related to interruption of supply lines falls squarely upon the shoulders of local law enforcement agencies. Thus, police in the United States play a role in both demand and supply reduction, although it is clear that different law enforcement agencies target and are responsible for different dimensions of these drug control strategies.

As Table 9-1 demonstrates, the structure of both the drug market and the law enforcement response is hierarchical. The leaders of the international drug cartels, the drug "kingpins," are clearly beyond the reach of local law enforcement agencies, and, in fact, seldom step foot on U.S. soil. The next rung down the drug distribution ladder is occupied by those involved in the importation or trafficking of drugs into the United States, and the subsequent distribution of large quantities of drugs throughout the country to mid-level wholesalers and dealers operating within specific cities or metropolitan areas. The producers, importers, and interstate traffickers are the proper focus of DEA, FBI, and other federal agencies (e.g., U.S. Customs at the border).

It is, however, the upper- and mid-level wholesalers and dealers to whom the traffickers supply drugs that often fall through the cracks between traditional federal priorities and local law enforcement abilities—the latter tending most frequently to identify visible, street-level dealers and drug possessors (users). It is to address the gap between federal and local police targets where the critical role of specialized drug units becomes apparent. This is not to say that local efforts always focus on the lowest rung of the drug distribution network or that specialized units focus exclusively on mid-level distributors; nor is it to say that some multi-jurisdictional drug task forces may be involved in investigations involving interstate traffickers in cooperation with the DEA or FBI.

Due to the nature of the drug market and the resources available to law enforcement agencies across the different levels of government, it is no surprise that the majority of arrests for drug-law violations are the result of local police efforts, and typically involve relatively minor (e.g., possession) offenses (Federal Bureau of Investigation, 2001). In reality, the largest groups of drug-law violators are the consumers of illegal drugs, and these are also the individuals most frequently arrested by law enforcement agencies. Of the more than 1.5 million arrests for drug-law violations in 2000, about 80%—a proportion which has remained fairly consistent since the early 1980s—were for drug possession (Federal Bureau of Investigation, 2001). In terms of the employees of the illegal drug industry, the largest in number and most visible to law enforcement, are the low-level street sellers. Local police departments in the United States made approximately 300,000 arrests for the sale/manufacture of illegal drugs during 2000 (Federal Bureau of Investigation, 2001). To put these local

TABLE 9-1   Hierarchy of the Illegal Drug Market and the Role of Specific Law Enforcement Agencies

Level of drug market	Place of activities	Law enforcement targeting them	Specialized drug unit/task force
Producer	Foreign countries	DEA, State Department	
Trafficker-importer into U.S.	U.S. ports and Borders	DEA, Customs, FBI	
Trafficker-interstate distribution	Transshipment points throughout U.S. (Large U.S. cities)	DEA, FBI	
Upper-level dealers	Large, but specific regional areas	Maybe DEA, FBI, or local police	
Mid-level dealers	Smaller, but specific regional areas	Maybe local police	Primary targets of specialized drug units
Street-level sellers	Specific neighborhood or out of specific residence	Local police	Secondary targets of specialized drug units
Low-level distributors/ assistants	Specific neighborhood or out of specific residence	Local police	
Consumers	Cars, homes, businesses, public places	Local police	

drug arrests into perspective, there were just over 40,000 domestic drug arrests made by the DEA during that same year (www.usdoj.gov/dea/stats/drugstats.htm). Thus, despite the perceptions by many regarding what constitutes "drug enforcement," the reality, for the most part, is that drug enforcement, or the identification and apprehension of drug-law violators by police departments, typically involve "low-level" possessors and sellers of drugs identified by beat-level police officers. This is not to say, however, that such enforcement activities are unimportant. Arresting drug possessors can produce specific and, sometimes, general deterrence for many users; it can potentially result in their referral or sentencing to substance abuse treatment. In the long run, this could have a positive impact on the demand for illegal drugs.

Exacerbating the problem of being able effectively to identify those supplying drugs to the street-level sellers has been the increased role that organized street-gangs have played in local drug markets. In what Johnson, Williams, Dei, and Sanabria (1990, p. 20) described as the "rise of vertically controlled selling organizations." the drug markets that evolved during the 1980s were much more likely to involve individuals from one criminal organization at several levels of the drug supply network. The street-level sellers have formal gang affiliations with those above them in the drug chain, which has resulted in a decrease in the willingness of these low-level sellers to provide the police with information in return for charge reductions. Obviously, this evolution has limited the ability of traditional law enforcement approaches to effectively penetrate the higher-levels of the organization, or work cases up the chain. The next section will examine what specifically law enforcement agencies can do to increase their odds of arresting drug-law violators, including the utilization of specialized drug enforcement units.

DRUG CONTROL STRATEGIES EMPLOYED BY LOCAL POLICE DEPARTMENTS

A great deal has been written over the past 20 years regarding the role of law enforcement in the United States' drug control strategy, which, as noted before, is usually categorized as either "supply" or "demand" reduction in focus. Kleiman and Smith (1990) analyzed and described law enforcement approaches to drug control efforts that also provide a basis for introducing the role and potential impact of specialized drug enforcement units. They described the "strategic bundles" which law enforcement agencies can develop and adopt for the purpose of reducing drug supply and demand. These strategies include:

1. Getting Mr. Big: High-level enforcement,
2. Sweeping the streets: Retail-level enforcement,
3. Concentrating on one market: Focused crackdowns,
4. Suppressing gang activity,
5. Controlling user crime, and
6. Protecting the youth (Kleiman & Smith, 1990, pp. 82–96).

The first three approaches can be considered more traditional, direct drug enforcement; whereas the latter three are intended to address some of the underlying problems associated with illegal markets, criminal behavior, and prevention strategies. "Getting Mr. Big" would tend to be associated with the strategy of targeting those at the higher levels of the drug distribution network, although who Mr. Big is can be relative. To the DEA or FBI, Mr. Big would have to be someone involved in the importation or distribution of many thousands, perhaps millions of dollars worth of illegal drugs. To a local police department, Mr. Big could be someone selling a few thousand dollars worth of cocaine or heroin. In fact, in Illinois, the sale of 15 grams of cocaine or heroin, which currently has a street value between $1500 and $2000, is an offense subject to a mandatory minimum of six years in prison. The important point is that the purpose behind getting Mr. Big, whomever that may be, is intended to have a negative impact on the local drug market by disrupting supply and driving up costs, albeit both for a limited amount of time.

The other two "traditional" policing strategies for drug control, "Sweeping the streets: Retail-level enforcement" and "Concentrating on one market: Focused crack-downs," which have been widely adopted in many cities in the United States, fueled much of the increase in drug arrests during the mid- to late 1980s. Implicit in these strategies is the development of a focused effort to achieve the goals of supply or demand reduction, although some of these efforts can be achieved without the development of a specialized drug unit or task force. For example, efforts to "sweep the streets" or "focused crackdowns" can be put into action through a change in general departmental policies. Officers can be directed to "focus" on specific neighborhoods (e.g., street-corners or parking lots where drug sales or use is suspected) or specific activities (e.g., visible drug sales), and rewarded for making drug arrests. Changes in patrol officer deployment so that resources are con-centrated in areas where illegal drug use and sales are considered to be prevalent, clearly can be employed without the need for specialized units.

Finally, additional training provided to patrol officers on what signs to look for during patrol or traffic stops to identify possible drug offenses (i.e., heightened aware-ness) has also been widely used and has contributed to the increase in drug arrests in the United States since the mid-1980s. However, the impact that these efforts have on retail drug markets is not well understood. For example, some have observed that the "obvi-ous problem with focused enforcement activity is the displacement of illicit transactions from one neighborhood to another" (Kleiman & Smith, 1990, p. 89). In a thorough review of research regarding the effectiveness of drug control efforts in the United States, the National Research Council (2001, p. 177) concluded that "little is known about the effectiveness of law enforcement operations against retail drug markets." On the other hand, Sherman et al. (1998) found evidence that a variety of police practices, including directed patrols and proactive arrests, reduced certain types of crime in the targeted areas.

However, what sweeps and crackdowns generally produce are large numbers of arrests for relatively minor offenses. With the crush of drug cases being funneled through the criminal courts, many of these cases wind up being diverted or dismissed; or, if prose-cuted and convicted, resulted in probated sentences. In some communities this may lead to an even greater dissatisfaction with the criminal justice system because the individuals who were selling drugs in front of concerned citizens' homes are often back in hardly any time at all. In the end, the available strategies or "strategy sets" that a local police department may adopt are influenced by numerous factors, which include: (1) nature of the drug prob-lem within a community, (2) available resources, (3) public demands, and (4) views of pol-icy makers and practitioners as to the cost-effectiveness of these law enforcement approaches, along with the cost-effectiveness of non-law enforcement strategies (e.g., pre-vention or treatment).

Obviously, resources will dictate which of these strategies can be adopted, as well as the diversity and scope of the efforts. Historically, only larger law enforcement agen-cies have had the resources and staff to support the operation of specialized units of any kind, and the same is true when it comes to specialized drug units or participation in drug task forces. Another factor influencing the approach a local police department may take regarding drug control initiatives is the extent and nature of drug abuse in the community served, that is, the nature of the illegal drug market. For example, large urban population centers, such as New York, Miami, San Diego, and Chicago have street-corner drug sales occurring in particular neighborhoods, and serve as ports of entry and transshipment points for illegal drugs ultimately distributed throughout the country. Thus, drug enforcement strategies in these jurisdictions usually include both supply interdiction efforts and street-level enforcement. On the other hand, smaller, suburban or rural communities are usually not faced with the issues of large drug shipments moving through their jurisdiction, or even open-air drug sales; rather, less visible drug sales and use. However, this is not necessarily a universal rule. For example, there are several drug corridors connecting the Mexico–U.S. border area in South Texas with supply routes to the North and Northeast that pass through many small towns and rural counties. These differences, along with a variety of other factors, influence and shape the strategies adopted and supported across different law enforcement jurisdictions.

It is here that we come to the critical role specialized drug enforcement units play in the portfolio of drug enforcement strategies available to local law enforcement agencies. In order for law enforcement agencies to make "better" drug arrests (e.g., those which will be prosecuted, rather than dismissed, and actually result in periods of incarceration for those involved in drug sales), it is necessary to employ a more formal, targeted effort. To do so, there is a need to have experienced officers dedicated to conducting the types of proactive investigations that will effectively identify the true "mid- to upper-level" dealers supplying drugs to cities and larger geographic regions. These officers can be either part of a specialized unit within a police department or members of a multi-jurisdictional drug task force.

DRUG ENFORCEMENT UNITS WITHIN LOCAL POLICE DEPARTMENTS

One of the most comprehensive national surveys done to gauge the prevalence of drug units and drug task force participation among local law enforcement agencies in the United States was conducted by the U.S. Department of Justice's Bureau of Justice Statistics through its Law Enforcement Management and Administration Survey (LEMAS), which has been conducted periodically during the 1990s. The degree to which local police departments operate their own specialized drug units, as well as the level of participation in multi-jurisdictional drug task forces was estimated as a result of this survey. With respect to "in-house" drug units, about 15% of all local police departments in the United States operated a drug unit with at least one full-time officer assigned during 1997, and the average number of full-time officers assigned to these units was six (Reaves & Goldberg, 2000, p. 10). The size of the jurisdiction served, whether or not local police departments operated their own drug units, and the size of these units are shown in Figure 9-1. Generally speaking, the larger the jurisdiction, the more likely the police department is to operate its own specialized drug unit, and the larger these units were. For example, more than three-quarters of departments serving a population of 100,000 or more had their own drug unit, compared to less than one-quarter of those departments serving fewer than 25,000 residents. All police departments serving populations in excess of 1,000,000 operated their own drug enforcement unit, and employed an average of 292 officers per unit. Among the small percentage of police departments serving populations of fewer than 25,000 residents, the average number of full-time officers assigned to these units was two.

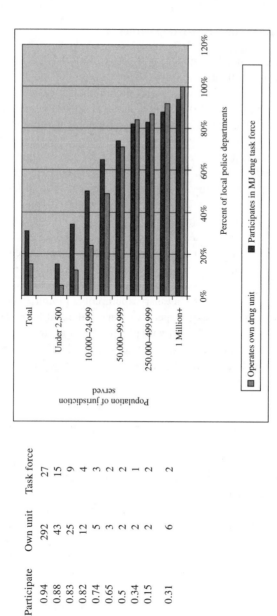

	Operates own drug unit	Participate	Own unit	Task force
1 Million+	1	0.94	292	27
500,000–999,999	0.92	0.88	43	15
250,000–499,999	0.87	0.83	25	9
100,000–249,999	0.84	0.82	12	4
50,000–99,999	0.71	0.74	5	3
25,000–49,999	0.49	0.65	3	2
10,000–24,999	0.24	0.5	2	2
2,500–9,999	0.12	0.34	2	1
Under 2,500	0.05	0.15	2	2
Total	0.15	0.31	6	2

FIGURE 9-1 Prevalence of in-house drug units and participation in multi-agency task forces.

PARTICIPATION IN MULTI-JURISDICTIONAL TASK FORCES

Participation in multi-jurisdictional task forces (MJTFs) followed a somewhat similar pattern to in-house drug unit operations. Departments serving larger populations tended to be more likely than smaller jurisdictions to participate. However, participation in drug task forces was consistently higher across the smaller to medium sized jurisdictions (those with populations under 50,000) than was the maintenance of individual drug units within departments. In larger departments, approximately 80% participated in a drug task force and operated their own drug unit as well. Another pattern evident in task force participation was that the number of full-time officers assigned to drug task forces tended to be less than the number assigned to in-house drug units. When participation in MJTFs was examined, almost one-third of all police departments had one or more officers assigned to these units, with participation rates varying across police departments serving jurisdictions of different sizes. In general, smaller jurisdictions (those serving populations of less than 50,000 residents) were more likely to assign officers to an MJTF than they were to operate a drug unit within their own department. For example, among those departments serving 25,000–49,999 residents, about one-half had their own drug unit, but two-thirds participated in a drug task force. Even among some of the smallest police jurisdictions in the country, those serving fewer than 25,000 residents, there was a 15–50% participation rate in drug task forces. Interestingly, it appears that during the 1990s there was a general decline in the prevalence of participation in multi-jurisdictional drug task forces by local police departments. Specifically, in 1990, 51% of all local police departments participated in a drug task force (Bureau of Justice Statistics, 1992), compared to 31% in 1997 (Reaves & Goldberg, 2000), and decreases were seen across almost all jurisdiction types (e.g., size of population served).

Assigning personnel to a drug task force, however, is only one way in which local departments can participate or receive the benefits of a task force's operations. For some small departments, it is still difficult to contribute personnel to a task force, since they may have only a few full-time officers at the outset. Thus, some task forces allow local departments to participate and receive the services of the task force, by making "in-kind" contributions, such as cash, equipment, and office space. One of the clear benefits of multi-agency drug task forces to smaller police departments that cannot operate their own drug units, therefore, is access to resources capable of conducting more comprehensive, covert, and/or long-term investigations.

Much involvement in drug task forces can be attributed to a considerable effort by the federal government to promote the development of these types of specialized drug units by providing funds needed to stimulate and facilitate involvement by agencies large and small in these efforts. Recognizing the limitations to traditional local drug control efforts, the federal government, through the State and Local Law Enforcement Assistance Act, and the Anti-Drug Abuse Acts of 1986 and 1988, provided grant money to states with a priority that these funds be used to establish and expand multi-jurisdictional drug task forces. With the grant funds authorized through these acts, there was increased emphasis and financial support directed at development of regional multi-jurisdictional drug task forces across the United States. Within the provisions of the Anti-Drug Abuse Act, also referred to as the Edward Byrne Memorial Block Grant Program, Congress authorized the distribution of more than $400 million annually in formula block-grant funds to state and local units of governments through State Administrative Agencies (SAA).

In general, these funds were to supplement existing drug enforcement efforts and had to be spent within specific purpose areas identified in the authorizing legislation, one of which was the creation and/or support of multi-jurisdictional drug task forces. By design and definition, they were intended to increase the capacity of local law enforcement agencies in targeting drug-law violators through increased information sharing, coordination of efforts, and having more time and resources to "work cases" up the distribution chain. Among the purpose areas that could be funded with the block-grant monies, multi-jurisdictional drug task forces were, and continue to be, very popular. Most SAAs allocated a considerable portion of their block-grant funds toward multi-jurisdictional drug task forces. For example, during federal fiscal years 1989 through 1994, more than $700 million of these federal block-grant funds were allocated to multi-jurisdictional task force efforts, accounting for 40% of the total block-grant distributions (Dunworth, Haynes, & Saiger, 1997). Between 1986 and 1993 alone, more than 700 drug task forces were formed with federal assistance (Coldren, 1993), and by 1998 more than 1000 drug task forces were in operation in the United States (Dunworth et al., 1997). It is estimated that in 1997 more than 6200 local police officers were assigned full-time to these drug task forces (Reaves & Goldberg, 2000), almost twice as many as the roughly 3300 agents employed by DEA (Bureau of Justice Statistics, 2001a). However, despite their popularity and pervasiveness, little research has been conducted to gauge the impact or effectiveness of these drug task forces. While there have been numerous analyses of task force characteristics, both nationally (Bureau of Justice Assistance, 1996, 1997) and within individual states (e.g., Draper, 1990; Farabaugh, 1990; Gilsinan & Domahidy, 1991), little has been done to fully understand the role these task forces play in specific regions, or how their efforts compare to traditional local enforcement activities. Also, little has been done to gauge, or assess the impact of these task forces. While these are difficult questions to answer given the problematic nature of measuring the drug problem, there are some methods that can be employed to begin answering basic questions regarding the efficacy of these units. In the following section, a specific methodology to evaluate drug task force operations in one state will be presented and the results discussed.

MULTI-JURISDICTIONAL DRUG TASK FORCES: THE ILLINOIS EXPERIENCE

Since the early 1970s, numerous local law enforcement agencies in Illinois have participated in MJTFs designed to identify and apprehend persons involved at the mid-levels of the illegal drug distribution network. Through shared resources and information, these task forces conduct overt and covert investigations within specific geographic regions covered by participating agencies. During 1999, there were 21 separate multi-jurisdictional drug task forces operating throughout most of Illinois.

Measuring the Performance of Illinois Drug Task Forces

Traditional performance measurements used in the field of policing are often times of limited value when gauging the operations of multi-jurisdictional drug task forces. First, unlike many police operations which tend to be reactive (e.g., calls for service, offenses reported to the police), drug task force activities are usually more proactive, "producing" arrests and drug seizures. There are a number of empirical questions that reveal the role and potential impact these task forces have in regional drug enforcement, including:

1. How do the numbers of drug arrests made by the task force compare with the numbers of drug arrests made by non-task force officers of local police departments participating in the task force?
2. In what ways, if any, do the drug offenses/offenders encountered by multi-jurisdictional differ from drug arrests by officers from local police departments participating in the task force?
3. Are task force arrests more likely to result in successful prosecutions and incarceration than similar arrests made by officers in local participating police departments?
4. Do these comparisons produce useful typologies of drug enforcement units?

The basic questions noted above regarding assessment of drug task forces raise additional questions regarding the appropriate bases of comparison and expose limitations with existing data. For example, and from both practical and political perspectives, who "gets credit" for arrests made by multi-agency drug task forces is problematic. Since arrests made by these task forces are the result of investigations involving the resources of numerous local governmental bodies, with individual mayors, police chiefs, sheriffs, and prosecutors involved, there are lots of people wanting recognition for their participation, particularly in high-profile or "big bust" cases. How task forces handle this can vary considerably. Task forces operating almost exclusively through covert means usually prefer that local police actually serve the warrants and make the arrests. This keeps involvement of the drug task force out of the public's and offenders' eyes. Other task forces may want more visibility and public awareness regarding their activities and efforts. In this case, when arrests are made, considerable press is sought to make it clear that the arrest was the product of an investigation by _____ Task Force (fill in the name/title).

From a research standpoint, how arrests by drug task forces are tallied and reported through Uniform Crime Reporting (UCR) statistics is another question. Since participating local police departments serving warrants may make many of the actual arrests resulting from drug task force investigations, drug task force cases can get lost in the reporting of the drug arrest by a local police department. Or worse, one arrest could be reported through the UCR by each of the local law enforcement agencies participating in the drug task force, resulting in a dramatic overcounting of these arrests. Another issue has to do with the reporting of drug task forces arrests through state criminal history record systems. In most instances, included on offender criminal history record is a list of previous arrests, which usually includes the offense and the arresting agency. However, given that task force arrestees are frequently processed (e.g., booked or fingerprinted) through local police departments, often times the local agency which processes the offender shows up as the arresting agency. The problem with this is that the fact that an individual was targeted and arrested by a multi-jurisdictional task force may be useful to criminal history record users (e.g., this individual may be involved in dealing at a relatively high level), but would be missed due to the reporting procedures.

Potential Typologies of Drug Task Forces

Although referred to generically, multi-jurisdictional drug task forces are not necessarily homogenous in terms of their organization, their operations, or the role they play in regional drug enforcement strategies. As noted by Coldren (1993), drug task forces can be organized or considered multi-jurisdictional in a number of ways. These organizational approaches include vertical, where agencies from numerous levels of government (e.g., federal, state, county and municipal police) participate and focus their efforts in a specific targeted area; or horizontal, where agencies from the same level of government collaborate to address specific aspects of the drug problem across their jurisdictions (e.g., municipal police departments which may be contiguous to one another) (see Phillips & Orvis, 1999, for a discussion of intergovernmental issues surrounding task forces). Another organizational structure includes task forces that are comprised of not only police agencies, but other criminal justice agencies as well, such as prosecutors, social service providers, drug treatment, or child welfare agencies (see Phillips, 1999, for a discussion of the organizational structure of task forces). Not uncommon are task forces that combine all of these organizational structures, including numerous local police departments, as well as representatives from county, state or even federal law enforcement agencies, prosecutors and non-criminal justice entities. Also, it is important to note that these multi-level/multi-agency organizational structures are not necessarily limited to those serving large urban areas, although involvement of some agencies, such as federal law enforcement, is more prevalent in more urban settings due to the higher number of agents assigned to these geographic areas. For example, in 1994, 46% of all prosecutors' offices indicated that they had staff as members of multi-jurisdictional task forces, with participation among large offices exceeding 80% (Bureau of Justice Statistics, 1996). Further, based on a survey done in 1995, it was found that federal law enforcement agencies served as members on nearly one-quarter of local drug task forces, with the DEA being the most frequently involved federal agency (Bureau of Justice Assistance, 1997). However, this is not to say that federal participation is limited only to task forces serving large, urban jurisdictions. There are numerous examples of where multi-jurisdictional drug task forces serving rural jurisdictions include, and may even be managed, by the DEA.

Another way to understand the unique nature of individual drug task forces is through an examination of the role that they play in the drug enforcement activities of their respective regions. One means by which typologies can be developed is by comparing the volume and nature of drug arrests made by drug task forces to those made during the course of traditional "drug enforcement" by the agencies which participate in the task force (e.g., see questions posed on page 17). Using Illinois task forces and data to illustrate this method of examining the role of drug task forces, it is clear that the role and activities of drug task forces vary considerably when looked across the broad array of jurisdiction types and task force models.

First, in terms of the volume of drug arrests, it is clear that the number of drug arrests made by task forces is low relative to those made by the local police departments which participate in drug task forces. Across all drug task force operations in Illinois, task force arrests accounted for less than 20% of all drug arrests in the regions that they served. This varies, however, from task force to task force, with some relationship to the size/type of the jurisdiction served (Figure 9-2). For example, task forces that serve predominantly rural jurisdictions tend to account for a larger proportion of regional drug arrests than do task

Mostly Rural   0.401606426
Mostly Rural   0.260162602
Mostly Rural   0.335135135
Mostly Rural   0.424920128
Mostly Rural   0.495412844
Mostly Rural   0.315525876
Mostly Rural   0.440677966
Mostly Rural   0.313235294

Urban/Rural Mix   0.07833733
Urban/Rural Mix   0.449288256
Urban/Rural Mix   0.302788845
Urban/Rural Mix   0.026584867
Urban/Rural Mix   0.176578786
Urban/Rural Mix   0.087557604
Urban/Rural Mix   0.258785942
Urban/Rural Mix   0.261131167

Mostly Urban   0.112840467
Mostly Urban   0.059069962
Mostly Urban   0.100307692
Mostly Urban   0.103566529
Mostly Urban   0.198198198

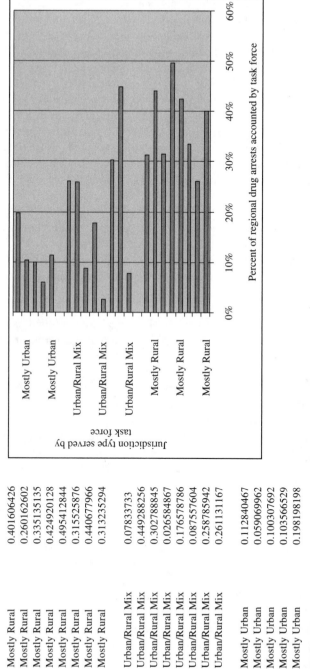

FIGURE 9-2   Percent of regional drug arrests accounted for by task forces, by unit, 2000.

193

forces operating in predominantly urban areas. This may be reflective of the pattern discussed previously: Smaller, more rural police departments are not likely to have in-house drug enforcement units, and therefore rely heavily on the drug task operations. It is also indicative of the nature of the drug markets in smaller, more rural jurisdictions where drug use and sales tend to be less visible, and are not conducted in ways that would attract the attention of the public or the police. In larger, more urban jurisdictions, most local police have their own drug units, and drug use and sales tend to be more visible to the public and police, resulting in larger volumes of drug arrests by these local agencies. Thus, some of the task forces in Illinois have account for a relatively large proportion of the drug arrests in the regions they serve, and thus have evolved into the primary, or at least substantial, drug enforcement efforts in those jurisdictions. Conversely, other task forces account for a relatively small proportion of the drug arrests within the covered jurisdiction.

The second, and perhaps more important set of criteria against which task force arrests can be compared to local arrests is the nature, or seriousness of the offenses for which arrests are made by drug task forces relative to those by local agencies. There are two different ways that this can be measured. First are the types of drugs involved in the arrests by task forces *vis-à-vis* local departments. One of the patterns revealed through analyses of the Illinois-specific task forces is that drug task force arrests were much more likely than local department arrests to involve what are perceived, by the public and reflected in the criminal punishments, to be the more dangerous drugs: cocaine, heroin, LSD, and methamphetamine. Arrests by local police departments were more likely to be for marijuana offenses. This predominance of marijuana arrests by local departments is not necessarily surprising, given that the most widely used, and most readily available illegal drug in Illinois is marijuana (Illinois Criminal Justice Information Authority, 2000). Local police department arrests in the jurisdictions covered by drug task forces tend to reflect the general patterns of drug use in the communities they serve. On the other hand, the fact that drug task force arrests are more likely to involve more serious substances, and are different than local arrests, would indicate that the targets of the drug task forces are more focused and planned than traditional local arrests. Another difference between drug task force and local arrests is the fact that drug task force arrests are most likely to be for drug sale or delivery, whereas the majority of local department arrests are for possession. Thus, task force arrests are not only more likely to involve substances viewed as more serious, but are also more likely to be for drug sale. Given these patterns, it is clear that most drug task forces are identifying and arresting individuals involved in different aspects of the drug market than become the focus local departments.

Another way to compare the nature of the arrests between drug task forces and local police departments is the likelihood of arrests leading to incarceration, and the degree to which task force arrests impact incarceration rates for drug offenses within a region. In general, felony-level task force arrests in Illinois were much more likely to result in a prison sentence than were local felony drug arrests. This further supports the view that drug task forces are more targeted in their enforcement activities (i.e., identifying and arresting more serious offenders), and may also support the hypothesis that the cases developed by task forces are stronger in court, since they tend to be more planned, often involve hand-to-hand/controlled buys, and also frequently involve video- and audio-surveillance of the transactions. In fact, some of the task forces in Illinois account for the majority of prison sentences for drug offenses within their respective regions, even though the task forces are small relative to the overall number of sworn police personnel within the region.

Combining the proportion of regional arrests accounted for by task forces with the proportion of prison sentences attributed to task force arrests, and the comparisons of these between drug task forces and local department arrests, can provide a better understanding of the unique roles that drug task forces can play in a region's drug enforcement strategies, and produces a number of different task force typologies that are useful in understanding the broad range of their activities. Keeping the comparisons to two dimensions, Table 9-2 summarizes the cells into which drug task forces can fall along the continuum of arrest volume relative to local agencies and prison sentences. As can be seen, there are two cells in the four quadrant typology which most task forces fall into: cell number 1, which includes those units which account for a relatively small percentage of the regional drug arrests and a relatively small proportion of the region's prison sentences. These types of units could be categorized as focused (low volume) and tend to cover more urban jurisdictions. The other quadrant into which a large number of units fall is the cell indicating that they account for a large percent of the arrests in the local area and a large percent of the individuals going to prison from the region served. These units arrest a relatively large volume of offenders, and account for a large proportion of the drug offenders that went to prison from the region served. Also, these units tend to serve rural areas.

The fact that the units are different in terms of their roles and activities is not necessarily problematic. It must be kept in mind that the extent and nature of the drug problem varies considerably across jurisdictions. Further, all drug task forces in Illinois, and indeed many across the country, are governed or directed in a broad policy sense by some type of oversight board. In Illinois, the composition of these oversight boards includes high-level representatives from the participating agencies (e.g., police chiefs, elected sheriffs, mayors, etc.), and may also include the elected county prosecutor(s) from the jurisdiction(s) within which the task forces operate. These policy boards set the overall task force goals and objectives, such as targeting street-level sales, targeting methamphetamine production, and so forth, and they leave day-to-day operations up to the task force supervisor.

TABLE 9-2   Potential Typologies of Multi-jurisdictional Drug Task Forces

Percent of regional arrests accounted for by task force	Percent of regional prison sentences accounted for by task force		
	Below median	Above median	Total
Below median	9 Task forces (5 urban and 4 urban/rural mix)	2 Task forces (1 urban and 1 rural)	11 Task forces
Above median	2 Task forces (1 urban and 1 rural)	8 Task forces (2 urban and 6 rural)	10 Task forces
Total	11 Task forces	10 Task forces	21 Task forces

MANAGEMENT ISSUES REGARDING SPECIALIZED DRUG ENFORCEMENT UNITS

Although there have been a number of things written regarding some of the management and administrative issues surrounding the operation of specialized drug enforcement units (e.g., see Coldren, 1993), given the unique nature of specialized drug enforcement units, there are also a number of management issues that should be pointed out and discussed briefly. In general, these issues can be broken down into three categories: (1) organizational structure, (2) "environmental" forces, including public sentiment, political considerations, and multiple "directors" or stakeholders, and (3) measuring the performance and effectiveness of unit efforts. While each of these issues could be a lengthy discussion in and of themselves, for the purposes of this chapter, only some of the more critical topics are addressed.

With respect to organizational structure, given that most specialized drug enforcement units are comprised of experienced law enforcement officers, often times these units are able to operate outside of the traditional tall hierarchies which are typical of police departments. This is particularly true when these units are made up from a number of individual law enforcement agencies (e.g., multi-jurisdictional units), since they are organizationally separate from any one specific agency. Thus, there may be a small number of unit directors or supervisors, and then a relatively or proportionately large number of agents or officers. There is usually little middle management between the task force officers and the director, and since most officers are experienced, there is little need for close supervision of daily tasks. Specialized units operating within a specific police department, on the other hand, are usually part of a much larger organizational hierarchy, which typically facilitates and promotes specialization.

When considering the environment within which specialized drug task forces or units operate, there are a few unique issues that need to be understood. One is that many of these units are designed to operate covertly, and therefore may not be well understood by citizens and some governmental agencies. Given the fact that these units tend to have little visibility (by design), it is imperative that unit directors maintain close contact and communication with stakeholders, funding organizations, and participating agencies. Some of this is accomplished through formal policy boards and period meetings. This is particularly critical given the limited resources available for governmental services. While having anonymity is critical in order to effectively perform covert investigations, it can also result in the public, practitioners, and policy makers not being cognizant as to what the unit is doing to address the drug problem. Balancing anonymity and ensuring that stakeholders are aware of what the unit is involved in is the role for adept unit or task force managers/supervisors.

Finally, measuring unit or task force performance has long been problematic and limited. As has been the case when gauging the performance of various criminal justice agencies or efforts, it is difficult to distinguish between quality and quantity. This is particularly important for specialized drug units, since their aggregate output (e.g., volume of arrests) is usually quite low relative to that of other law enforcement strategies. However, it is clear that the removal of someone higher up in the drug distribution network, which these specialized units tend to focus on, has more of an impact on the overall drug economy than arresting numerous low-level sellers or users. Slowly, criminal justice policy makers are beginning to recognize the limitations of relying simply on raw numbers of arrests or prosecutions, and are beginning to see the utility of examining the quality of the arrest. Also, while there has been a move toward

increased use and consideration of public views/opinions of police as a measure of performance (e.g., citizen surveys), as was noted previously, given the covert nature of many of these
units, this may not necessarily be useful to assess specialized drug unit operations or activities.

CONCLUSION

Given that drug markets are continuously changing, partly in response to changes in consumer demands and partly in response to the reactions of the criminal justice system, law
enforcement agencies must have the capacity to respond and address these changing markets. At the local level, traditional patrol, and more recently community-policing initiatives,
will continue to generate substantial arrests for drug-law violations. However, the nature of
these arrests and the behavioral characteristics of arrestees identified through these arrests
are relatively minor, frequently involving drug possession only or, at most, low-level street
sales. While there are benefits to these approaches in order to have a greater capacity to
interrupt the availability of illegal drugs and disrupt the drug market in a substantial way,
law enforcement needs the capability to conduct more long term, planned investigations,
which often times requires going across traditional jurisdictional boundaries and requires
more time and resources than traditional approaches. DEA fills this role at the highest levels of the drug distribution chain, identifying and arresting individuals importing drugs into
the United States and moving large quantities of drugs across the country. What often times
is missed, however, are those individuals that fall into the middle of this continuum: the
mid-level drug dealers and suppliers within specific jurisdictions. The strategy that has been
employed to identify and apprehend these individuals is the specialized drug unit within
police departments and multi-jurisdictional drug task forces, the latter combining the assets
of numerous agencies. While the development of drug task forces was not necessarily new
during the mid-1980s, when the focus on drug enforcement in the United States increased
dramatically, support by the federal government to establish and maintain MJTFs resulted
in a dramatic increase in their prevalence and membership.

One of the most significant issues which may face drug task forces in the future is
the ability to continue operation if federal funds which have provided the "glue" to hold
these units together is eliminated. It is clear from the data collected through a variety of
surveys and research that, for the most part, federal funds do not account for a large proportion of the overall operating budgets of most drug task forces. Specifically, a 1994 survey conducted in 1994 found that only about 17% of all multi-jurisdiction drug task force
budgets were supported with federal funds (Bureau of Justice Assistance, 1996).
However, for smaller, local agencies that historically have not had their own drug units,
and for which participation in drug task forces has been limited, it is often these federal
funds that allow their participation, and, concomitantly, produce the only real drug
enforcement effort possible in their respective communities. Further, almost two-thirds
(65%) of multi-jurisdictional task forces responded that without the federal support
provided through the Byrne Program (discussed earlier) they would be forced to shut
down, citing that the federal money provides "the glue that keeps a task force together
and operating" (Bureau of Justice Assistance, 1996, p. 12).

Another potential issue facing drug task forces is competition from other drug task
forces, and how the proceeds of asset forfeiture are distributed. For example, one of the

primary reasons given for the disbanding of the drug task force that served the suburbs of Chicago for nearly 30 years was the removal of local officers from that task force to allow them to participate in the High-Intensity Drug Trafficking Area (HIDTA) task force operating in the same general region. Given the larger scale of operations, and the promise of higher levels of asset seizure sharing, many departments opted for participation in this higher-level (in terms of targets) task force, leaving a gap in terms of addressing the mid-level drug traffickers/dealers operating across suburban communities.

Given the division of labor extant in American drug enforcement—again referring to Table 9-1—the specialized drug enforcement unit may be viewed as an essential building block between the efforts of individual, local street officers and the "Mr. Big" focus of federal agents. It is not reasonable to expect the average patrol officer responsible for undifferentiated calls for service to develop the specialized acumen of officers more highly trained and allowed to specialize in drug enforcement (or any other specialized form of criminal investigation for that matter). Nor is it reasonable (or wise; see Chapter 11) to expect federal resources to be focused on local law enforcement problems. That's not the mission of the DEA or the FBI—DEA doesn't have enough agents and, especially since September 11, 2001, that's not what we want the FBI to be doing anyway. The specialized drug enforcement unit, whether within a single-agency unit or as a multi-jurisdictional task force, appropriately managed, is the most logical police response to a major segment of the controlled substance distribution network.

RECOMMENDED READINGS

COLDREN, J.R. (1993). *Drug control task forces: Creating and implementing a multijurisdictional unit*. Washington, DC: U.S. Department of Justice.

HAYSELP, D., & WEISEL, D. (1992). Local level drug enforcement. In G.W. CORDNER, & D.C. HALE (Eds.), *What works in policing? Operations and administrations examined*. Cincinnati, OH: Anderson Publishing, pp. 35–48.

KLEIMAN, M.A., & SMITH, K.D. (1990). State and local drug enforcement: In search of a strategy. In M. TONRY, & J.Q. WILSON (Eds.), *Drugs and Crime*. Chicago: University of Chicago Press, pp. 69–108.

PHILLIPS, P.W. (1999). De facto police consolidation: The multi-jurisdictional task force. *Police Forum*, 9(3), 1–6.

PHILLIPS, P.W., & ORVIS, G.P. (1999). Intergovernmental relations and the crime task force: A case study of the East Texas Violent Crime Task Force and its implications. *Police Quarterly*, 2(4), 438–461.

REFERENCES

BUREAU OF JUSTICE ASSISTANCE (1996). *A report to the Attorney General: Multijurisdictional task forces and use of overtime and related issues FY1994*. Washington, DC: U.S. Department of Justice.

——. (1997). *Multijurisdictional task forces: Ten years of research and evaluation*. Washington, DC: U.S. Department of Justice.

BUREAU OF JUSTICE STATISTICS (1992). *Drug enforcement by police and sheriff's departments, 1990, special report*. Washington, DC: U.S. Department of Justice.

——. (1996). *Prosecutors in state courts, 1992 and 1994*. Washington, DC: U.S. Department of Justice.

——. (2001a). *Drugs and crime facts*. Washington, DC: U.S. Department of Justice.

——. (2001b). *Probation and parole in the United States, 2000*. Washington, DC: U.S. Department of Justice.

COLDREN, J.R. (1993). *Drug control task forces: Creating and implementing a multijurisdictional unit*. Washington, DC: U.S. Department of Justice.

DRAPER, G. (1990). *Arrests and dispositions of persons arrested by the Texas anti-drug abuse task forces*. Austin: Criminal Justice Policy Council.

DUNWORTH, T., HAYNES, P., & SAIGER, A.J. (1997). *National assessment of the Byrne formula grant program*. Washington, DC: U.S. Department of Justice.

FARABAUGH, D. (1990). *Evaluation of drug task forces in Maryland*. Towson, MD: Maryland Office of Justice Assistance.

FEDERAL BUREAU OF INVESTIGATION (2001). *Crime in the United States*. Washington, DC: Government Printing Office.

GILSINAN, J.F., & DOMAHIDY, M. (1991). *Multijurisdictional drug enforcement task forces in Missouri: What works and what doesn't work*. Saint Louis, MO: Department of Public Policy Studies, Saint Louis University.

ILLINOIS CRIMINAL JUSTICE INFORMATION AUTHORITY (2000). Statewide strategy to control drug and violent crime. Chicago: State of Illinois.

JOHNSON, B.D., WILLIAMS, T., DEI, K.A., & SANABRIA, H. (1990). Drug abuse and the inner city. In M. TONRY, & J.Q. WILSON (Eds.), *Drugs and Crime*. Chicago: University of Chicago Press, pp. 9–67.

KLEIMAN, M.A., & SMITH, K.D. (1990). State and local drug enforcement: In search of a strategy. In M. TONRY, & J.Q. WILSON (Eds.), *Drugs and Crime*. Chicago: University of Chicago Press, pp. 69–108.

NATIONAL RESEARCH COUNCIL (2001). *Informing America's policy on illegal drugs: What we don't know keeps hurting us*. Washington, DC: National Academy Press.

OFFICE OF NATIONAL DRUG CONTROL POLICY (2000). *Drug Related Crime*. Washington, DC: Author. [NCJ 181056]

PHILLIPS, P.W. (1999). De facto police consolidation: The multi-jurisdictional task force. *Police Forum, 9*(3), 1–6.

PHILLIPS, P.W., & ORVIS, G.P. (1999). Intergovernmental relations and the crime task force: A case study of the East Texas Violent Crime Task Force and its implications. *Police Quarterly, 2*(4), 438–461.

REAVES, B.A., & GOLDBERG, A.L. (2000). *Local police departments 1997*. Washington, DC: U.S. Department of Justice, Bureau of Justice Statistics.

SHERMAN, L.W., GOTTFREDSON, D., MACKENZIE, D., ECK, J., REUTER, P., & BUSHWAY, S. (1998). *Preventing crime: What works, what doesn't, what's promising*. Washington, DC: U.S. Department of Justice.

10

Horse Mounted and Bicycle Units

Mitchel P. Roth

INTRODUCTION

Since the advent of policing, it has been obvious that officers on horseback or bicycle have distinct advantages over their bipedal counterparts. There are no accurate statistics for the number of mounted police forces or bicycle units in the United States today. In the early 1980s there were between 70 and 82 horse mounted units (Carfield, 1982). In an updated survey in the late 1980s these numbers ranged between 100 and 120 (Carfield, 1982). By one estimate there were 650 mounted units nationwide in 1997 (Mounted Police Units, 1997). Most recently, according to the International Police Mountain Bike Association (IPMBA), approximately 5600 (43%) of all local police departments use bike patrol on a routine basis (www.ipmba.org/facts.htm, 2003). This number includes 90% of police departments serving 100,000 or more residents, as well as 13% of all sheriff's departments, including 50% serving populations of 500,000 or more citizens (www.ipmba.org/facts.htm).

It is clear that the numbers of both units have multiplied in recent years as concerns for public safety and better public relations became a paramount concern for city managers and county administrators. As modern policing continues to emphasize constructive interaction between officers and members of the public, the initiation of horse mounted and bicycle police units has contributed significantly to community-policing efforts.

Over the past decade bicycle and mounted patrols have become a familiar presence in most large cities, from the shopping malls to the vice districts. This chapter examines the emergence of bicycle and horse mounted police patrol as highly effective community police strategies in the late twentieth century. While a considerable amount of research has focused on police officers that patrol by foot or automobile, until recently there has been little research directed at the more community friendly officer on bicycle or horse—traditions that have a rich history dating back more than a century. Unfortunately, there is a dearth of printed material published on the this topic. Any overview of these units includes a perusal of the increasing number of websites heralding developments in specialized police units.

A survey of websites for mounted police forces indicates a growing revival of these units. The same holds true for bicycle units. These department websites demonstrate the renewed interest in these forms of mounted policing. Most of the units listed in Tables 10-1 and 10-2 offer similar information on their web pages, highlighting these specialized units.

Exemplifying the information available on mounted police departments, the Portland Police Mounted Patrol Unit, for example, patrols primarily the downtown area, combatting street crime. It is also used for special events, crowd control, and park problems. Officials favor this unit due to its high visibility factor and quick response

TABLE 10-1 Mounted Police Forces Cited on the Internet

	Founded
Albuquerque Police Department, New Mexico	1977
Bergen County Mounted Patrol, New Jersey (volunteer)	1986
Boston Police Department, Massachusetts	1883
Daytona Beach Police Department, Florida	1995
Fayette County Sheriff's Office, Ohio	1997
Fort Lauderdale Police Department, Florida	1983
Fort Worth Police Department, Texas	1977
Lake County Sheriff's Office, Florida	(no date)
Monterey County Sheriff's Office, California	1939
New Mexico Mounted Patrol (volunteer)	1937
New York Police Department	1871
Orlando Police Department, Florida	1982
Pembroke Pines Police Department, Florida	1987
Pennsylvania State Police	1905
Providence Police Department, Rhode Island	1997
Rochester Police Department, New York	1887–1932; 1997–
Saint Paul Police Department, Minnesota	1994
Stillwater Police Department, Oklahoma	1986
Toledo Police Department, Ohio	1989–1991; 1995–
U.S. Park Police, Washington, DC	1936
Vallejo Police Department, California	1995
Ventura County Sheriff's Department, California	1978
Virginia Beach, Virginia	1985
Waterloo Police Department, Iowa	1990
Waukesha Police Department, Wisconsin	1987
Wichita Police Department, Kansas	(no date)

TABLE 10-2   Bicycle Units Cited on Internet

	Founded
Abilene Police Department, Texas	1995
Boise Police Department, Idaho	1989
Canton Ohio Police Department, Ohio	1994
Concordia University Police Department, Montreal	1996
Escondido Police Department, CA	1992
Fort Lauderdale Police Department, Florida	1985
Governors State University Police Department	1997
Hendersonville Police Department, Tennessee	(no date)
Horseheads Police Department, New York	1994
Indianapolis Police Department, Indiana	1991
Kenosha County Sheriff's Department, Wisconsin	1995
Marshall University Police Department, WV	1993
Moorehead Police Department, Minnesota	(no date)
North Yorkshire Police, England	2002
Port Hope Police Department, Ontario, Canada	1994
Rochester Police Department, NH	1991
Romulus Police Department, Michigan	(no date)
Salem Police Department, MA	(no date)
San Angelo Police Department, TX	(no date)
San Bruno Police Department, CA	(no date)
Santa Monica Police Department, CA	(no date)
Seattle Police Department	1987
University of Akron Police Department, Ohio	(no date)
University of Kansas Police Department	1991
University of Oklahoma Police Department	1990
University of Wisconsin Police Department	1992
Vancouver Police Department, Canada	1991

capability in congested areas. Currently seven officers staff the unit. Formed in 1983, the Fort Lauderdale Mounted Police Unit was established to control college revelers during Spring Break and for ceremonial purposes. This unit consists of one sergeant, seven officers, and one civilian groom. They also patrol shopping areas during peak hours and have seen service in crowd control situations such as protests and union strikes. The San Jose Police Department Mounted and Parks Enforcement Unit was created in 1986 to add to the police presence downtown. They work during festivals and patrol parks and other downtown areas.

The horse has proven an excellent law enforcement tool. Horse patrols can negotiate various geographical barriers to patrol cars, and since they move slower make a more visible presence for a longer period. According to the San Jose Mounted Police, one officer on horseback is as visible in a crowd as 32 officers on foot (www.sjpd.org/mounted.html#welcome).

It is clear from this survey that the most common reasons for the existence of horse mounted units are riot control, community relations, and high visibility. However, mounted units are rarely absent from parades and state funerals, adding a touch of historical pageantry to the proceedings. Horses, like a K-9 partner will defend their riders. They are also environmentally sound and can work in parks, forests and beaches where vehicles either cannot operate or would cause great damage. According to a survey of a number of mounted police unit websites, it is generally accepted that mounted officers are six times more visible than foot officers (www.geocities.com/Heartland/Ranch/992).

The main disadvantages associated with horse patrols are inclement weather, speed over long distances, the vulnerability of the horse, its limited carrying capacity, and its tendency to litter the street (although it is biodegradable). Other detriments should be considered before starting a mounted unit. Horses are very expensive to select, train, and maintain, and are potential liability problems.

Bicycle units, for the most part, are simply extensions of normal police patrol units. Members of these units are typically officers on the force who are motivated enough to want to try something new. Most bike units were instigated at the behest of officers already on the force instead of being created as separate units. Officers in bike units are almost universally volunteers from other units in the department.

While many cities introduced bicycle units in the late 1890s and early 1900s during the heyday of the bicycle, most if not all were soon disbanded, following the introduction of motorized vehicles in the early twentieth century. Bicycles enjoyed a revival in the 1970s and 1980s as law enforcement across the nation experimented with alternative methods of crime control.

In 1987, Seattle, Washington established the first official bike patrol in the modern era. By the 1990s a number of forces introduced their first bicycle units. The use of bicycles in law enforcement continues to gain momentum across the United States and in other countries as an effective and economical method of police patrol. Bike patrols give police departments often hindered with budget cuts the ability to respond to any given situation quickly and in an economic manner. Bicycle patrols have proven especially effective in pedestrian areas that are not accessible to automobile traffic.

Somewhere between the horseback police officer and the patrol car-bound policeman, bicycle police are making an impact on current police patrol strategy. Unlike horse mounted officers, which are impossible to miss, bicycles are fast, quiet, and can maneuver without being observed. Like horses they are also good for community relations. Bikes have proven effective in apprehending a particular milieu of stealthy malefactors, including shoplifters, burglars, drug dealers, and graffiti sprayers.

Officers on bike offer the community a number of advantages. It is estimated that these officers have 4–5 times as a many contacts with the public during their shift compared to police officers in vehicles—many are self-initiated by public relations-oriented officers. Police mounted on bikes are deceptively fast in urbanized areas. They can get to spots that cars cannot and are able to respond to medical emergencies quickly while crowds prove a barrier to cars. The mere presence of bike officers raises the awareness of bikes as legitimate road users and encourage other cyclists to follow road rules. Cheap to buy and maintain, bikes give the public a better value for the tax dollar. Ten to fifteen bike officers can be fully outfitted for the cost of one patrol car.

Bicycle patrols vary in size, from a single officer assigned in a secondary function on a small-town police force to a large department, such as the Houston Police Department, which devotes 513 bicycle patrol officers (237 primary and 276 secondary) (*Telemasp*, November 1998). At universities and colleges they have been used in large crowd situations such as football games, concerts, and commencement ceremonies.

As with other forms of mounted policing, bicycle patrols also face certain disadvantages. Officers frantically pedaling to a crime scene can arrive too tired to physically assist the situation and, except for congested traffic conditions, it takes longer to arrive at the scene (Siuru, 1996, p. 53).

Bicycle units are hampered by a number of factors that limit their hours of operation, including weather, terrain, and trained personnel. Close to 85% of these units operate at night, but only 50% operate year-round (Sharp, 1999). In addition, horses can move faster than bicycle cops and can carry more emergency equipment.

Several forces are now experimenting with "electrically-assisted" bikes, which allow officers to arrive at a crime scene ready to handle physically demanding confrontations. Virtually noise-free, at the same time they maintain the element of surprise. Supporters of the new technology claim it is the fastest way to maneuver through congested traffic and is less intimidating than a patrol car.

HORSE MOUNTED AND BICYCLE UNITS

History of Horse Mounted Units

Although it is unknown when the first horse was used in a police action, most historians trace the utilization of mounted units as peace keepers to King Charles' *Articles of War*, published in 1629 (Carfield, 1982). While Texas and Australia implemented mounted police units by the 1820s, credit should be given to London's 1805 Horse Patrol as the earliest mounted police force (Campbell, 1968). Consisting of 52 men and animals, the Bow Street Horse Patrol patrolled main roads leading into London up to 20 miles outside the city limit (Campbell, 1968). Like most twentieth-century mounted police officers, these early peace keepers were recruited from the military, usually the cavalry (until 1939). Similar to the Pennsylvania State Police, the London mounted was housed in cottages near their patrol routes. Each man was armed with a sabre, pistol, handcuffs, and a truncheon. The Bow Street Mounted diverged from their modern-day counterpart in one important area—they were not involved in dealing with crowds and rioting, but mainly road patrol.

As rioting and social disorder increased in England during the industrial revolution of the nineteenth century the mounted police played a more crucial role in crowd control. During the early years of the twentieth century London's Mounted saw their duties expand to include controlling suffragettes, other horsemen, patrolling common lands, and escorting members of the monarchy.

During the early 1820s an American mounted policing tradition developed in frontier Texas. The Texas Rangers had an on-and-off existence during the nineteenth century. Following annexation in 1845 Federal troops took over border patrol and the Rangers were disbanded. But, by the 1850s they were reorganized for a brief stint to protect the wide-ranging Texas frontier from Comanche raids (Webb, 1965). After several cycles of reorganization the Texas Rangers made the transition to the twentieth century and a new set of challenges including labor strikes, border raids, prohibition violations, small town violence, and Ku Klux Klan activities. By the 1930s the Ranger tradition was in dire need of updating. Traditionally undermanned, their numbers were trimmed and the force was in danger of obsolescence because of urbanization and modern science. With the introduction of the automobile and the train, their days as an effective mounted police force became a nostalgic memory (Roth, 1998).

New Mexico and Arizona took similar steps toward improving law enforcement at the turn of the twentieth century based on the frontier ranging tradition best exemplified by the Texas Rangers (Prassel, 1981; O'Neal, 1987). However neither force would last as a mounted force into the 1930s.

During the nineteenth century most major urban centers in the United States had implemented mounted patrol as part of its developing public safety strategy. As early as 1805, New Orleans established the Gendarmie. Half the force was mounted, a costly proposition in any time period. This force was considered essential for hunting down runaway slaves. But, within just a couple of years the unit was disbanded because of the high costs of maintaining a horse patrol.

During the heyday of post-Civil War Reconstruction, New Orleans reorganized a mounted police force for the first time since 1806 (Rousey, 1996). But a much larger budget allowed for 5% (about 36 officers) of the force to be mounted. Primarily employed "in the suburban periphery of the city" (Rousey, 1996, p. 130), three years later the mounted force was doubled in size to 70.

In 1871 New York City became the first American municipality to formally adopt a mounted police unit. San Francisco initiated a mounted force three years later and in 1883 Boston became the third American city to introduce such a unit. The Boston Mounted Unit proved invaluable in hunting down criminals in the winding alleys of the city's Back Bay area, where beat patrol officers faced high brick walls and narrow passageways. Here a mounted officer, astride his horse, could peer over the walls and observe suspicious activities that an officer on foot would miss (see Appendix 1).

The New York City Mounted was formed with the express purpose of controlling a certain class of people, "who, since the avenues to and beyond Central Park had been laid with smooth pavements and become crowded with vehicles, had become fast and reckless driving . . . at rates of speeds perilous" to citizens (Campbell, 1968, p. 87). Mounted Squad No. 1 was originally composed of a sergeant, 12 patrolmen, a hostler, and 15 horses. Their initial success led to a larger force which by the turn of the century consisted of more than 200 mounted men.

In 1904, the Street Traffic Regulations Bureau was organized in an attempt to relieve vehicular congestion, particularly in the vicinity of Wall Street. The Mounted officers were so successful in expediting traffic that the Merchants Association of the City of New York accorded the Mounted Unit with ceremonial flags at the annual Police parade.

Although the New York City Police Department owned even more horses by the next decade, the emergence of the automobile signaled the decline of the mounted police. During the social unrest following World War I, horses once again were used for crowd control. They were used sporadically between the wars, but their numbers clearly diminished after World War II, partly because it became increasingly vexing to find appropriate stable facilities in America's largest city. During the financial crisis in New York City in 1976, the unit was threatened with being disbanded. However, the creation of the Police Foundation launched a major program to urge the donations of horses and related supplies. Before the year ended more than 30 horses and over $5000 had been contributed from donors around the country (Campbell, 1968).

Between 1915 and 1921, 23 states established some incarnation of the state police model. With the loss of National Guard units to the war effort in 1916, Colorado, Georgia, Maine, Michigan, New York, and Rhode Island organized temporary units primarily composed of large mounted units stationed in rural areas and small towns. Their mandate gave them statewide police powers (Bechtel, 1995).

In the following years, union unrest and other social changes led to varying attitudes toward mounted state police units. While the Pennsylvania State Police were vilified as the "Cossacks" and "Black Hussars," the New York State Police were heralded and even romanticized as the "Grey Riders" (Reppetto, 1978). The popularity of the New York force over its Pennsylvania counterpart can be explained by the reluctance of New York State officials to use the force as strike breakers (Van de Water, 1922).

As transportation problems became more complex in the early twentieth century and cities became more crowded, alternatives to mounted policing were considered. A pattern soon became apparent where by the 1930s most mounted units had been replaced by automobiles and mechanized scooters. Horses and motor traffic were considered antithetical, with horses becoming a traffic hazard on busy city streets.

However, the pattern changed in the 1970s as cities rediscovered the advantages of mounted police officers. But, having discarded the mounted units many decades earlier, few police forces had experienced riders. No organizational pattern of mounted policing has emerged in the past 30 years. Some cities provide horses and tack, while others, because of budgetary constraints, require officers to supply their own mounts.

In Rochester, New York, the city provides the saddles, tack, and other necessities for the horses. Downtown corporations, who support the reintroduction of the units, donate many of the horses. A private fund was also set up to allow individuals to contribute to the "Friends of the Mounted Police" fund. An outside corporation manages this on behalf of the patrol. In Rochester and other cities these tax-free donations have provided the funds to increase the number of horses.

The Fort Worth Police Department inaugurated its Mounted Patrol in the late 1970s in response to transient problems taking place in the historic Stockyards area. Like other western communities, there was also a desire on the part of the citizens in the historic section of the city to see an officer on horseback both as a working police officer and as a reminder of the city's western heritage. Often a figment of imagination, or the invention of the local chamber of commerce, such was the case in "historic" Fort Worth, which never had a formal mounted patrol unit in its early days.

BICYCLE MOUNTED UNITS

There is little detailed history on bicycle units. Most discussions of these units in department histories and websites offer a photograph and a brief caption with an estimated date, but little else. Most information on the evolution of bicycle policing prior to the late twentieth century is anecdotal. What is clear is that in the late 1890s traditional methods of police patrolling had become inadequate for the changing nature of urban America. In 1896 New York City implemented America's first bicycle patrol (Jeffers, 1994).

During the mid-1890s, America witnessed a "bicycle craze." There were more than three million bikes on the streets by 1896 (Jeffers, 1994). So-called wheeling clubs sprung up simply to organize races or share the fun of bike riding. However, the problem of careless bicycling in New York City became such a hazard by 1896 that the acting chief of police issued a warning to "wheelers." Acting Chief of Police Moses Cortright stated, "There are many thoroughfares in the city where the traffic on vehicles of all kinds, and especially bicycles, is so great that it is necessary for the protection of life and limb that such traffic should be regulated" (quoted in Jeffers, 1994, p. 206).

A year earlier, New York City Police Commissioner Theodore Roosevelt inaugurated a "scorcher squad" composed of four bicycle officers to regulate traffic in the most dangerous sections of the city. Roosevelt also implemented an eight-mile-an-hour speed limit. Violators would be stopped by bicycle cops who then inspected their driver's license (Jeffers, 1994). During its first year of service, Chief of Police Conlin noted that the bike squad "[h]as been of great service to the public, that it has much increased the efficiency of police service on the thoroughfares patrolled by the bicycle squad, and has been instrumental in the accomplishment of police work" (quoted in Jeffers, 1994, p. 208).

Within seven years, the squad was boosted to 100 members and credited with making 1366 arrests in 1902 alone (Jeffers, 1994). By the end of the decade, horseless carriages had become affordable to the masses and the Bicycle Squad was overwhelmed in an attempt to control the city's traffic maelstrom.

As bicycles became popular in the late 1890s, the Indianapolis Police Department introduced the "Flying Squadron" in 1897. The Indianapolis story is fairly unique because bicycles were used before horse patrols, which were not implemented until the turn of the century (Indianapolis Police Department History, 2003). It so happens that these discontinued units were reinstituted in reverse order, the horse in 1983, and the bike in 1991 (Indianapolis Police Department History, 2003).

While patrolling on horseback or by foot had certain advantages, the introduction of the "steam carriage," a precursor to the automobile, demanded better and faster traffic control because the increasingly crowded streets were filled with horse-drawn trucks, steam carriages, and other forms of transportation. This made city streets unsafe for pedestrians in major cities.

By 1912 it was decided that using automobiles instead of bicycles would be a better way of controlling traffic and New York City established a Traffic Division. The Bicycle Squad was not disbanded until 1934. Instead of consigning the bicycles to auction or the garbage, the police department commendably sold them to local children for as little as 25 cents each.

Today, bicycle patrols have witnessed a grand reemergence in urban American policing.

STRUCTURAL VARIATIONS: HORSE MOUNTED UNITS

There is a variety of mounted horse patrol units. Horses can be found in law enforcement in a wide range of categories, including sheriffs' posses, park police, search and rescue teams, campus police, state police, and city forces. Some cities and counties offer volunteer mounted units. There are also variations of volunteer units. Some cities, such as Los Angeles, have volunteer units composed of actual police officers who volunteer the use of their own horses and tack. The first step in creating a mounted police unit is to decide the unit's mission. Seasons and weather patterns, as well as the physical terrain (urban or wilderness) will influence the effectiveness of the horse patrol.

Typically, full-time mounted units are used in the same manner as a foot or radio car police officer beat. Part-time units are utilized for specialized patrol. This model uses horses and riders on a seasonal basis for special events, crowd control, or directed enforcement missions (Mounted Guide Homepage). Reserve police units are another model, one that assists full-time patrols. Specialized Wilderness patrol is employed in more environmentally sensitive areas. Less frequently, horses are used by corrections officers who supervise and guard prisoners on outdoor work assignments. Other units are used for search and rescue, tracking lost persons.

It is difficult to assess the comparative popularity of various horse units due to the dearth of information provided by law enforcement on mounted units. However, the following horse mounted models exemplify the diversity of horse mounted units.

Park Police

Horse patrols have proven effective for patrolling park and recreational areas, some covering thousands of acres. With limited road access mounted units can patrol environmentally sensitive areas with little negative impact. Mounted police are better suited for patrolling a geographically diverse area or one that is heavily wooded.

Of the many park police units, the U.S. Park Police Horse Mounted Unit is one of the oldest. Created in 1934, the unit patrols the streets and wooded areas of Washington, DC. Through the years its mandate has expanded to include federal parks across the United States, including San Francisco's Golden Gate Park. However, this unit is most in evidence in the nation's capital during numerous parades, celebrations, and civic activities. During the unrest of the 1960s and 1970s, the U.S. Park Police Unit faced large-scale violence that necessitated new training and new strategies for managing and dispersing crowds.

Volunteer Organizations

Maintaining a mounted police force is an expensive proposition. The New Mexico Mounted Patrol (NMMP) is an unfunded unit that can be called on by the Governor, the State Police, or other local police forces to assist professional police officers. Members of this unit are required to complete 268 hours of classroom and practical training. Each potential trooper must complete a 100-hour ride-along with a full-time officer, who will evaluate his or her fitness to continue as a mounted patrolman. The NMMP is reportedly the only known volunteer law enforcement organization originating from action by a state legislature (www.dps.nm.org/nmmp/mphist.htm).

The NMMP model fills an important void for police agencies that cannot maintain a sufficient number of full-time police in emergency situations. This mounted unit has been involved in drug raids, public security functions, traffic control, and DWI road blocks, and has even aided U.S. Secret Service in protecting President Bill Clinton on a campaign stop in Las Cruces in 1996 (New Mexico Department of Public Safety, 2001).

The Los Angeles Police Department (LAPD) has maintained a volunteer mounted unit since 1981. Unlike the New Mexico model, the LAPD unit is composed of full-time officers who volunteer their own equipment, provide their own transportation to various demonstration sites, and conduct their own group training. Trained in riot and crowd control, as well as in counter-sniper and advanced arms training, the LAPD mounted has been used as a tactical support unit on a number of occasions, most notably during the 1992 Los Angeles riots (Conley, 1994, p. 70), although on this occasion they provided tactical support driving conventional vehicles. The LAPD unit was also used to patrol for looters during an earthquake in the early 1990s.

　　　Some mounted units combine volunteer civilian riders with sworn police officers. The Lake County Sheriffs Office (Florida), for example, consists of two sworn deputies and their horses and between 10 and 16 Civilian Volunteers and their horses. This unit performs a variety of patrols, civic functions and searches for missing persons and criminals. Each member of the civilian element must complete 32 hours of training prior to their first assignment and then must attend monthly training sessions.

Mounted Posses

In many communities mounted posses of volunteer members assist official law enforcement efforts. Posse members do not have to be actual police officers, but can be local ranchers or other horse people who receive specialized training and assist primarily in Search and Rescue operations. They also might take part in parades, special events, or other activities that need a bit of "pomp and circumstance." Unlike their more official police counterparts, posse horses do not need to be a special breed, height, or color, but still need to be calm and people friendly by nature.

　　　The Monterey County Sheriff's Posse, Inc. (MCSP) fits this category. Created in 1939, today the MCSP has 69 members and maintains a 50-acre ranch in Salinas, California. The posse is a private non-profit corporation and is not an agency or department of the County or Sheriff's Department. The unit sponsors various community activities and rides in a number of parades each year.

Mounted Police Search and Rescue

Some police departments maintain horses just for patrol and search and rescue operations. Due to a history of hastily improvised rescue attempts in the southern California desert that resulted in the loss of equipment and injuries to peace officers, in 1949 the Palm Springs Police Department formally created a volunteer citizen's search and rescue unit. In the half century since its inception, the Palm Springs Mounted Police Search and Rescue (PSMP-SAR) unit has performed more than 1000 rescues using horses, helicopters, and hikers. Each member must be a sworn reserve officer, a trained and certified technical rope rescuer, and must provide their own personal gear, which can cost almost $2000 per member.

Campus Mounted Units

The University of Wisconsin at Madison established a mounted police unit in 1989. It works in conjunction with mounted officers from other units in the county in a variety of settings. On campus from three to five mounted police are used during each college football game. This mounted unit handles everything, including crowd control, vehicle extractions, assisting ejections from the stadium, providing information to fans, and as good community relations for children and parents. Beginning in 1998, the unit began handing out police officer trading cards featuring members of the horse unit. When not working sports events the mounted officers escort VIPs, ambulances, and buses. Once again they prove their usefulness with their height advantage and mobility in situations where crowds prevent squad cars easy access.

Collateral Duty Units

Some police departments have organized mounted special units that are used sporadically for collateral duties. For example, the Vallejo Police Department in California boasts a Mounted Unit Color Guard. Established in 1995, this unit is composed of VPD members with three or more years experience. In order to join officers are required to pass a Basic Assessment Test, which measures the equitation skills of both horse and rider as a team. Prospective members must own their own mounts and agree to rent the horse to the Police Department whenever the unit receives an assignment. In addition, officers are responsible for transporting mounts to the assigned site.

In order to join the unit each officer must first attend a 40-hour certified school. After graduation, they must attend mounted enforcement training once a month, and pass an extensive, in-service test annually. Besides acting as a color guard, the unit provides crowd control and security during special events and holidays. While on assignment each officer is responsible for taking care of the horse's feeding, watering, grooming, first aid, and must assure that their mounts stay current with vaccinations, worming, and shoeing.

Also considered a collateral assignment unit, the Ventura County Sheriff's Office Mounted Unit was established in 1978 and is now often used to train other California mounted units. This unit has been utilized in a number of specialized capacities, including high profile gang patrols and anti-burglary patrols.

Some departments engage a mixture of full-time officers and reserves. The Haledon, New Jersey police, for example, use six reserve officers and five full-timers to patrol a community of under 5000 (Weinblatt, 1996).

STRUCTURAL VARIATIONS: BIKE UNITS

Today bikes are used not only by city and county police agencies, but also in military installations, colleges, shopping malls, security companies, and park and wildlife law enforcement. Despite the number of agencies using bike patrols, their duties and modes of operation are quite similar, not nearly as diverse as the variety of mandates given to horse mounted units. Most bicycle officers operate in a similar manner as officers in patrol cars. They answer calls, enforce laws, and even issue tickets to bicyclists.

According to IPMBA, which maintains a directory of units with statistics on more than 650 bicycle police patrols, the average size of a bicycle unit is nine officers, although several have more than 50 officers.

Typical in its mission statement, the mission of Bicycle Patrol of the Canton Police Department is clearly inspired by community policing. The unit emphasizes the alleviation of quality of life issues by attempting to handle rowdies, loiterers, deteriorating housing, graffiti, prostitution, etc. (Canton City Police, www.neo.Irun.com, 1).

On the other hand, the Abilene Texas Police Department bike unit, formed in 1995, is considered unique in that it is made up of volunteers who work bike detail primarily on off-duty time. Similar to other early police bike units, five officers initially offered to provide their own equipment and work for compensated time if allowed to form their own unit. Officers work four day/10 hour shifts with three days off. The idea here was that officers would not be taken off regular patrol allowing compensated time to accumulate and be "paid back" during the day scheduled for directed assignments.

Abilene Police Chief Martin has claimed that the use of bikes by his department has "been the most effective public relations tool in the department's history" (Abilene Police Department, 2003). The Abilene bike unit was given the opportunity to establish itself as an important crime fighting tool when the department began getting complaints of problems in popular six-block downtown area, home to coffeeshops, nightclubs, a bus depot, and an ice cream shop. The introduction of bike patrol minimized the problem in a short time and returned the area to the public.

University and College Bike Units

A number of universities and colleges have found bicycle units to be a perfect fit for their environments. The use of bikes has allowed officers greater patrol access in the academic and residential areas not accessible by car, or too large to cover efficiently on foot. Created in 1992, the University of Wisconsin Police Department bike unit covers a wide range of activities. Patrols are used during University of Wisconsin home football games and Green Bay Packer preseason games held in Madison. The bike unit patrols parking lots and side streets, enforcing laws ranging from ticket scalping to felony drug possession and sales. They also work crowd control at concerts and football games (www.uwpd.wisc.edu/bikepat.htm).

The University of Kansas Police bike unit, created in 1991, found that when faced with budgetary restraints, it could make up for budget cuts with bike patrol. This economical solution to manpower cuts allowed the department to maintain the ability to still respond quickly in certain situations (www.ukans.edu/cwis/units/kucops/department/bicycle.html). The University of Northern Iowa established its bike unit in 1994 as part of a commitment to community-based policing. It has two trained officers assigned to each shift. Recently the University of Northern Iowa bike unit joined the local Cedar Falls Police Department in patrolling college neighborhoods by bike "particularly in the early evening" (www.uni.edu/pubsaf/ops/bp.html).

In Akron, Ohio, the University of Akron Police Department Patrol Division has two officers trained on mountain bikes for each shift. In a college environment bike patrol members can speed quickly from one building to another without disrupting pedestrian traffic. They can also patrol between buildings to check isolated areas (www.uakron.edu/police/division). Since 1996 the Concordia University Bike Patrol Unit in Montreal, Canada has routinely patrolled streets, alleys, and indoor and outdoor parking lots. Officials at the university claim that these officers can respond to calls quicker and cover four to six times more ground than a foot patrol officer (htttp://members.tripod.com/~Puc7Tyr/index.html, p. 1).

OPERATIONAL CONSIDERATIONS

Horse Mounted Units

Horse mounted officers have seen their duties expand in recent years. For example, in Fayetteville, North Carolina, mounted officers have begun using radar guns to clock speeding cars from horseback (news-journalonline.com/2002/Feb/2002/NOTE6). Unless on regular patrols, mounted units are deployed according to specific needs, usually for ceremonial purposes such as parades, funerals, and other special events. Other units are utilized for crowd control at civil disturbances, sporting events, concerts, and large gatherings of people such as the annual New Orleans Mardi Gras celebration.

Police officials favor mounted police units for a variety of reasons. Chief among them is the high visibility of an officer on horseback, and the capability to quickly respond to an emergency in congested areas. In cases of riots and civil unrest, research indicates that one mounted police officer is equal in effect to 10–20 police officers on foot. Mounted officers have proven an excellent tool in law enforcement. They can negotiate various geographical barriers to patrol cars, and since they move slower, make a more visible presence for a longer period.

The deployment of mounted officers should be considered according to the situation. Routine patrol functions may include riding horses a minimum of five hours per shift, five days per work-week (8-hour shifts), for a total of at least 25 hours per week. Popular patrol locations include shopping centers and other shopping areas during peak hours, parks, and downtown areas during normal business hours. Where car burglary is a significant problem, as in movie theater parking lots and shopping mall parking areas just before Christmas, because of the additional height perspective provided by the mount, horse mounted units provide extraordinary surveillance, apprehension, and deterrence advantages.

Horse mounted units are occasionally called on for crowd control situations such as strikes and protests. The Fort Lauderdale Mounted police are used to control college revelers during Spring Break, and for ceremonial functions.

During the 1990s, the Los Angeles Police Department Mounted Unit was deployed following the 1992 riots. Mounted patrol teams can control crowds efficiently by moving them, preventing them from moving, or splitting two opposing factions. The LAPD mounted unit was also called into action following several major earthquakes to patrol affected areas for looters. The Los Angeles unit has close to 30 members, including three sergeants. Officers are dispatched daily to different sections of the city, often in a drug enforcement capacity.

In New York City, the mounted unit utilizes specialized vehicles such as 9-horse tractor trailer combos, 6-horse trucks, and 2-horse combos. When horses are injured they are transported in the 2-horse combos. Evidence suggests that horses seem to prefer being transported in these familiar and reassuring surroundings. However, there are instances when ambulances are required from the American Society for the Prevention of Cruelty to Animals (ASPCA).

Bicycle Units

The major advantage enjoyed by bicycle patrol officers is their face-to-face contact with citizens. These units typically engage in proactive patrol in an effort to reduce crime by high visibility, presence, and enforcement activity. These special units have proved successful at addressing such community problems as loud parties, loitering, and car break-ins. Reserves pedal mall parking lots and golf courses, and have even reportedly seen action at special events such as Civil War reenactments.

As far back as 1977, cities such as Richardson, Texas, with a population of 67,000, implemented a bicycle patrol made up of five two-officer teams (Yarbrough, 1977). Considered innovative for its time, each team worked a six-hour tour of duty encompassing a five to six block area. Area of patrol was based on the analysis of statistical offense data. In this early prototype, bicycle teams were transported to their patrol areas by a team supervisor in a departmental van. The supervisor remained available in order to assist the team if needed to transport suspects. However, due to budgeting, this unit was eventually disbanded, despite its success rate.

Today bicycle police patrols are commonly deployed in urban settings. Larger metropolitan areas offer more challenges to bike patrol strategies than smaller towns. Cities such as Phoenix, Arizona have implemented a plan, in which the city is divided into six precincts, each patrolled by bike squads of approximately ten officers. Typically working in pairs, precinct bike officers are often used to support officers in patrol cars. Otherwise they can be seen throughout the city, working problem neighborhoods, housing projects, parks, and the downtown.

Deployment naturally varies according to the needs of each precinct. Its effectiveness has led the Phoenix Police Department to equip and train school resource officers to apply bike patrol to school policing. Some fire personnel are even bike-trained, allowing the fire department to make paramedics available for large parades and celebrations, where people may face dangers from the omnipresent Arizona sun. Paramedics on bicycle have proven most effective when streets are too crowded or congested for traditional vehicles.

The Phoenix police have augmented the precinct bike patrols with a Quick Response Team (QRT). Members are available when needed to work whichever part of the city they are required in. Each member of the QRT team keeps personal equipment in his car to use as needed. This allows officers to peddle their bikes back to their cars, put them on racks, and report directly to the required emergency situation (Stockton, 1999).

In New Mexico, the Rio Rancho police department serves a community of 60,000 and covers almost 100 square miles. Created in 1993, the seven members of the bike patrol are required to live within a 25-mile radius. Each officer has a take-home car outfitted with bicycle racks and prisoner cages. In order to save money bike officers perform their own routine maintenance on the their equipment. Because of limitations posed by terrain and distance, officers are deployed to handle special problems that require the low-profile and stealthy bicycle (Lesce, 1999).

Police officers who have patrolled by both car and bike have found that public relations are improved. One officer noted that he would engage at least five citizens in long conversations while on bike patrol, compared to none while behind the wheel of a police car (Weinblatt, 1996). Bicycle patrol has been well received in Walnut Creek, California. A city of more than 60,000 people, and covering 15.1 square miles, which includes 2000 acres of open space, a large retail district, and an extensive paved trail system, Walnut Creek is considered an ideal size for such a unit (Weinblatt, 1996, p. 13). Bicycle patrols are used in a variety of community environments—from the inner city to upper-middle-class communities. Each has its own crime problems. But, wherever they have been used, the response by the business community has been quite favorable.

In many police departments, bicycle patrols are dispatched according to calls-for-service. Some departments place restrictions on the types of calls handled by bike officers. In Texas, there are several departments that prohibit bike officers from working major collisions and from working on major highways. One department prohibits these units from responding to calls where a paddy wagon is required.

Bicycle officers are faced with a number of obstacles that rarely hinder officers behind a car wheel or on foot. No matter how fit an officer is, many large cities pose special problems for bike-bound police. In New York City for instance, bicycle officers take special training courses from expert cyclists on avoiding hazards including metal plates, sewer grates, and the opening of car doors.

According to the Boston Police Department, its police officers on bike "have fewer accidents and injuries than those in cars" (www.transalt.org/blueprint/chapter16/chapter16c.html, 2, 2002). Police officials attribute this record to the fact that bicycle police are in better shape and more alert on duty. One Boston official noted that, "Being in a cruiser, you have all that metal around you, you become complacent. On a bike, you see and hear better. Also you are going slower" (www.transalt.org, 2002).

ADMINISTRATIVE CONSIDERATIONS

Horse Mounted Units

A number of issues need to be considered before determining whether to establish a horse mounted unit, particularly during times of budget tightening. These issues include analyzing the type of areas to be patrolled, the location of the area and its primary use, and the activities it serves.

Before seeking administrative approval for starting a horse mounted unit, several questions require definitive answers, including: Who will be the constituency for the service? Who will benefit from the creation of this special unit? And, how will the benefit be measured?

Another area of consideration is the cost of horses, equipment, and care. Annual operating budgets can easily exceed $2000 per horse. Included in this estimate is the initial prorated cost of the horse, tack, grain, hay, bedding, and blacksmith and veterinarian fees. Other variables can also impact the total expense. One of the more controllable expenses would be the selection of stable facilities. For example, standing stalls are more expensive to build, equip, and maintain than box stalls. Horses housed in box stalls utilize more bedding than one in a standing stall. On the other hand, an idle horse in a box stall receives more exercise than a standing stall. However, if the horse is a police mount it should receive the requisite amount of exercise regardless of the type of stall.

Chief among the more uncontrollable expenses are illness, thrown shoes, and the price and quantity of feed. In any case, a well-planned stable management program can keep these expenses and variables to a minimum. No matter how much planning goes into starting mounted police unit, costs can be prohibitive for smaller, less-capitalized forces. For example, when Santa Fe, New Mexico began its pilot program in 1995, the state legislature allocated $32,000 for the program, which was mainly targeted at providing bridles, saddles, special horseshoes, and insurance for nine officers who provided their own mounts (Mounted Police Units, 1997).

According to the Fort Lauderdale Police Department, the budgetary costs for its mounted unit in 1998–1999 was $710,456. At that time the unit was composed of one sergeant, seven officers, and one stable attendant. Almost 90% of the budget ($636,266) was devoted to personnel expenses. The balance was used for operating expenses such as animal care and related equipment. In addition, in 1994, with input from mounted officers, the city built a $525,000 state-of-the-art stable (Fort Lauderdale Police Department, 2001).

Budgets for mounted units should consider per horse program costs for veterinary care (average $50 per month); stall shavings ($3 per day per stall); stall and barn maintenance ($50 per horse per month); shoes every four to six weeks per horse (in California this runs $30–40 per shoeing); and new horse tack ($1150–1600 per horse). Also included under expenses are insurance coverage, officer training, and other miscellaneous costs such as fence and gate repairs, pasture watering, grounds equipment, special uniforms, and safety gear (Mounted Guide Homepage, 2001).

Every mounted unit should devise a yearly budget justification, which should include the number of officers and mounts, patrol duties, shifts, and special assignments. Most units have at least one sergeant for each shift. A sergeant's duties is responsible for performing all duties and assignments required by Operations Support Captain, Patrol Commanders and shift captains. Some units require the sergeant to:

1. Ensure that daily work schedules and special assignments are completed and distributed;
2. Maintain statistics of activity and files on stable and horse management;
3. Handle day-to-day functions of the unit;
4. Assist in budget preparation;
5. Ensure training requirements are met for both officers and horses;
6. Coordinate related stable management activities, such as blacksmith, veterinarian, stable tours, etc.; and
7. Inspect horses, tack, and related equipment to ensure it is in proper condition.

Routine patrol functions usually consist of officers riding a horse a minimum of five hours per shift five days a week. Officers groom their horses prior to beginning the workday for one hour, and weather permitting, bathe the horse for one hour after their shift. The rest of the officers' workday can include patrol assignments, necessary breaks, and tack and equipment cleaning and maintenance.

Today, the New York City Mounted Unit is a component of the Traffic Control Division of the Department's Patrol Bureau. They patrol a diverse territory, which can include Coney Island's boardwalk, City Hall, Times Square, and Yankee Stadium. While much of their equipment is standard with the rest of the department, as with other mounted units many parts of the uniform are unique. Most distinctive are the riding breeches for the yellow stripe on the outside seams. Unit members have the choice between wearing English-style black leather riding boots or placing leather leggings over high-top black shoes. Members are also required to wear helmets.

There are a number of reasons for retiring police horses besides age concerns. Police horses are taken out of service because of injuries or developing bad habits or "attitudes." One police horse that had worked on paved services developed a foot inflammation problem. However, during convalescence the horse's behavior changed and he began to buck and rear. Unsuitable for police work, the force decided to sell him at its stolen property auction.

In another situation, the Chicago Police Department announced the retirement of a 35-year old horse after a 22-year career. During this time he had developed severe arthritis and was unable to stand very long. However, in this case the horse's trainer reluctantly chose to euthanize the horse, explaining, "I bought him, I trained him, and I trained the riders on him. The arthritis just got worse, and we were using him less. He couldn't get his hind legs up and out of the stall. We can't take the chance of a horse going down on the street" (Mounted Police Units, 1997, p. 13). In other cases, where the horses are in better health, the Chicago police sends them to one of three pastures it uses for retired mounted horses.

According to Naber Technical Enterprises, a training company devoted to mounted units, there are certain protocols that should be followed to insure the proper retirement of police horses (http://members.nbci.com/_XMCM/mntdpolice/Adopting.htm). Although the majority of mounted police units use horses that are owned and cared for by individual officers, many larger cities, including Boston, New York, and Chicago, maintain their own "stable" of horses. In order to find suitable adoption for these horses, one suggestion is to contact the closest city over 300,000 and ask the police for their police horse retirement policy. Almost every state has some incarnation of a mounted officer's association that also can be queried.

Procuring Horses

Depending on funding a number of alternatives exist for procuring horses. For example the Fort Lauderdale mounted unit established in 1983, was initially started with confiscated funds taken from drug dealers and other continuing criminal enterprises. Here was a case when no tax dollars were required, yet the unit was able to purchase all of its gear, including horse trailers, and was capitalized well enough to construct stables as well. In subsequent years, private citizens have donated all of the horses.

During the 1990s, a number of police departments took their campaign to recruit horses to the public. In 1997, the Los Angeles Police Department began its efforts to recruit horses. With 31 officers and only 28 horses the department papered the city with "wanted" posters for donated horses (Mounted Police Units, 1997). The department's stringent requirements result in only 1% of every 600 applicants being accepted annually. Even the vaunted Pennsylvania State Police has used recruitment campaigns to find adequate mounts. Pennsylvania remains one of the few state police forces to maintain mounted units.

Horses are acquired in a number of ways by mounted patrol units. The most popular in Doeren's 1989 study involved either donation and purchase. Other programs acquired horses on loan, others through trade, using an officer's privately owned horse, and in one case acquired a horse by trade. Most programs use some type of combination of these methods of procurement (Doeren, 1989, p. 15). In any case, it is important that a trial period be given each horse before the purchase is finalized. A contract stipulating a 30-day trial period should be required, in order to have enough time to evaluate the nature and potential of the horse, as well as to expose it to the various conditions that could affect it as a police mount.

Selection of Mounted Patrol Personnel

Applicants for mounted horse patrol should be selected on their past performance, experience, health, attitude, and most importantly, basic knowledge of, and experience working with horses. The U.S. Park Patrol, for example, requires a minimum of three years' patrol experience prior to being assigned to the special unit. In order to maintain stability and continuity, and to hone their equestrian skills, mounted officers should be willing to serve terms of between two to five years in the unit.

Typically, the vast majority of horse mounted officers have joined the unit at their own volition. Nearly all police agencies that use mounted units select officers from the ranks after they have proven themselves on patrol. Once a mounted candidate passes the initial requirements of having an outstanding and discipline free police record, he or she is assessed according to riding skills. It is not mandatory that a candidate is an expert rider. All that is required is that the officer is not afraid of horses, displays good balance, and has the potential to become a good rider.

Training for Horse Mounted Units

One of the major problems confronted by any mounted police unit is knowing the limitations and capabilities of the unit. Continuous training and instruction should be mandatory in order to fulfill the unit's potential. Both officers and horses should graduate from recognized mounted horse training programs, no matter how much experience the horse and rider already have. This is particularly true because of potential liability issues that could arise from on the job incidents with the public.

The amount of time required to fully train a police horse depends on the horse's previous training, its ability and desire to learn new skills, and its ability to accept new situations. In the initial training phase, the horse should become accustomed to having every part of the body touched, from the sensitive ears down to the hooves.

The patrol horse should then be taught the skills necessary to easily work in an urban environment. As the horse becomes more proficient, training becomes more intense, subjecting the mount to a nuisance training. Horses should undergo obstacle course training, that can include a number of scenarios. Some training methods have required horses to walk over an old mattress, go through plastic streamers, negotiate a foot bridge, and to walk across discarded tires. Other training includes introducing horses to the police cruiser's lights and sirens, shooting off firearms, and using firecrackers to stimulate holiday reveling. Other demonstrations should imitate crowd conditions, with actors shaking saddle cloths in the horse's faces and yelling.

Selecting Horses

Several criteria need to be considered when selecting police horses, including gender, height, weight, age, and breed (Doeren, 1989, p. 10). The majority of the horses in one 1989 study were geldings (castrated males), which were preferred to mares and stallions, because of their quieter nature, and thus are considered safer, presenting less liability issues. In the same study that examined eleven police agencies representing Chicago, Baltimore, Detroit, Boston, Albuquerque, New Orleans, Richmond, Virginia,

Sacramento, California, U.S. Park Service, and the Connecticut Governor's Horse Guards, all favored larger horses. A variety of reasons were offered, including the ability of larger horses to carry heavy officers for longer time periods; larger horses give greater visibility; they are better for crowd control and command more respect (Doeren, 1989, p. 11). Height is also important because a tall officer on a small horse presents a poor appearance.

In addition to size and gender, younger horses were preferred between the ages of three and seven, which reportedly insured a certain maturity, adequate term of service, and an ability to accept training. With the average life span of a horse at about 20–25 years, the average service for police work should be 10–15 years, depending on the horse's conditioning at an early age. Exceptions can of course be made as far as age range provided the horse is above average in conditioning and physical conformation. Veterinarians are qualified to estimate a horse's age, as can an experienced horseman by examining the front teeth.

Other considerations go into horse selection, some unrelated to peacekeeping duties. For example the U.S. Park Service uses horses that are only black in order to maintain conformity in parades and other public events.

Among the biggest problems faced by mounted units is adequate funding. An expensive investment, horses require a great deal of financial support. Horse trailers, trucks, and tack are part of the expenses that should be expected. Other problems involved inclement weather, injury to horses on duty, sanitation problems associated with biological functions, and the limited hours that a horse can work each day. The day-to-day care of horses is also very time-consuming.

Bicycle Units

There are a number of concerns that need to be considered when introducing a bike patrol. Much thought must go into purchasing the right bikes (see below). Just as important for budgetary reasons is having some type of bicycle fleet management plan. Keeping inventory for a bike fleet is challenging, since unlike motor vehicles, bikes are easily stolen or misplaced. It is essential to mark the bike frames with some emblem that identifies it as department property, hoping that it might deter theft. Bicycle frames have serial numbers, usually located near the pedals at the bottom of the frame. Numbers should be recorded. Additionally, the bike should be given a control number for inventory/maintenance purposes (Richardson, 2002). Other suggestions for maintaining inventory include etching number on wheels to prevent wheels being changed from one bike to another. This is a safety hazard since the drive train and the brakes are fine-tuned to each wheel. Finally, it is best for accounting concerns to assign bikes to specific personnel rather than generically to a unit or office.

Since police departments may have a bike fleet ranging 5–500 bicycles, storage concerns need to be considered. Although motor vehicles can be left out in the elements, bicycles need to be stored in an environment that will guard against rust and dry rot. Among the devices that are available to maximize storage space are simple bicycle storage hooks that can be easily obtained at hardware stores and professionally manufactured storage racks. Ceiling hooks should be mounted at least two feet apart to protect handlebars. Storage access should only be accessible to authorized personnel (Richardson, 2002).

Selecting Bike Patrol Equipment

When selecting a helmet, consider the construction type, comfort (venting and fit), and safety. Vendors offer a wide range of safety helmets for prices ranging from $40 to 200. This brings up the question, "Is a more expensive helmet safer than a less expensive one?" Consumer Reports ratings in 1999 suggest that a higher price does not mean greater safety protection. Helmets are currently made from expanded polystyrene (EPS) foam. This is considered the best material for absorbing impact and energy from crashes. An added benefit is that it can hold up under a wide range of temperatures and weather conditions. There are two common types of construction. The in-mold construction model (average cost of $50) is considered superior to the tape-on-shell style (King, 2003). These are exceedingly resistant to separation on impact or in a crash, because of the molding procedure and rigidity of the one-piece construction (King, 2003). It is recommended that a helmet should be replaced after impact or every 3–5 years. Of course this depends on use (Bicycle Helmet Safety Institute, www.bhsi.org, 2002).

Most bike patrol officers ride in every weather conditions except when it is icy or in hard rain. It is recommended that officers learn to layer clothing items since temperatures can vary widely during the day in some regions of the country. Uniforms run the gamut from slightly altered police uniforms to custom made bike apparel. The Albany police bike unit wears a yellow and black colored uniform, hence their nickname "the Bumblebees (www.albany.edu/police/units)."

Procuring Bicycles

Compared to horse mounted units, it is easier to budget for bike patrols. According to one observer, bicycle budgets "are often transparent to agencies' overall budgets," which accounts for their growing popularity (Sharp, 1999, p. 76). However, there is a wide disparity in the amount of funding accorded this unit. Obviously, expenditures often reflect the importance given the special unit by its department. In New York City, bicycle units are considered important, and officers receive the best training and equipment. Although not all departments must require top-of-the-line equipment, top-of-the-line training is a must for liability reasons.

Due to some of the disadvantages associated with bike patrols, some police departments are loathe to direct budget money directly to the bicycle units. Administrators often will seek funding sources that are not direct budget charges but do provide the availability of these units for specific functions. This may include securing donations or outside funding. Some departments even refurbish abandoned bikes.

By one estimate it costs $1200 per year for one bike, plus annual maintenance cost of about $200 (Sharp, 1999, p. 76). Smaller departments may find this price prohibitive. Although the average bicycle unit is composed of nine members, some agencies have at least 50 unit members. For comparative purposes, it costs between $23,000 and 30,000 to purchase a new police cruiser, and another $4000 per year for maintenance.

One way to reduce costs for a bicycle unit is for officers to learn how to perform bike maintenance. According to one source, "an in-house maintenance program is the best cost-saving program a department can initiate (Wideman, 1995, p. 35)."

In order to fund its bike patrol, the Fair Lawn, New Jersey, Police Department acquired new equipment, training, and winter gear with donations and grants. In order to start up the operation, the department paid for start-up costs and training with confiscated drug funds through county prosecutor's office. Such budget creativity can defray start-up costs for under-funded departments.

In Murphysboro, Illinois, an innovative approach allowed it to fund a bicycle patrol in 1997. Adopting the community-oriented vision of policing, the department received the donation of two mountain bikes from the local Wal-Mart. To purchase the four remaining bikes, the department used money collected through confiscated cash and property. Doing its part in the community-policing relationship, the bicycle store offered to service the bikes for free.

Types of Bicycles

Before an agency begins a bicycle program one of the most important issues to consider is what type of bicycle to purchase. The selection of bicycles for patrol fall into several categories, which includes standard bicycles, those specially designed for police work, those modified for police work by a bicycle shop, and those modified by the police department. The majority of bike patrols use some type of mountain bike, due to their versatility, durability, and performance. But, according to one source "the components that the mountain bike is fitted with—the drive train, brakes, and wheels—are more important than the frame itself" (Richardson, 2002, p. 2).

Most police departments have adopted mountain bikes. Equipped with packs over the back wheels to hold first aid kits, bike repair tools, routine paperwork, and ticket books, the bikes are easily carried on police car racks until they are needed for special patrols. The Santa Clara County Sheriff's office in California raised money for its bike program through a Citizens Options for Public Safety (COPS) fund. The money raised allowed the force to purchase six bikes, which cost $1000 per bike to be completely outfit for duty (Alaimo, 2001).

The variety of accessories used by bike patrols reflects the diversity of the units. Many bicycles are equipped with the same items found in police cars, including first aid supplies, automobile accident reports, flares, and other objects needed to perform law enforcement duties. Some departments use special emblems, specialized racks, lighting systems, electronic equipment, and special saddlebags. Others have experimented successfully with electric motors that are used for uphill riding. Some bicycles are even equipped with sirens, radar units, and computerized speedometers and odometers.

There is little diversity in personal equipment. Standard equipment usually includes handguns, holsters, radios, helmets, uniforms. Also common are riding gloves, pepper spray, batons, whistles, and special riding shoes.

Selection of Bike Patrol Officers

The overwhelming majority of bike patrol officers are volunteers from the police force. Officers who volunteer for assignment in the bike patrol unit are expected to meet certain criteria as established by the department to qualify. Following an officer's acceptance, the next requirement is satisfactory completion of a training course, often provided by the prestigious International Police Mountain Bike Association (IPMBA) Police Cyclist Course. This course is utilized by departments around the world. The course covers such varied topics as bicycle handling skills, night operations, bicycle maintenance, emergency maneuvers, nutrition, and defensive and pursuit cycling. Before attending such a course an officer should be physically capable of:

1. Sustained pedaling on flat and elevated surfaces;
2. Bicycle sprints of up to a half mile or more;
3. Bicycle maneuvering in motor vehicle and pedestrian traffic;
4. Riding over obstacles such as curbs, parking blocks, up and down stairs, and off-road type surfaces;
5. Carrying of bike and equipment weighing up to 40 pounds for short climbs and descents of less than 100 yards;
6. Defensive and pursuit cycling; and
7. Dismounting bike to a running transition (Marshall University site).

Bike Patrol Training

Similar to their horse mounted counterpart, proper training is essential for any bicycle patrol officer, whether they are full-time, part-time, or volunteers. The Seattle Police Department is considered one of the pioneers of bike policing. Some departments have used this force to train their new officers. No matter the training, it should include how to use the bike in crowd control, how to draw your weapon on the bike, and other areas.

IPMBA training consists of basic bicycle techniques of braking, maneuvering up and down steps, how to take down subjects using the bicycle, and being conscious of road hazards. One of the leading advantages of using bicycles in police work is the stealth factor. Most criminals keep watch for marked patrol units. The almost silent bicycle can pull up on crimes in progress unnoticed and in virtual silence.

The IPMBA Police Cyclist Course is the most utilized and state-recognized training program. It covers handling skills, night operations, maintenance, emergency maneuvers, nutrition, group riding, etc. In addition to the Police Cyclist Course, the IPMBA offers other courses at annual conferences. These include the Maintenance Officer Certification Course (MOC), the Advanced Police Cyclist Course, the Emergency Medical Services Bicycle Operations Certification Course, and the Police Cyclist Instructor Development Course (open for current instructors and candidates for instructor). Most forces and state training boards consider it necessary in the current liability-conscious society.

Some police forces have brought bike training in-house. For example, in 1996, the Richmond, Virginia police department used the experience and knowledge of its own officers to maintain control over its curriculum and to be cost-effective. It also offers the program to other law enforcement agencies. This course is unique in its education of officers on state bike laws, which historically had not been enforced.

Other training courses can range from 16-hour courses to four-day training programs. In California, the Newport Beach and Santa Monica Police departments take three-day courses. The Santa Monica force serves a population of more than 100,000. About 5% of the 196 full-time officers are assigned to bike patrol (Weinblatt, 1996). According to one source training for this unit includes strategic riding, safe operation, obstacle riding, and rapid dismount.

There is evidence that not all officers are willing to undergo the rigorous training program. In fact, indications are that many officers feel that this duty should be left to younger officers, since riding shifts often include riding up to 25 miles per day.

Liability Issues for Mounted and Bicycle Units

Liability policies should be carefully considered when using mounted or bicycle police officers. Legal problems can arise when volunteer officers that are not actual law enforcement officials are injured on the job. For example in a 1996 case in Ohio, a volunteer member of the mounted unit was injured and was denied compensation (Mounted Police Units, 1997). In this case the Ohio Court of Appeals ruled that the rider was not entitled to worker's compensation benefits for injuries that occurred when her mount slipped on shale in the creek. The court found that posse members, who lack arrest powers, could not be considered law enforcement officers or other employees under Ohio law.

In order to guard against injury and other liability issues involving both officers and civilians, it is of paramount importance to consider the disposition of the horse. The consensus is that horses should have quiet, even-tempered dispositions to prepare them for dealing in close proximity with crowds. Bad habits to be wary of include biting, kicking, rearing, bolting, and other characteristics that might threaten the public or the officer. It is also important that the horse does not mind petting or touching by children and adults (Doeren, 1989, p. 14). They also must not object to saddling, grooming, nor should they balk, rear, shy, or be hard to catch when turned out. Horses should also be evaluated for stable vices, such as cribbing, halter pulling, tail rubbing, weaving, bolting, and other habits that can lead to more serious problems.

No matter which special unit a police officer works in, there is a tendency to develop a "too intense" identification with the unit, whether it is SWAT, K-9, or mounted. There is anecdotal information suggesting that horse mounted officers develop a "cowboy" or "mountie" alterego, that if one source is correct, leads to the occasional lassoing of agitators in angry crowds (Fairburn, 1989, pp. 56–57).

Liability issues are also a concern for bike patrol units, and topics such as injuries and lost work time should be considered while developing a strategic plan. Most injuries are the result of falling off the bicycle or colliding with fixed objects. In hot, humid climates, bike officers are encouraged to develop good hydration habits, keeping water with them at all times. Departments usually provide sunblock, and skin cancer has not become a problem for bike officers.

Recognizing the importance of physical fitness, the Phoenix, Arizona bike unit permits officers to work out on duty for two hours each week. Several local gyms also offer special rates to officers. In order to avoid excessive injuries, bike training should focus on scenarios from previous incidents, do a lot of night training, and work on uneven terrain that could present a hazard. Other training aspects that can protect departments from liability include emergency maneuvers, nutrition, group riding, and bicycle maintenance. Learning to fall properly should be an important aspect of this training.

SUMMARY AND CONCLUSION

Demonstrating the cyclical nature of crime control, police departments throughout the United States have resurrected several police innovations (bicycles and horses) from the nineteenth century. The oldest recorded mounted police force was established in London in 1805. Over the next century the idea of mounted police spread to the larger cities of Britain and the United States. The growing influence of community policing in the 1980s and 1990s has led to the ready adoption of these people-friendly patrol options. Both alternatives invite constructive community contact between police officers and community residents.

The benefits of mounted policing cannot be measured in dollars alone. More accessible to the community than an officer in a patrol car, there has been little resistance to this type of patrol, whether on horseback or bike. Using these special units creates opportunities to interact with the public that do not exist with the traditional patrol car approach. Bike officers can often be spotted giving lectures at schools on bike safety and the importance of helmet-wearing. Some departments have "bicycle safety rodeos" to enhance community awareness of safety issues.

Mounted police officers, whether on horseback or bike, have the same jobs as officers in patrol cars, but feature some advantages and disadvantages. Among the most frequently cited reservations about creating horse mounted units are the expense, training, and maintenance burdens, as well as possible liability and insurance issues. Although horses offer advantages of height and visibility in crowd conditions, mounted police also offer larger targets for beer bottles and other hand thrown objects. However, the more positive attributes associated with horse units outweigh the negative. Horses offer a superior field of vision in crowds, parking lots, and uneven terrain. In addition, horses can go where motor vehicles cannot, carry more emergency equipment than a bicycle and are faster. One instructor compared a mounted officer to "an eight-foot tall, 1,400 pound police officer" (Fairburn, 1989, p. 55). Horses will defend their riders, are environmentally sound, and can work in wilderness areas without damaging the environment.

Bicycle police units have gained wide acceptance by the public and police forces over the past 20 years. Bicycle units have made an impact in drug enforcement, campus patrols, neighborhoods, tourist assistance, and special operations. Similar to the horse, the bicycle offers a proactive and flexible tool for crime control and efforts to suppress crime through high visibility, presence, and enforcement activity. Inclement weather poses special problems for bicycle officers, and in regions with severe weather problems, these patrols do not operate year-round. Technological advances and improvements in training and tactics have made the bicycle an important, if not indispensable addition to police operations. By 1999, more than 150 law enforcement agencies had taken the next step by adopting electric bicycles, or electrically assisted bikes (Siuru, 1996). According to one supplier of electric bikes, in 1999 they had become "the hottest product on the law enforcement market today" (Siuru, 1996, p. 81). The silence of the electric drive system insures a stealthy approach in areas of high crime activity. The adoption of horses and bicycles for police patrol has contributed to the popularity and success of community policing as law enforcement enters the twenty-first century. There is little doubt that this age-old answer to proactive crime control will continue to make an impact on police strategy in the next decades.

RECOMMENDED READINGS

CAMPBELL, J. (1968). *Police horses*. New York: A.S. Barnes.

CARFIELD, W.E. (1982). *Comparative analysis of twenty-five mounted units in the United States*. William E. Carfield: Eastern Kentucky University.

DOEREN, S.E. (1989). Mounted patrol programs in law enforcement. *Police Studies*, *12*(1), 10–17.

FAIRBURN, D. (1989). Mounted police training. *Law and Order*, March, 55–57.

FULTON, R.V. (1993). Community policing on horseback. *Law Enforcement Technology*, May, 32–34.

ROTH, M.P. (1998). Mounted police forces: A comparative history. *Policing: An International Journal of Police Strategies and Management, 21*(4), 707–719.

SHARP, A.G. (1999). Bicycle budgets: How to reduce wear and tear on your bicycle budget. *Law and Order*, April, 75–79.

TEXAS LAW ENFORCEMENT MANAGEMENT AND ADMINISTRATIVE STATISTICS PROGRAM (TELEMASP). (November 1998). *Bicycle Patrols, 5*(8), 1–11.

REFERENCES

ABILENE POLICE DEPARTMENT BICYCLE UNIT (2003). [Available at: http://www.abilenepolice.com/spe.htm.]

ALAIMO, M. (2001). *Mounted police debut this week in Saratoga.* www.metroactive.com/papers/saratoga.news/07.02.97/Mounted Police.html.

BECHTEL, H.K. (1995). *State police in the United States: A socio-historical analysis.* Westport: Greenwood Press.

BICYCLE HELMET SAFETY INSTITUTE (2002). www.bhsi.org.

CAMPBELL, J. (1968). *Police horses.* New York: A.S. Barnes.

CARFIELD, W.E. (1982). *Comparative analysis of twenty-five horse mounted units in the United States*, William E. Carfield: Eastern Kentucky University.

CONLEY, C. (September 1994). Los Angeles police department mounted unit. *Law and Order*, 70–73.

DOEREN, S.E. (Spring 1989). Mounted patrol programs in law enforcement. *Police Studies, 12*(1), 10–17.

FAIRBURN, D. (March 1989). Mounted police training. *Law and Order*, 55–57.

FORT LAUDERDALE POLICE DEPARTMENT (2001). *Mounted Patrol Unit.* http://ci.ftlaud.fl.us/police/mounted.html.

INDIANAPOLIS POLICE DEPARTMENT HISTORY (2003). [Available at: http://www.indygov.org/ipd/aboutipd/history.htm.]

JEFFERS, H.P. (1994). *Commissioner Roosevelt: The story of Theodore Roosevelt and the New York City Police, 1895–1897*, New York: John Wiley & Sons.

KING, K. (2003). *Helmet safety: Separating fact from fiction.* www.ipmba.org/newsletter-0308-helmetsafety.htm.

LESCE, T. (1999). Riding with the Rio Raucho Bike Patrol. *Law and Order*, April 1999, pp. 70–72.

MOUNTED GUIDE HOMEPAGE (2001). http://members.nbci.com/_XMCM?mntdpolice/Newmntdguideindex.htm.

MOUNTED POLICE UNITS (1997). www.law.utexas.edu/dawson/theme/polic_1997.htm.

NEW MEXICO DEPARTMENT OF PUBLIC SAFETY (2001). www.dps.nm.org/nmmp/impact.htm, February 10.

O'NEAL, B. (1987). *The Arizona rangers.* Austin: Eakin Press.

PRASSEL, F. (1981). *The Western peace officer.* Norman: University of Oklahoma Press.

REPPETTO, T.A. (1978). *The Blue Parade.* New York: Free Press.

RICHARDSON, T.J. (2002). *Bicycle fleet management: Keeping track of the company bikes.* www.ipmba.org/newsletter-0308-fleetmgmnt.htm, pp. 1–4.

ROTH, M. (1998). Mounted police forces: A comparative history. *Policing: An International Journal of Police Strategies and Management, 21*(4), 707–719.

——. (2001). *Historical dictionary of law enforcement.* Westport, CT: Greenwood Press.

ROUSEY, D.C. (1996). *Policing the southern city: New Orleans, 1805–1889.* Baton Rouge: Louisiana State University Press.

SHARP, A.G. (1999). Bicycle budgets: How to reduce wear and tear on your bicycle budget. *Law and Order*, April, 75–79.

SIURU, W.D., Jr. (1996). Electric-assisted bikes: Fast, silent, and environmentally friendly, *Law and Order*, 44(7): 53, 55.

STOCKTON, D. (1999). Phoenix Bike Patrol battles the heat. *Law and Order*, April 1999, 66–69.

TEXAS LAW ENFORCEMENT MANAGEMENT AND ADMINISTRATIVE STATISTICS PROGRAM (TELEMASP). (November 1998). *Bicycle Patrols*, 5(8), 1–11.

VAN DE WATER, F.F. (1922). *Grey riders: The story of the New York state troopers*. New York: G.P. Putnam's Sons.

WEBB, W. (1965). *The Texas rangers*. Austin: University of Texas Press.

WEINBLATT, R.B. (1996). Reserves patrol on bicycles: This new breed is cutting a wide path as they pedal forth. *Law and Order*, 44(3), 13–14.

WIDEMAN, M. (July 1995). Cyclist training in Georgia. *Law and Order*, 34–35, 40.

YARBROUGH, K. (1977). Plainclothes bicycle patrol: Silent—preventive—effective. *FBI Law Enforcement Bulletin*, May, 10–13.

APPENDIX 1
Origins of Some Horse Mounted Units in Major American Cities
Those in Continuous Service:

New York City, 1871 to the present,

San Francisco, 1874 to the present,

Boston: 1883 to the present,

Baltimore: 1890 to the present,

Cleveland: 1906 to the present,

New Orleans: 1920 to the present.

Those that have been reactivated:

	Deactivated	Reactivated
Chicago	1945	1974
Detroit	1870	1893
Miami Beach	1940	1979
Portland, OR	1928	1980
Richmond, VA	1925	1941
St. Louis	1948	1971
Wilmington, DEL	1928	1980

11

Federalism, Federalization, and Special Units: An Issue in Task Force Management

Peter J. Nelligan
Peter W. Phillips
W.A. Young, Jr.

Special units that combine law enforcement agencies horizontally across the same level of government or vertically across different levels of government, or both, multiply the opportunities and challenges of managing task force units created and operating totally within a single agency. While these multi-jurisdictional task forces (MJTFs) offer the promise of combined expertise, expanded resources, and geographical reach, they also raise issues of accountability, autonomy, coordination, communication, and control. If a federal agency is part of the mix, they also raise important issues concerning the United States' historical and Constitutional commitment to federalism. When MJTFs proliferate under the dominance of federal law enforcement agencies, they may constitute a nearly invisible force toward the federalization of criminal law, criminal investigation, and prosecution that historically have been the responsibility of state and local law enforcement agencies. The phenomenon may have adverse impacts on local law enforcement agencies as well as important principles concerning the appropriate roles of different levels of government.

Multi-jurisdictional task forces are discussed in several other chapters of this volume because they represent a significant proportion of the special units deployed today in law enforcement. In each of the other chapters, the authors have made recommendations for the management of the special units they are describing. None of the other chapters, however, emphasizes the political aspects of MJTF management.

In this chapter, therefore, the authors examine a specific case for the purpose of demonstrating the potentially adverse impact on local law enforcement of the proliferation of MJTFs under federal control. In particular, it examines events in one jurisdiction where the Federal Bureau of Investigation, eager to formally establish an MJTF, exerted extreme pressure on a municipal police department to join, despite the police department's concerns that such an association would constitute an improper surrender of its autonomy as well as have other harmful effects.

METHODOLOGY

The methodology employed in this research is that of the descriptive case study. Relying on interviews, official documents, and press reports, the authors reconstruct and analyze events surrounding a discrete phenomenon. Descriptive research, is defined by Best (1959) as

> The description, recording, analysis, and interpretation of the present nature, composition, or process of phenomenon; [focusing] on the prevailing conditions, or how a person, a group or thing behaves or functions in the present. (p. 12)

Further, the value of descriptive research is to tell the "what is" of a specific problem. "Descriptive studies serve many very important functions. Under certain conditions, it is of tremendous value just to merely know what the current state of the activity is" (Borg & Gall, 1967). Descriptive case studies permit researchers to investigate a phenomenon in much more detail than more large-scale or tightly designed methodologies. Case studies are particularly useful in the early stages of the examination of a phenomenon when it already has not been fully described and analyzed. They can identify the important elements and issues of a phenomenon that will provide direction for larger-scale studies.

Case studies have, however, inherent limitations. Due to their complexity, they involve a certain level of subjectivity. Different researchers examining the same phenomenon may draw different conclusions. More importantly, the findings of case studies are of unknown generalizability. The question of whether the properties of a single case are idiosyncratic or shared across members of the class can only be answered by further research.

We begin our discussion by presenting the historical context of federalism and criminal law.

FEDERALISM AND THE FEDERALIZATION OF CRIMINAL LAW

Nearly 95% of the nations of the world have unitary governments in which all political power is exercised by the central political authority (Edwards, Lineberry, & Wattenberg, 1999). Under such arrangements there are no issues of intergovernmental relations; the central government has complete formal control and may dictate to, or indeed abolish, any lower-level governmental entity. The United States is one of a handful of nations that adopted a federal system in which power is shared among different levels of government. While this arrangement was intended to preserve liberty and to place the relative duties of government at the level of greatest competency, it also gives rise to constant controversies over the boundaries of the duties, powers, and prerogatives of the national and state governments. James Q. Wilson (2000) comments, "Since the adoption of the Constitution in 1787, the single most persistent source of political conflict has been the relations between national and state governments" (p. 34).

The choice by the founders of a federal form of government was not happenstance. Given their recent history with England, the colonies were not inclined to completely surrender their sovereignty to a powerful central government. Rather, they wished political power and government operations to remain close to and accountable to the people. What emerged from the Constitutional Convention was a document that substantially limited the powers of the federal government and reserved to the states the power and duty to regulate the conditions of life in the communities of the new nation.

Although the Constitution empowered the federal government to make and enforce criminal law, the founders clearly contemplated that most regulation of conduct by use of criminal law would take place at the state and local level. The jurisdiction of the United States courts was limited to distinctively federal business and to provide an impartial forum to resolve disputes between states and between the residents of different states. The Constitution did not confer on the federal government general criminal jurisdiction. Federal criminal law was limited to those realms in which the federal government had a specific duty, such as coining money, operating a post office, regulating conduct on its property, and regulating interstate commerce. According to legal historian Lawrence Friedman (1993), in the early decades of the nation's history:

> The federal government was a bit player, a spear-carrier in the drama of criminal justice. The state courts were the exclusive venue for ordinary cases, ordinary offenses; the federal courts handled only, special, "federal," crimes. (p. 71)

Federalism has operated in criminal justice substantially as expected by the founders. For all of the nineteenth-century criminal justice was almost exclusively a local concern (Friedman, 1993). Nearly all criminal prosecutions were brought under state criminal statutes, based on charges brought by local police and prosecutors in state courts before state judges; and the guilty fell under the authority of state penal systems. By the mid-twentieth century, however, the trend in criminal justice, as with other areas of governmental activity, was away from federalism and toward federalization.

Despite an initial framework that severely limited the powers of the national government in relation to the states, the long-term historical trend has been a flow of power away from the states and to the federal government. As the population of the nation increased and became mobile; as communities, states, and regions integrated economically; as transportation and communication became easier; as newspapers addressed mass markets; problems increasingly came to be perceived as national in scope, calling for the uniform and centralized solutions that only a national government could deliver. The process proceeded slowly but methodically in criminal law and criminal justice.

Beale (1996) identifies three major periods of expansion of federal criminal jurisdiction. The first period, the decades just after the Civil War, was Congress's attempt to respond to new problems associated with economic growth and particularly increased commerce, transportation, and communication across state lines. The first broad extension of federal criminal law was a mail fraud statute passed in 1872. Other notable statutes were the White-Slave Traffic (Mann) Act of 1910; the Harrison Act of 1914, taxing narcotic drugs; and the Dyer Act of 1919, making it a federal crime to transport stolen vehicles across state lines. Congress claimed authority to extend the criminal law into areas that previously had been left to the states in its Constitutional authority to regulate interstate commerce. In *Champion v. Ames* (1903), the Supreme Court ruled that Congress's prohibition of the transportation of lottery tickets across state lines was a Constitutional exercise of its power to regulate interstate commerce.

The second period of expansion was an era inclusive of prohibition, the Great Depression, and the New Deal. Prohibition signaled the government's willingness to aggressively regulate moral conduct, and brought a flood of cases into the federal courts. The conclusions of a 1937 congressional committee that the states could not adequately respond to a growing criminal activity spurred Congress to create even more federal crimes (Beale, 1996). New Deal regulatory schemes included new federal crimes and penalties. Again Congress claimed and the Supreme Court generally upheld the interstate commerce rationale for its enactments.

Beale (1996) divides the most recent period of federal criminal expansion into two sub-periods: during the 1960s and 1970s the federal government focused on organized crime and control of drugs, and during the 1980s and 1990s more specifically on violence. Two notable enactments of this period were the laws making automobile hijacking a federal offense and the law creating gun-free school zones. The cumulative effect of the federalization trend during these three historical periods is that the number of federal crimes has increased from 17 about two centuries ago to somewhere in the vicinity of 3000 today (Beale, 1996). Indeed the pace has increased markedly in recent years. The American Bar Association (1998) concluded that of all the federal crimes enacted since the Civil War, more than 40% were enacted since 1970, and that the pace has been the fastest since 1980. More important than the numbers is the fact that many of the most recent federal enactments overlap state statutes. However, that trend may be changing.

As mentioned previously, since the federal government has no constitutionally established power to define, investigate, and prosecute crimes, Congress has justified the federalization of criminal law under one of its enumerated powers—the power to regulate interstate commerce. During nearly all of the twentieth century the federal courts showed great deference to the Congressional assertion that nearly any act affected interstate commerce.

At the end of the twentieth century a much more conservative U.S. Supreme Court signaled a change in direction in considering the relative powers and duties of the federal government and the states. In 1994, in a five to four decision in *U.S. v. Lopez* (1995), Justices Rhenquist, Kennedy, O'Connor, Scalia, and Thomas struck down struck down Section 922 (q) of the Gun Free Schools Zone Act of 1990 (18 U.S.C. sec. 922(q)(1)(A)) making it a federal offense to knowingly possess a firearm in a school zone. The majority held that possession of a firearm in a school zone does not "substantially" affect interstate commerce and was therefore beyond Congress's power to prohibit. Five years later, in *U.S. v. Morrison* (2000) the same justices went even farther, striking down the Violence Against Women Act (1994) that prohibited victims of gender-related violence to seek civil remedies in the federal courts. Extending its principles in the *Lopez* decision, the court strongly asserted that generalized police powers belong to the states and that there is ". . . no better example of the police power . . . than the suppression of violent crime and vindication of its victims" (p. 1754). It emphatically rejected the idea that ". . . Congress may regulate noneconomic, violent criminal conduct based solely on that conduct's aggregate effect on interstate commerce . . . " (p. 1754).

Thus, in the *Lopez* and *Morrison* cases, the Supreme Court indicated that most violent crime is a matter for the states and that it will not tolerate federal usurpation of the states' rights and duties to suppress it under vague connections to interstate commerce. The Supreme Court clearly wants federal law enforcement out of the violent crime business. However, the proliferation of or restriction of federal criminal statutes is only part of the story. Statutes require an administrative apparatus for enforcement. It is to the growth of this apparatus we now turn.

Although the Revenue Cutter Service was established in 1789 and the Secret Service emerged during the Civil War, it wasn't until the creation of the Bureau of Investigation in the Justice Department in 1909 (renamed Federal Bureau of Investigation in 1935) that a broader federal law enforcement presence began to emerge (Morgan, 1983). It grew rapidly as Congress created new federal crimes. Arguably the most important event in its history was the appointment of J. Edgar Hoover as its director in 1924. Hoover cleansed it of corruption, professionalized it, and spent nearly 50 years expanding its influence and burnishing its image. In addition to the routine enforcement of the increasing number of federal criminal statutes, the FBI and its predecessor agency responded to real and imagined threats to the nation that waxed and waned during most of the twentieth century—espionage and subversion during and after World War I, gangsterism during the 1920s and 1930s, espionage and domestic communism after World War II, civil rights and political dissent in the 1960s and 1970s, and organized crime today (Budget of the United States, 2000; Morgan, 1983).

Over the years the FBI was viewed by most Americans in and out of law enforcement as a model of professionalism and expertise. Its crime laboratory, fingerprint identification system, National Crime Information Center, and National Academy (NA) are all major resources for local law enforcement.

Morgan (1983) further points out that "Doubts were voiced by some about the steady expansion of the FBI, but they were ignored" (p. 770). Indeed, if one examines only the budget and number of employees of the FBI for the past decade, one can see the results of the most recent expansions. As shown in Table 11-1, from Fiscal Year 1990 to 2000 the number of authorized special agent positions in the bureau increased approximately 19%; total personnel, 28%; and appropriations, 95% (Congressional Research Service, 2000). In the last five years, the FBI has been building agent strength at a rate of approximately 300 per year. One of the uses of this manpower is the movement of the FBI into the suppression of violent crime, often through the mechanism of MJTFs.

We arrive, then, at something of a paradox. Ending a long historical trend, the U.S. Supreme Court, has signaled a halt to the federalization of the criminal law, at least the proliferation of statutes. On the other hand, it seems that federal law enforcement agencies, particularly the FBI is moving aggressively to expand its role. We turn now to consideration of the MJTF in this endeavor.

DEFINING AND COUNTING MULTI-JURISDICTIONAL CRIME TASK FORCES

Multi-jurisdictional task forces are not easy to either define or count. One source reports that there are approximately 700 MJTFs employing almost 10,000 officers funded by federal discretionary grant monies as part of the Anti-drug Abuse Act of 1986 and 1988

TABLE 11-1   FBI Authorized Personnel and Budget Appropriations, Fiscal
              Years 1990–2001

Fiscal year	Special agents	Support personnel	Total personnel	Congressional appropriation**
1990	9,851	12,729	22,580	1,684,444
1991	10,314	13,369	23,683	1,697,121
1992	10,479	14,118	24,597	1,927,231
1993	10,273	14,118	24,107	2,007,423
1994	9,875	13,538	23,323	2,038,705
1995	10,067	13,675	23,742	2,138,781
1996	10,702	14,336	25,038	2,530,045
1997	11,271	16,039	27,310	2,837,610
1998	11,545	16,311	27,856	2,937,000
1999	11,677	16,755	28,432	2,987,000
2000	11,733	17,080	28,813	3,002,876
2001	11,828	17,375	29,203	3,277,562

*2001 estimated.
**in Millions.
Sources: For 1998–2001, The Budget of the United States, 2000, pp. 640–642, 2001, pp. 648–651.

(Lyman & Potter, 1997). That number is probably low when all types of MJTFs are concerned, since it does not consider the number established with only state and/or local funds, or those established to specialize in types of crime not related to the drug problem. Although the formation of an MJTF is often considered as a possible solution whenever a particular crime problem appears to transcend normal jurisdictional boundaries, one is hard put to find a formal definition of the term.

An MJTF is often defined by its function, its advantages, and/or its goal. For example, Lyman and Potter (1997, p. 430) note that crime task forces have the advantage that "they can legally cross over mutual jurisdictional boundaries." Government grant programs that promote the use of crime task forces may take one of several directions in their definition. The Bureau of Justice Assistance (1996) described the crime task force through its attributes or advantages:

> The Byrne Formula Grant Program has led to improved cooperation among agencies, enabling the agencies to work as a single unit across jurisdictional boundaries. Smaller departments have been able to engage in undercover activities that they could not perform solely with their own resources. Task forces have enabled agencies to dedicate personnel full time to such activities as drug enforcement, gang abatement, and major financial investigations. The use of such dedicated personnel permits task forces to increase the size of caseloads and obtain better equipment. Undercover operations are improved by the facilitated exchange of undercover officers among agencies. These officers are able to

operate more effectively because they are unknown to local drug dealers and members of the target community. Task forces have been able to adopt a problem-solving approach that includes targeting and apprehending higher-level criminals, deterring other distributors from entering markets, and making undetected movement across jurisdictional boundaries more difficult. (p. 1)

For the purposes of this chapter, a non-crime-specific definition of an MJTF is applied:

An MJTF is a special law enforcement organization with multijurisdictional authority created by the agreement of several government bodies to more effectively combat a delineated crime problem and using the combined resources, both human and logistical, of several law enforcement agencies to more efficiently combat the stated problem for the term of the agreement. (Orvis, 1999; Phillips & Orvis, 1999)

EVOLUTION OF THE CRIME TASK FORCE CONCEPT

Documentation regarding the beginnings of the task force concept is sparse. Robert F. Kennedy, Attorney General during the early 1960s, is generally recognized in the law enforcement community as the "father" of the Organized Crime Drive Task Force. He theorized that a combination of representatives from federal, state, and local law enforcement agencies, answerable to and directed by a strong federal central authority and insulated from area politics, would have the greatest chance of eliminating the organized crime threat posed by the Mafia (Hyatt, 1991). The task force concept proved so successful as an organized crime enforcement strategy that its application to other types of criminal investigations soon became abundantly clear, with variations in name only (Phillips & Orvis, 1999).

In the *Multi-Agency Investigative Team Manual*, Brooks, Devine, Green, Hart, and Moore (1988), following the serial murder episodes of the late 1970s and early 1980s, developed procedures to assist agencies and task forces in managing complex homicide investigations and to assist them in preparing submissions to VICAP, the FBI's Violent Criminal Apprehension Program. This manual contains one of the first acknowledgements that the management of intergovernmental relations is different than the management of a single law enforcement agency's actions (Phillips & Orvis, 1999).

The mid-1980s through the mid-1990s was a period when federal grants were often awarded to local and state law enforcement agencies with the purpose of promoting the use of task forces. The Bureau of Justice Assistance (1996) counted 493 of these formula-grant funded task forces in 29 states. Among the respondents, city police officers and sheriff's deputies, at 88%, were those cited most often as members. Federal law enforcement officers were participants in less than 25% of the task forces. It has been federal funding since 1988 that has resulted in the most rapid proliferation of crime task forces. Dunworth, Haynes, and Saiger (1997) report that there were between 900 and 1100 MJTFs within the next decade, with the overwhelming majority of these being drug task forces.

Within the Justice Department's Office of Justice Programs, the Executive Office of Weed and Seed now supports efforts at approximately 200 sites nationwide with grants averaging about $225,000 annually. Interagency collaboration is a central component of the weed and seed strategy and the law enforcement approach at most sites includes multi-agency task forces. Dunworth and Mills (1999) report that:

> The sites developed varying degrees of increased local, State, and Federal coordination, whether in targeting offenders, narcotics operations, prosecution, or probation/parole. Local responses ranged from increasing communication through monthly meetings to creating formal interagency and multijurisdictional task force operations housed at the same facility. (p. 6)

For FY 1995, the U.S. Attorney General's Office received a one-time Congressional appropriation of approximately $15 million in no-year funds (balances may be carried forward through subsequent years until the funds are totally disbursed) for the purpose of developing and enhancing Violent Crime Task Forces and similar task force operations (e.g., Violent Crime/Fugitive Task Forces). This was different from monies appropriated to the FBI for its Safe Streets Task Forces and could supplement, not supplant the FBI appropriations. Subsequently 142 task forces were approved to receive varying degrees of support, primarily to be used for state and local officer overtime, limited clerical support, and some special equipment such as cell phones and secure radios. According to an administrator in the U.S. Attorney's Office, funds are now depleted (McDonough, 1998). The number of FBI Safe Streets Task Forces, however, continues to multiply.

FBI SAFE STREETS TASK FORCES

After the breakup of the Soviet Union and the diminished need for FBI resources allocated to national security, and commensurate with an increase in violent crime in the U.S., on January 9, 1992, the FBI initiated the Safe Street Task Force (SSTF) program to focus on this domestic problem. The main focus of any SSTF is on the primary violent crime problem in the geographic area served by the task force. In general, the focus of SSTFs has followed national crime patterns; since 1992, they have emphasized fugitives, gangs, and enterprise crime where racketeering statutes can be applied. Some SSTFs are organized to deal with specialized crimes in particular geographic areas, for example, auto theft on the Mexican border, cargo theft in port areas, and high-tech theft in Silicon Valley (Neu, 1999; Rachlin, 1993).

> Safe streets is an initiative designed to allow the Special Agent in Charge of each FBI field office to address street, gang and drug related violence, through the establishment of FBI sponsored, long-term, proactive task forces focusing on violent gangs, crimes of violence, and apprehension of violent fugitives. (FBI, 1997)

Funding for local law enforcement involvement in SSTFs comes from the FBI's budget. The data entered at Table 11-2, gathered from several sources, shows the growth of SSTFs from 1995 to 1999. SSTFs presently total 164 (FBI, 1999). It is within this context that controversy emerged regarding the conditions set forth in a newly promulgated Memorandum of Understanding (MOU) concerning the establishment of regional violent crime task forces.

TABLE 11-2   Growth of Safe Streets Task Forces, 1995–1999

Source	Year	Total SSTFs	FBI Field offices involved	Violent crime	Other specia- lized	FBI agents assigned	Other fed. agents	State and local officers
Shur, D.	1995	119	52	80	39	n.r.	n.r.	350
FBI	1997	152	54	111	41	n.r.	n.r.	n.r.
FBI	1998	162	54	124	38	781	182	1207
FBI	1999	164	52	126	38	785	142	1255

Note: Depending on source and year, Violent Crime SSTFs may be labeled: "Violent Crime," "Violent Crime/Fugitive," and/or "Violent Crime/Gang." For the purposes of this table, those separate designations have been grouped simply as Violent Crime SSTFs. None of the reports reviewed revealed the exact number of FBI Resident Agencies (subordinate to the field offices) that are the actual sites of the SSTFs.

A CASE IN POINT

The East Texas Violent Crimes Task Force (ETVCTF) was organized in 1993 through a "Gentleman's Agreement" among the chief executives of the following agencies: the Tyler, Longview, and Marshall Police Departments; the Gregg and Smith County Sheriff's Offices; the FBI and Bureau of Alcohol, Tobacco and Firearms resident agencies in Tyler; the U.S. Attorney's Office for the Eastern District of Texas and Smith County District Attorney's Office. The western edge of the geographic area served by the task force is approximately 100 miles due east of Dallas, the eastern edge being close to the Louisiana border. The area is bisected by Interstate Route 20 and has a population of approximately 275,000.

In a study of the ETVCTF, Phillips and Orvis (1999) found that for the first six years of its existence, the task force operated on a strictly informal basis. No MOU or other written document outlining mutually agreed upon operational standards was executed by the participating agencies. Although the participating agencies designated particular officers as the operational members of the task force, they came together as a unit only when the task force was activated. Activation was initiated when one of the participating agencies opened a violent crime investigation and requested assistance from the task force. When the case was solved or leads at least temporarily exhausted, the officers went back to performing their regularly assigned agency duties until again "called out" (Phillips & Orvis, 1999). Within its suite of offices in Tyler, the FBI did provide meeting space for task force officers, a modest amount of clerical support, and any agent who participated in a task force investigation had access to superior Bureau resources. Particularly significant in retrospect, however, is the fact that the local police agency in whose jurisdiction the violent crime was committed became the lead agency in that investigation and directed the efforts of the task force when it was activated. The FBI did neither more nor less than it has done historically to assist local law enforcement. As it operated, the ETVCTF was not significantly different from any other mutual aid pact of the type commonly encountered in the police, fire, and emergency medical services.

Beginning in October 1998, however, relationships began to change. The Supervisory Senior Resident Agent (SSRA) in charge of the Tyler FBI office promulgated a draft document titled, "Memorandum of Understanding, East Texas Violent Crime Task Force" (dated at Tyler, TX, October 1). When this document was received by the largest of the participating organizations in the heretofore informal task force—the Tyler Police Department—certain provisions of the proposed, formalized, and contractual agreement raised red flags.

Based on concerns raised by the Tyler police chief by telephone, the Tyler SSRA sent an undated letter in which he attempted to explain certain provisions, particularly the kinds of crimes for which the FBI felt it should be notified (Wilkins, n.d.). In that letter to the police chief, the Tyler FBI SSRA stated: "There are many state crimes wherein the FBI has concurrent jurisdiction. New Federal laws have come into effect over the course of the 1980s and 1990s as a result of congressional concerns" (p. 1). He also provided:

> a list of crimes that should be used to determine when to call the FBI in Tyler for assistance and/or because of our investigative jurisdiction:
>
> - Carjackings
> - Drug related homicides
> - Drive-by shootings
> - Armed robberies of commercial establishments
> - Kidnappings
> - Extortions
> - Bank robberies/burglaries
> - Sexual exploitation of children
> - Robberies/burglaries with intent to obtain controlled substances
> - Cases with the potential for Interstate Transportation in Aid of Racketeering, e.g. murder/robbery for hire when the arrangements may have necessitated crossing state lines
> - Murders wherein the motive is insurance proceeds. (Wilkins, n.d., p. 1)

The point here is not whether the FBI does or does not have concurrent jurisdiction—it is that these are all crimes completely covered by Texas criminal statutes. As will be discussed later, a local law enforcement chief executive cannot abdicate his or her responsibility, either to the governance structure or to the citizens of the municipality, by delegating authority over a case to an outside agency. The chief executive is accountable for the conduct of investigations of crimes defined by the penal laws of that state as committed in the jurisdiction and reported to his or her agency.

This was followed by a second draft of the MOU sent to the police department by the SSRA in January 1999 (Wilkins, 1999). Meetings were then held among the following participants in the informal East Texas Violent Crimes Task Force: the Tyler Police Department, the Smith County Sheriff's Department, the Smith County District Attorney's Office, and the FBI. This resulted in a third draft of the MOU being distributed in April (Dobbs, 1999).

Subsequently, the FBI summarily rejected the April revisions to the MOU, which satisfied the major concerns of the Tyler Police Department. In this case, the rejection came not from the Tyler SSRA, who had apparently approved of the latest changes, but rather from the regional supervisor of the Tyler SSRA. In a letter to the Tyler police chief dated May 20, 1999, Edward Leuckenhoff, Assistant Special Agent in Charge of the Dallas Division of the FBI said:

> I stress that the standard MOU language, including the requirement of day-to-day administration by the FBI, is set by DOJ [the Department of Justice] in Washington. Neither the Tyler Resident Agency nor the Dallas Division of the FBI have any discretion or latitude in this regard. (p. 2)

The version of the MOU attached enclosed with Leuckenhoff's letter was exactly as it had been proposed originally, making the official FBI position both unalterable and unacceptable to the Tyler Police Department. For the reasons described below, the Tyler police chief refused to sign the MOU and did not remain an official participant in the task force.

In its analysis of the draft MOU, the Tyler Police Department perceived that the FBI was seeking to expand its influence over local violent crime investigations. A local agency that signed such an agreement would be obligated to adhere to stipulations that, for all intents and purposes, gave administrative and operational control over specific local cases to the Tyler office of the FBI. The proposed East Texas Violent Crimes Task Force Agreement required that:

1. The day-to-day operation and administrative control of the East Texas Violent Crimes Task force be the sole responsibility of the Supervisory Special Agent of the FBI.
2. The investigative methods employed be consistent of the policies and procedures of the FBI and certain unspecified guidelines from the Attorney General's Office.
3. The FBI Supervisory Senior Resident Agent assigned to the East Texas Violent Crimes Task force oversee the prioritization and assignment of accepted cases and related investigative activity in accordance with unstated or otherwise nebulous objectives of the task force.
4. All investigative reporting would be prepared in compliance with existing FBI policy.
5. Investigative files of the ETVCTF be maintained in a locked file room within the Tyler FBI Resident Agency and all case files would be prepared using standardized FBI forms.
6. The FBI maintain all evidence and original tapes acquired during the course task force investigations. All task force personnel adhere to FBI rules and policies governing the submission, retrieval, and chain of custody.
7. Matters designated to be handled by the East Texas Violent Crimes Task force would not knowingly be subject to non-task force law enforcement efforts.
8. There be no unilateral action taken on part of any participating agency relating to East Texas Violent Crimes Task force investigations.
9. Informants and witnesses be managed according to FBI and the Attorney General's guidelines. Informant and witness files would be maintained in the Dallas FBI office.
10. No unilateral press release be made by any participating agency without the prior approval of other participants. No information pertaining to the East Texas Violent Crimes Task force itself will be released without approval of all participants.
11. Any assets, awards, or seizures of property will be divided on the equal basis [rather than a formula based on personnel-hours contributed or a like variable].

12. Participating agencies agree to provide the full-time resources of its personnel for the life of the operation.
13. Subject to funding availability and legislative authorization, the FBI reimburses the participating local and state agencies the cost of overtime worked by task force members assigned full-time to the task force.
14. Continuation of the MOU, and, by implication, the task force, be subject to the availability of necessary funding (FBI, 1998).

It is important to note that local law enforcement administrators adopted differing strategies for coping with these possibly objectionable changes in policy. For example, one administrator told the Tyler police chief, "I'll sign anything to get the money, then do as I damn well please." Apparently, this approach is acceptable to the FBI because the Special Agent in Charge of the Dallas Division is reported as having said that in practice, the task force policies are not that rigid, and quoted as saying:

> This is a collective agreement. Only the management and overall running of the task force has to be supervised by someone. Many times, what you read in black and white, when dealing with the federal government, some standards have to be reviewed. (Gillaspie, 2000)

The Tyler police chief and executive staff members found the ETVCTF proposal to be unacceptable. They did not believe it was appropriate for a local law enforcement agency to relinquish responsibility to the FBI for the administration and operational oversight of investigations of local major violent crimes occurring within the jurisdiction of the city. They also felt that to contribute four full-time officers to the task force, as was demanded by the FBI, was an excessive drain on the department's resources.

In a letter to FBI Assistant Special Agent Leuckenhoff dated May 27, 1999, the Tyler Police Chief Bill Young stated:

> The MOU as written places all human resources, housing, administration and operational responsibility and control directly under the authority of the FBI, applying FBI investigative processes, policies, practices, and procedures. This approach to a taskforce does not in my opinion depict what I consider a "partnership" [a term used by the FBI's Leuckenhoff in earlier correspondence]. (p. 2)

Another matter of concern to the Tyler police administrators was the prohibition from discussing violent crime issues with the media without the approval of all task force participants. An obligation to be forthright with the media—and through them to the public regarding serious, violent, and sometimes spectacular crimes occurring within Tyler—was perceived by the police chief. This obligation could be neither abdicated to an outside authority nor postponed until all task force participant leaders could be polled.

Likewise, the language prohibiting any enforcement efforts relating to violent crime cases adopted by the task force by non-task force officers was tantamount to telling all other officers that they should not actively seek investigative leads because they would be disallowed from following up on them anyway. This would not only perpetuate the common rift between uniformed and plain-clothes officers that has been a perennial police problem (see, e.g., Guyot, 1991), it would risk driving the wedge of elitism between groups of investigators themselves.

The most perplexing stipulation of all was the continuation of the task force being contingent upon funding availability versus the actual issue of "violent crime." It was the opinion of the Tyler police administrators that the continuation of a Violent Crimes Task Force ought to be contingent upon the existence of violent crimes—not the availability of federal dollars. To them, the question was: What is the objective: the elimination of a major crime problem or simply spending federal money?

Even FBI headquarters offers differing opinions as to what the MOU actually mandates. Kenneth Neu (personal communication, December 15, 1999), Violent Crimes Section, FBI Headquarters, stated that there is no standardized, overall Memorandum of Understanding (MOU). There is, however, standard language related to the:

a. Federal Tort Claims Act in reference to deputizing local officers as federal agents;
b. Handling of informants based on regulations applicable to all federal law enforcement agencies;
c. Prosecution of enterprise crime in relation in assuring adherence to RICO-case guidelines from the Attorney General;
d. Overtime and reimbursement, in that appropriate language must be included to refer specifically to the current year's appropriation bill or authorizing public law.

With regard to the controversy in Tyler, Texas, wherein the police chief had refused to sign the MOU, Neu (personal communication, December 15, 1999) noted that the FBI had recognized problems before, referring to misunderstandings of language in the document. He said that the FBI was trying principally to (a) assure a truly coordinated effort and (b) provide financial tools, that is, for overtime reimbursement to local departments. In terms of media releases, for example, the FBI wants to (a) minimize leaks that could compromise complex investigations and (b) reduce the possibility of releasing conflicting information. The language in the MOU is rather explicit regarding the FBI's role in making news releases.

When asked what the FBI's position would be if a participating State or local law enforcement agency head deviated from any of the conditions of the MOU, Neu (personal communication, December 15, 1999) said,

> If one of the agency heads deviates from conditions of the agreement, it should be up to the entire group to get together and decide what to do about the "variation." We [the FBI] know what the history of civil litigation has been and we know that we'll be party to it in the future. We have an obligation to protect the government from future lawsuits.

When the Dallas Division of the FBI informed the Tyler police chief that the MOU was unalterable because the Department of Justice had prescribed its content, they also sent a copy of that letter to the Tyler mayor (Leuckenhoff, 1999). The chief's refusal to sign the MOU led him into a confrontation with the mayor and the city manager, wherein he was accused of being "uncooperative" with other area law enforcement agencies, in particular, the FBI.

The chief remained steadfast in his position that the conditions set forth in the MOU usurped local prerogatives and legal mandates. Further, to sign the agreement without the intent to abide strictly to the contractual language was professionally unethical and professionally abhorent.

The chief's resignation was then requested and he complied. Approximately six months later, an internal candidate was chosen by the city manager as Tyler's new police chief. To fulfill a "campaign pledge," and within only a few weeks of his official appointment, the new chief signed the MOU in a highly publicized meeting.

It is interesting to note that the Tyler Police Department is a signatory to an MOU defining participant obligations in the Drug Enforcement Administration/Tyler Multi-Agency Task Force. (This MJTF covers a six-county area of East Texas from the DEA Resident Agency in Tyler.) The DEA task force was formalized in 1991 and has been reauthorized by all participants each year thereafter, including the last reaffirmation, August 18, 1999 for the federal fiscal year 2000. Of 17 sections specifying participating organization obligations, only one provision of this MOU contains language similar to that found in the FBI MOU. That one provision relates to the day-to-day supervision of the Tyler police officer while acting as a special DEA officer (DEA/Tyler Multi-Agency Task Force Agreement, 1991; amended 1999, p. 2). Is DEA not a Department of Justice bureau, too?

As mentioned above, the Tyler police chief did not sign the MOU. The quotation earlier from the Dallas FBI supervisor that the Justice Department has dictated the contractual terms under which any federal support for a regional violent crime task force shall take place identifies the crux of the issue. *Who is, or ought to be, responsible for the resolution of local crime problems, the FBI and DOJ officials in Washington DC or the local police and sheriffs?*

Heymann and Moore (1996), in their article "The Federal Role in Dealing with Violent Street Crime: Principles, Questions, and Cautions" said:

> Our basic assumption is that federal engagement in the control of violent street crime should be guided by principles that reflect the basic principles of federalism: That, for the most part, governments that are smaller and closer to the people will make better judgments about what is publicly valuable to do than governments that are larger and more remote, and that the decisions of the former should be left undisturbed unless issues of fundamental rights, coordination across states, or opportunities for social learning are engaged. If, however, political reality requires the federal government to be given a role, the role should emphasize financial rather than operational assistance. (pp. 104–105)

This statement sums up the perspective of the Tyler Police Department. The participation and financial support of the FBI would be sincerely appreciated, but it should not be necessary to relinquish local control and responsibility to the FBI or the DOJ to be the recipients of federal support. The fact remains that the pre-existing Violent Crimes Task Force, wherein the lead agency in a particular criminal investigation was the agency in whose jurisdiction the crime occurred, was efficient, effective, and more Constitutionally appropriate than the structure now being mandated unilaterally by the FBI.

The memorandum from FBI Headquarters to all field offices states that the wording in certain sections is mandatory and in other sections, only preferred (Neu, 1999). It appears that the Dallas Division Office made the decision simply to impose the preferred language as mandates, too, and to tell the municipal and county agencies that such imposition had come from higher-ups. This may or may not have had a substantive effect on the Tyler Police Department's decision not to participate, but it does tend to indicate that the FBI consciously blocked efforts to tailor the agreement to the needs of local law enforcement agencies.

It is important for municipal, county, and state law enforcement agencies to work well with federal agencies. It is likewise extremely important that federal agencies, in their haste to assist local governments, observe Constitutional principles. Walter Williams (2000) expressed this notion when he wrote:

> In the name of one social objective or another, we are creating what the Constitution's Framers feared – concentration of power in Washington and the creation of a Super State. The Framers envisioned a Republic. They guaranteed it in Article IV, Section 4 of the Constitution, making a State's authority competitive with, and in most matters exceeding, federal authority. Now it's precisely the reverse.
>
> In the pursuit of lofty ideals like health care, fighting crime, and improving education, we Americans have given up one of our most effective protections against tyranny – dispersion of political power. (p. 6; emphasis ours)

Using changes in the federal law enforcement agencies' budgets as the principal index, some authors (e.g., Oliver, 2002) have suggested that an unintended consequence of the 9/11 attack and the War on Terrorism is a change in federal crime control policy. Federal law enforcement agencies—the FBI in particular—in pursuit of Al Quida, no longer have the resources (especially personnel resources) to investigate crimes of the type discussed earlier in this chapter. Miller (2001) (cited in Oliver, 2002) notes that both the Attorney General Ashcroft and the FBI Director Mueller "have spoken . . . about the need for the FBI to retool by shedding its role in areas where the FBI's jurisdiction overlaps with another agency, such as carjacking cases, auto thefts, bank robberies, weapons violations, child support matters, and drug investigations."

Has this happened? Data on federal prosecutions collected and analyzed by the Transactional Records Access Clearinghouse (TRAC) suggest *not*. TRAC (2003) reports that "in the immediate months after 9/11 criminal referrals and prosecutions were down but that the volume rebounded fairly quickly" (http://trac.syr.edu/tracreports/terorrism/fy2002.html). Further, for fiscal year 2002, terrorism prosecutions remained a minor portion of all federal criminal prosecutions (http://trac.syr.edu/tracreports/terrorism/fy2002/pctfilterrorrismG.html).

The main issue here, however, is not whether federal crime control policy has changed to somewhat curtail federalization of the criminal law. The issue is whether or not the FBI is willing to abandon its role in the management of local crime task forces—power gained is neither frequently nor easily abdicated. Furthermore, the issue highlighted by this case study relates directly to the politics played by the FBI to gain control over a segment of local law enforcement from which it derived certain power. Is there any evidence that the FBI, or certain of its higher-level bureaucrats, will not employ the same tactics in the future when there is a perceived Bureau advantage to be gained? Despite 9/11, that evidence has not yet been forthcoming and the authors believe that the lesson of this case study will be relevant far into the future of federal-local task force operations.

SUMMARY AND CONCLUSION

Federalism bespeaks both a separation of powers between the federal government and the states and a certain predominance of states' rights over federal usurpation of their authority— a founding notion of our country and preserved until relatively recently. The late twentieth century, however, brought considerable expansion of the power and operations of the federal government in all areas of American life, including criminal law and the criminal justice system. Much of this expansion was simply a response to the imperatives of economic and population growth, mobility, and the increasing integration and complexity of all aspects of national life. Other aspects of federalization arose less out of national imperatives than from a tendency of Congress to respond to the passions and fears of the moment, organizational entrepreneurship by federal criminal justice agencies, or a tendency of the American public to seek the power of the federal government to provide solutions to local problems. The result has been an increasing overlap of federal and state criminal law.

Until recently, also, there has been little opposition to the federalization of criminal law and the criminal justice system. Lately, however, legal scholars, the organized bar, and the federal judiciary have expressed great concern about this tendency. The U.S. Supreme Court, which has customarily acquiesced to federalization, recently has shown a willingness to slowdown or even reverse the trend by striking down criminal statutes it regards as exceeding Congress's authority under the commerce clause. It is possible that there may be a decline in accumulation of federal criminal statutes that address essentially local problems.

Another related process, much less visible, has been the expansion of the organizational reach of some federal criminal justice agencies. In order to pool resources and expertise to address specific crime problems, law enforcement agencies have created interagency task forces. In their early form, task forces consisted of federal agencies only. Next, federal agencies sought cooperative alliances with state and local law enforcement agencies as equal partners. Most recently the Federal Bureau of Investigation through the use of federal funds and standardized "Memoranda of Understanding" sought to create and exercise substantial control over violent crime and anti-gang task forces consisting of local agencies. Horizontal co-operative arrangements become hierarchical, with the FBI at the top.

The pressure on local agencies to enter these agreements is enormous, as is evidenced by events surrounding the refusal of the Tyler Police Department to sign an MOU to join an FBI-sponsored violent crime task force in East Texas. Local law enforcement executives who regard such arrangements as an improper surrender of local autonomy and responsibility risk at a minimum being branded as "uncooperative" and at worst risk losing their positions. If, as FBI officials asserted, that at the operational level control was not to be heavy handed, the insistence on such an overreaching MOU is difficult to comprehend. Why say it if you don't mean it? Those who value the federalist idea that essentially local crime is best addressed by local agencies may be justified in their concern that this is simply one more manifestation of the federalization of criminal law and the criminal justice system.

Events concerning the East Texas Violent Crime Task Force and the Tyler Police Department raise a number of further research questions. What are the varieties of formal agreements among the members of MJTFs across the country, especially when a federal agency is involved? Notwithstanding the varieties of formal agreements, how do MJTFs actually behave at the operational level? How many other local law enforcement agencies found the FBI's standardized MOU to impose onerous or improper conditions? If any have objected, what have been the consequences? And finally, does the proliferation of MJTFs under federal control threaten Constitutional principles by contributing to the federalization of criminal law? Clearly, much more research attention must be devoted to the federalization of the criminal law and to the management of MJTFs.

RECOMMENDED READINGS

AMERICAN BAR ASSOCIATION (1998). *The federalization of criminal law*. Washington, DC: Author.

HEYMANN, P.B., & Moore, M.H. (1996). The federal role in dealing with violent street crime: Principles, questions, and answers. In A.W. HESTON, & N.A. WEINER (Eds.), *The Annals of the American Academy of Political and Social Sciences, 543* (January), 103–115.

OLIVER, W.M. (2002). 9–11, federal crime control policy, and unintended consequences. *ACJS Today, 22*(3), 1, 3–6.

PHILLIPS, P.W. (1999). De facto police consolidation: The multijurisdictional task force. *Police Forum, 9*(3), 1–6.

PHILLIPS, P.W., & ORVIS, G.P. (1999). Intergovernmental relations and the crime task force: A case study of the East Texas Violent Crime Task Force and its implications. *Police Quarterly, 2*(4), 438–461.

TRANSACTIONAL RECORDS ACCESS CLEARINGHOUSE [Providing a constant watch on federal law enforcement and the courts.] Available at: http://trac.syr.edu.

REFERENCES

ABRAMS, N. (1983). Federal criminal law enforcement. In S. Kadish (1983). *Encyclopedia of crime and justice* (Vol. 2, pp. 779–785). New York: Free Press.

AMERICAN BAR ASSOCIATION (1998). *The federalization of criminal law*. Washington, DC: American Bar Association.

BEALE, S.S. (1996). Federalizing crime: Assessing the impact on the federal courts. *The Annals of the American Academy of Political and Social Science*, 543, 39–51.

BEST, J.W. (1959). *Research in education*. Englewood Cliffs, NJ: Prentice-Hall.

BROOKS, P.R., DEVINE, M.J., GREEN, T.J., HART, B.L., & MOORE, M.D. (1988). *Multi-agency investigative team manual*. Washington, DC: U.S. Department of Justice.

BORG, W.R., & GALL, M.D. (1967). *Educational research*. New York: David McKay.

BUREAU OF JUSTICE ASSISTANCE (1996). *Multi-jurisdictional task forces: Use of overtime and related issues*, FY 1994. Washington, DC: Author.

——. (1998). *Edward Byrne memorial state and local law enforcement assistance* [On-line]. Available: http//www.ncjrs.org/txtfiles/ebmfs98.txt.

CHAMPION v. AMES 188 U.S. 321 (1903).

CONGRESSIONAL RESEARCH SERVICE (2000). [Excerpts from] The Budget of the United States 2000 (pp. 640–642); [and] 2001 (pp. 648–651).

DOBBS, D. (1999). Correspondence to Captain Gregg Grigg, Tyler Police Department, 6 April.

DRUG ENFORCEMENT ADMINISTRATION (1999; 1991). DEA/Tyler multi-Agency task force agreement. Dallas, TX: Author.

DUNWORTH, T., HAYNES, P., & SAIGER, A.J. (1997). *National assessment of the Byrne formula grant program*. Washington, DC: National Institute of Justice [NCJ 162203].

DUNWORTH, T., & MILLS, G. (1999). *National evaluation of weed and seed*. Washington, DC: National Institute of Justice [Research in Brief]. [NCJ 175685].

EDWARDS, G.C., LINEBERRY, R.L., & WATTENBERG, M.P. (1999). *Government in America* (4th ed.). New York: Longman.

FEDERAL BUREAU OF INVESTIGATION (1997). *FBI violent crime initiatives* [On-line]. Available: http://www.fbi.gov/kids/gang/safest.htm.

——. (1998). Memorandum of understanding: East Texas violent crime task force. Tyler, TX (1 October).

——. (1999). FBI violent crime initiatives. Washington, DC: Author (April).

FRIEDMAN, L.M. (1993). *Crime and punishment in American history*. New York: Basic Books.

GILLASPIE, A. (2000). Tyler police rejoin task force. *Tyler Morning Telegraph*, 2 March, pp. 1, 6.

GUN FREE SCHOOL ZONES ACT, 18 U.S. C. Sec. 922(q)(1)(A) (1990).

GUYOT, D. (1991). *Policing as though people matter*. Philadelphia, PA: Temple University Press.

HEYMANN, P.B., & MOORE, M.H. (1996). The federal role in dealing with violent street crime: Principles, questions, and answers. In A.W. HESTON, & N.A. WEINER (Eds.), *The Annals of the American Academy of Political and Social Sciences* (1996), *543*, pp. 103–115.

HYATT, W.D. (1991). Strike forces and fighting organized crime in the United States. *Corruption and Reform*, 61–73.

LEUCKENHOFF, E.H. (1999). Correspondence to Chief Bill Young, Tyler Police Department, 20 May.

LYMAN, M.D., & POTTER, G.W. (1997). *Organized crime*. Upper Saddle River, NJ: Prentice-Hall.

McDONOUGH, M. (1998). Personal communication. 29 July.

MILLER, B. (2001). Crime on the back burner? *The Washington Post National Weekly Edition*, *19*(7), 30.

MORGAN, R.E. (1983). Federal Bureau of Investigation: History. In S.H. KADISH (1983). *Encyclopedia of crime and justice* (Vol. 2., pp. 768–774). New York: Free Press.

NEU, K.E. (1999). Safe Streets Violent Crimes Initiative, Memorandum of Understanding (MOU), Safe Streets Task Forces (SSTFs). Washington, DC: FBI. [Memorandum to all field offices] (12 March).

OLIVER, W.M. (2002). 9–11, federal crime control policy, and unintended consequences. *ACJS Today*, *22*(3) 1, 3–6.

ORVIS, G.P. (1999, March). "The evolution of the crime task force and its use in the twenty-first century." Paper presented at the Annual Meeting, Academy of Criminal Justice Sciences, Orlando, FL.

PHILLIPS, P.W., & ORVIS, G.P. (1999). Intergovernmental relations and the crime task force: A case study of the East Texas violent crime task force. *Police Quarterly*, *2*(4), 438–461.

RACHLIN, H. (1993, May). Making the streets safer. *Law and order*, *41*(5), 59–62.

SHUR, D. (1995, April). Safe streets: Combining resources to address violent crime. *FBI Law Enforcement Bulletin*, *64*(4), 1–8.

TRANSACTIONAL RECORDS ACCESS CLEARING HOUSE, 2003. [Available: http://trac.syr.edu/ tracreports/terrorism/fy 2002.html]

U.S. v. LOPEZ 514 U.S. 549 (1995).

U.S. v. MORRISON 120 S. Ct. 1740 (2000).

VIOLENCE AGAINST WOMEN ACT, 42 U.S.C. Sec. 13981 (1994).

WALKER, D.B. (1981). *Toward a Functioning Federalism*. Cambridge, MA: Winthrop.

WILKINS, J (n.d.). Correspondence to Chief Bill Young, Tyler Police Department.

WILKINS, J (1999). Correspondence to Chief Bill Young, Tyler Police Department, 15 January.

WILLIAMS, W. (2000). Have we learned about government murders? *Tyler Morning Telegraph* (6 January), p. 6.

WILSON, J.Q. (2000) *American government* (5th ed.). Boston: Houghton Mifflin.

YOUNG, W.A. (1999). Correspondence to Assistant Special Agent in Charge, FBI, Dallas (27 May).

12
Summary and Conclusion

M.L. Dantzker

Patrol is best described as the "the backbone of policing" and Criminal Investigation is generally considered the second major component in a police agency (Dantzker, 2003). However, neither entity can address every issue or problem that may arise within the community and thus, needs support from other components or units. These support units often may be described or designated as special units whose role or function is to address or focus on specific issues that may drain too much time and resources from patrol and criminal investigations. This chapter serves as a general examination of the concept of special units as support units and as a capstone to this text.

THE CONCEPT OF SUPPORT UNITS

Although police agencies primarily rely on patrol and criminal investigations to handle a large proportion of the agency's requests for services, neither can completely or effectively perform the communities' requisite needs without assistance from other areas. As noted by the 1967 Presidential Commission on Law Enforcement and the Administration of Justice,

> In addition to improving the competence of top and middle management and applying known principles of organization, police departments must establish special units whose function is the continual planning, administration and assessment of police practices and procedures. A chief of police, particularly as his force grows larger, cannot alone effectively administer a department, devise policy, or evaluate performance. He needs supporting staff, administrative personnel and management specialists who can supervise and train personnel, assist in policymaking, fulfill administrative functions, and assess the soundness of existing practices and procedures. (Task Force Report, 1967/1990, p. 48)

The special units referred to are more commonly recognized as support units or support services (Dantzker, 2003).

> A support unit can simply be defined as a unit of the police agency that lends technical or investigative support to the patrol and criminal investigation divisions. Support services are critical to the efficient operation of any police line function, such as identifying a criminal suspect, searching a building, or effecting an arrest. (Dantzker, 2003, p. 129)

On the heels of the 1967 President's Commission's report, the 1973 National Commission on Criminal Justice Standards and Goals suggested that,

> Support services are critical to the efficient operation of any police line function. The failure of a vital support service can severely impair the effectiveness of an operational police unit. All police agencies must take measures to insure that their support units are organized and operated to promote the most effective accomplishment of basic agency goals. (Police, 1973, p. 292)

As noted by Phillips in Chapter 1, support units can be either specialized (assigned to complete certain tasks) or designated as a "special unit" (formed to address a persistent or contingent problem). The number of support units formed in an agency is often contingent on the size of the size of the police agency and whether there is a true need for such a unit (Chapter 1).

> The number of support units an agency has is limited by its size. It could be argued that in smaller agencies too many support units burden the agency by depleting resources from the more crucial units (e.g., patrol). However, some may argue the same thing is true for larger agencies. (Dantzker, 2003, p. 133)

A second issue, discussed in all the previous chapters is personnel selection. Deciding how to staff these units can be problematic for a police administrator. Who will work in these units (non-sworn or sworn officers) often depends on the type of unit being formed (Dantzker, 2003). For example, a crime scene search unit could be staffed by non-sworn experts while a fugitive retrieval unit would require sworn officers. Agencies that want to staff these units without depleting the ranks of sworn officers will hire non-sworn personnel, which as noted in several chapters, can be a cost benefit and effective for the police agency.

Finally, while the use of non-sworn personnel in certain support units may save the agency money in personnel costs, it does raise a concern for liability.

> The courts are holding police agencies liable for their shortcomings (e.g., improper or poor training of officers) and their officers' activities. Therefore, because the agency is responsible for all its employees, the question of liability is moot as long as the civilians' acts are governed by agency policies, agency policies do not violate any constitutional rights, and the civilian is properly trained. (Dantzker, 2003, p. 134)

Because the concept of support or special units is to provide additional resources and assistance to the patrol and criminal investigative functions, recognizing what types and how they are to be utilized is important (Loveday, 1998; Woods, 2002). This text provides a much needed insight into this particular area of policing.

THE NOTION OF SUPPORT

To better understand the concept and existence of special units as support units, this text has provided a well-rounded perspective with each chapter offering clarity to the uniqueness and use of such units. As previously noted, Phillips (Chapter 1) began the exploration by examining specialization versus special units. He clarifies that specialization refers to officers who are trained to do certain tasks while the special unit is designated to address a particular problem. The key point to his chapter is that an agency must have an actual need for a special unit before forming one. To support the development, Phillips offers suggestions for planning and decision-making discussing all the various required elements.

Aptly following Phillips' point on whether there is an actual need for a special unit is Rubenser in Chapter 2. The main focus of her chapter is addressing whether a special unit is functional (serves a real purpose) or legitimacy maintenance (creating public support for the unit by making it look like it's an appropriate, useful unit). In making this determination, she suggests that three questions be answered:

1. Whether this special unit engages in activities related to its official mandate,
2. Whether officers assigned to the unit believe they are engaging in a legitimate police function, and
3. Whether community residents, as consumers of the unit's services, are aware of and have positive attitudes about the unit's activities.

To test the debate, she examined the Omaha Police Department's (OPD) special unit, Nuisance Task Force (NTF), a unit that was originally created as a six-month program but because of the high volume of complaints that continued needing attention became a permanent unit.

The study of OPD's NTF required Rubenser to examine several different sets of data which included the unit's activities, interviews with officers, interviews with other agencies and with consumers. Her analyses found that:

1. The unit did engage in activities closely related to its official mandate.
2. Officers have a positive attitude, and
3. Principal consumers are aware and value the unit's work.

Ultimately, she concludes that the NTF does perform a functional purpose but that her findings cannot be generalized to all special units within the OPD or to policing in general. Obviously, each agency would have to conduct a similar study to determine whether its special units were either functional or simply maintaining legitimacy.

With community policing being the current trend in policing, whether special units are compatible to this approach is important. Shernock (Chapter 3) investigates this issue by examining the relationship between special units and community policing. His discussion begins by noting that in order to understand the relationship or compatibility between special units and community policing that it is important to:

1. Understand the meaning of a special unit as a distinctive type of unit within the police organization;
2. Realize that there are three entities interacting—the special unit, the larger police organization, and the community; and that special units have effects on the role that officers in other units play regarding the community;

3. Distinguish between two very different interpretation or models of community policing; and

4. Whether special units can be incorporated into the roles of the generalist COP officer.

Shernock then addresses each of these issues indicating the positives and negatives of their application. For example, regarding the meaning of a special unit as a distinctive type of unit within the police organization Shernock notes that the special unit ". . . is not only a part of the organization, but an entity that tends to institutionalize itself in the organization in a way which affects the organization's structure in relating to the community or task environment."

Despite the broadness of the beginning of his chapter, Shernock settles in to a specific discussion of two types of special units, volunteer auxiliaries and paramilitary police units (PPUs). His discussion explores the existence and use of such units and how they can both have positive and negative impacts on the police organization and in the relationship with the community. Areas of interest he covers include looking at two ways of integrating special units with COP as co-optation and problem-solving; identifying with and alienating the community; expansion into and usurpation of other police jurisdictional domains; the effects these two special units have on professionalism and stratification within the police organization; and the cooperation and coordination of such units in the police organization.

Overall, Shernock concludes that little has been written about community policing and specialized units. His chapter while examining only two such units, offered a necessary and useful insight as to how special units can be viewed within the larger scheme of community policing. Of particular interest, he clarifies that community policing should become part of the special units rather than special units becoming part of the community-policing organizational structure.

While the first three chapters establish the foundations for special units, Chapter 4 begins placing attention on particular types of special units beginning with the Edler Abuse Unit. Here Sapp and Mahaffey-Sapp first establish that the population in the U.S. is getting older creating more possibilities for elderly victims and thus a need for attention by policing and law enforcement. Interestingly, they note that, to date, little research has been done on the elderly but what is available is discussed. From here they address the important elements surrounding such a unit beginning with identifying the types of elder abuse (physical, psychological, material, and medical); statutory response of which they indicate has been limited (Adult Protective Services' legislation); and how the training for such a unit is very important and that a team approach should be taken.

The next part of their discussion offers operational considerations for such a unit that include knowledge of abuse reporting requirements, legal constraints of professional action, and the degree of competency of the elderly. Administrative considerations suggested include management and supervision issues along with selection and training. Their conclusions or future directions for such a unit include the developing of programs to help reduce elder victimization, educational programs, mandatory reporting laws, public policy, and more research. In all, they conclude that because of the growing trend of elderly people in the U.S., police departments will have to recognize how this change will affect their ability to function and that a special unit may be necessary for effectiveness.

Although a long time part of our society, it has only been within the past 15–20 years that gangs have required specific attention from policing, to the point that many police agencies have formed special gang units to address the problems. In Chapter 5 Rush and Orvis explore the concept and need for a gang unit. They begin by addressing one of the most important issues, whether there are signs of a gang problem and whether it requires a special unit to handle the problems. This is followed with a discussion on defining the gang problem and definitions of a gang. They note that while many definitions may exist, ultimately a gang problem exists when there is evidence of three or more individuals whose cohesive existence is based on criminality. When evidence exists, then a gang unit is probably prudent and they define how the gang unit can be a solution.

Rush and Orvis continue the discussion by identifying the different types of gang units (youth services, gang detail, and the gang unit) and demonstrating how each type has a particular role and function depending on the level of gang activity. As with any special unit how the unit is managed is important and they address this issue followed by suggesting that the duties of a gang unit officer include profiling, tracking, intelligence, assisting, interacting, operations, and investigations. Who should be in such a unit and its supervision are addressed along with special problems the unit may face such as training and cultural awareness. To demonstrate how such a unit might work Rush and Orvis discuss The Westminister, CA., Police department's unit as a model for other gang units. Overall, they conclude that police agencies should not wait until after gangs become a problem before they address them and that the type of unit formed is dependent upon the level of gang activity.

There are those who may argue that any crime committed against a person or another's property is not done out of love nor necessarily a result of hate. Yet, due to an increasing amount of crime that appears to be reflective of the offender's bias toward the victim because of the victim's race, gender, ethnicity, or sexual preferences laws have been created to specifically categorize such crimes as hate crimes. The result of this new classification is an ever increasing amount of a particular crime police agencies must investigate. In an effort to assign resources specifically to these types of crime police agencies have begun establishing Bias crime units. Vaughn (Chapter 6) explains to us the elements associated with such a unit.

Vaughn begins by noting that bias crime units investigate hate crimes that are motivated by bias and defines bias as put forth in the law. As to why such a unit is necessary he offers a discussion on hate crime statistics and the statutes and laws. While indicating that there are bias units at both the federal and state/local levels, he acknowledges that these units are still rare in the U.S. Any agency considering the creation of such a unit needs to consider administrative and operational issues. Vaughn indicates that administrative considerations should include unit structure and authority, training, zero tolerance among its officers regarding their own biases, developing accurate data, and establishing community partnerships. Operational considerations he offers include recognizing indicators of bias crime (victim and offender characteristics and modus operandi), preservation of evidence, offender profiles, and victim assistance. His conclusions are consistent with previous chapters' authors in that a bias unit shouldn't be created just for the sake of creating such a unit but to address a legitimate problem.

Volunteerism has long been a staple of society, especially as it relates to policing. The examination of volunteerism and its application to special units is accomplished by Phillips (Chapter 7) who sets out to identify and explain the varied roles of volunteers in contemporary American policing and to examine some of the special considerations important to police agencies in the management of volunteer units. In accomplishing his goals, Phillips begins by defining a volunteer unit and identifying rationale for using volunteers. His rationale includes cost-effectiveness, better services, citizen participation, source of recruits, and co-production. A key point is that there is a cost benefit to police agencies by using volunteers as the examples Phillips offers demonstrate how agencies saved money.

The use of volunteers is not all positive, as Phillips notes that there is opposition to such units. Internal opposition can be from sworn officers who believe that volunteers may take jobs away, are not always available when needed, and that there is a high turnover creating difficulty in establishing any kind of continuity. Phillips also suggests that external opposition comes from citizens themselves who may feel that they are being policed by non-police or being short-changed for their tax dollars. Despite the opposition, volunteer units continue to exist in policing and to such an extent that Phillips claims that it is difficult to quantify the types of volunteer units because of the variation, variety, and diversity of such units. Furthermore he advises that sources of volunteers are broader than perhaps believed identifying alumni of citizen police academies as one of the best sources.

Phillips closes his discussion by addressing the selection, training, and supervision of volunteers and the importance of incentives and rewards for these individuals who generally volunteer for the intrinsic value but being acknowledged as valued by the agency is important too. He concludes that developing a qualified volunteer corps is an important, continuing innovation in and for policing.

One of the most recognized special units in policing today is SWAT, primarily because of all the media attention this unit has received in both the popular press and the entertainment arena. To better understand this unit, Taylor, Turner, and Zerba in Chapter 8 describe the historical development of this type and similar police paramilitary type units to its current existence. They begin with the historical development noting that the first modern full-time SWAT unit was created by the Los Angeles police department in 1967 but that the first specialized weapons teams was actually created in the 1880s in New York City. With its initial role being assaultive in nature that is created for tactical response, this unit would eventually become more oriented toward negotiation starting in NYC in 1973. Because of the change in its nature the SWAT unit would soon give way to crisis intervention by means of a Critical Incident team.

Taylor, Turner and Zerba continue with the Critical Incident theme noting that this type of unit may be composed of two units, one that negotiates and the other tactical. As with previous chapters these authors explore the selection issues, policy, training, and command and structure. They conclude their discussions with a look at the multi-jurisdictional teams and alternatives for smaller agencies that cannot create such a unit nor have access to enough other agencies to form a multi-jurisdictional unit. In closing, they advise that police agencies must change with the teams and emphasize how non-violent responses are growing in popularity for these paramilitary type units.

Perhaps one of the most popular units in today's policing is that which deals specifically with the drug problems. Specialized drug enforcement units are very popular and Olson (Chapter 9) examines the role that these specialized units play in the array of strategies used by police departments in their drug control efforts. He begins by identifying two types of drug units, those belonging to individual departments and those that are multi-jurisdictional and follows this up with a brief history of the drug problem in the United States. In light of all the problems, Olson indicates that there are really two types of control strategies employed, supply reduction and demand reduction. However, he advises that the biggest violators are the consumers of illegal drugs.

Because of the two types of drug enforcement units, Olson discusses their prevalence in local police departments, which apparently continues to grow despite their competition from multi-jurisdictional drug task forces (MJDTF). To demonstrate how the multi-jurisdictional task force operates, Olson offers the "Illinois experience" which highlighted measuring performance, identified potential typologies, and recognized management issues such as organizational structure and environmental forces. All the cited issues are applicable to any level's MJDTF. Ultimately, it appears that Olson favors the MJDTF approach suggesting that individual agency drug units must be able to address changes in society, operate with less and less federal funds, and expect competition from other drug task forces or units.

The popularity of community policing has brought with it a return to earlier days of policing, particularly the use of horses and bicycles. Therefore, the growth of Mounted and Bicycle units has been inevitable and requires recognition as they receive from Roth in Chapter 10. He begins with a brief overview of the Mounted and Bicycle units establishing pros and cons of both. Then beginning with the Mounted Unit, Roth offers a brief historical tour claiming the first official horse patrol begin in London (1805) followed up with a similar discussion about the Bicycle unit. Roth continues his exploration by identifying a number of variations of the Mounted units which included park police, volunteers, posses, search and rescue, campus, and collateral duty.

As has been noted by previous authors there are operational and administrative considerations for the establishment of any type of special unit. Roth continues this trend by noting that operational considerations for both types of units are response, presence, accessibility, and citizen contact. The administrative considerations include areas to be patrolled, the location of the areas and their primary uses, and the activities of the area. Additional administrative considerations included benefits, costs, selection and obtainment of horses and bicycles, personnel selection, liability, and training. He concludes that by creating Mounted and Bicycle units policing is reviving an old means of patrol but a means that has its place in today's policing.

To provide a missing component from previous chapters that focused on the multi-jurisdictional task force (MJTF), in Chapter 11 Nelligan, Phillips, and Young focused on the political aspects of managing an MJTF. To accomplish this, they examined a specific case of events in one jurisdiction where the FBI exerted pressure on a municipal police department to join into an MJTF, despite concerns by the department as to what effect the association may have on its autonomy.

Prior to discussing the particular case, Nelligan et al. offered a discussion on federalism and the federalization of criminal law following up with defining and counting MJTFs. Included in this latter discussion is a look at the evolution of the crime task force concept and the introduction to the FBI's Safe Streets tasks force, which is ultimately the focus of this chapter. The remainder of the chapter is the results of a descriptive case study of the FBI's attempt to implement such a task force in Tyler, Texas, despite serious concerns and reservations by the administration of the Tyler Police Department (TPD). A detailed description of its development and the processes were offered. Furthermore, the authors provided the requirements for the proposed task force agreement between the FBI and TPD, which included 14 points. Unfortunately, the push by the FBI to have things accomplished more its way than TPD's lead to the resignation of the TPD Chief. This result conjured up a major concern of how much influence a federal agency could employ over a municipal agency, thus creating the issue of federalism and its influence in the creation and management of a MJTF. As Nelligan et al. conclude, "much more research attention must be devoted to the federalization of the criminal law and to the management of multi-jurisdictional task forces."

CONCLUSION

Obviously, patrol and criminal investigations may be the primary functions of a police agency, but cannot necessarily address all the problems of the community without assistance from units within the agency whose primary function is to address a particular problem or set of problems. This is accomplished by special units. While the number of special units is limited by available resources and determined by needs analysis, this text has offered valuable insight as to the role and function of such units through the examination of specific types, units that seem to be growing in popularity in policing today, each with its own particular purpose and role in the scheme of policing.

REFERENCES

DANTZKER, M.L. (2003). *Understanding today's police* (3rd ed.). Upper Saddle River, NJ: Pearson Education Inc.

LOVEDAY, B. (1998). Improving the status of police patrol. *International Journal of the Sociology of Law*, *26*(2), 161–196.

POLICE (1973). Washington, DC: National Advisory Commission on Criminal Justice Standards and Goals.

TASK FORCE REPORT: THE POLICE (1967/1990). (Originally published by the U.S. Government Printing Office in 1967.) New Hampshire: Ayer Company, Publishers, Inc.

WOODS, T. (2002). Starting a bike patrol. *Law and Order*, *50*(4), 78–85.